T0398688

CREATIVITY FROM SUBURBAN NOWHERES

Rethinking Cultural and Creative Practices

Edited by Ilja Van Damme, Ruth McManus, and Michiel Dehaene

Looking at suburbs as places of creativity gives rise to novel and thought-provoking narratives that typically run counter to the idea that suburbs are sites of "ordinary," "mundane," and "everyday" practices. Far from being geographies of "nowhere" – dull, materialistic, and monotone – suburbs are unpacked as being heterogeneous and historically layered places of living, work, and creation. Situating creativity in place and time, *Creativity from Suburban Nowheres* displaces mainstream understandings of creativity and widespread stereotypes commonly associated with the suburbs. Contributors explore the particular forms of creativity that suburbs elicit both in the process of their making, materialization, and community construction, and in the myriad ways in which suburbs are inhabited and experienced.

Anchored in historical and geographical research, this volume highlights how and in what forms creativity should be understood in the suburbs, why and when creativity can be found, and how the notion of suburban creativity overthrows ingrained and dominant normative viewpoints. Rather than seeing creativity arise *despite* its suburban location, *Creativity from Suburban Nowheres* illuminates the emancipatory potential of suburbs for creativity.

(Global Suburbanisms)

ILJA VAN DAMME is an associate professor of urban history at the University of Antwerp.

RUTH McMANUS is an associate professor of geography and associate dean for teaching and learning in the Faculty of Humanities and Social Sciences at Dublin City University.

MICHIEL DEHAENE is an associate professor in the Department of Architecture and Urban Planning at Ghent University.

GLOBAL SUBURBANISMS

Series Editor: Roger Keil, York University

Urbanization is at the core of the global economy today. Yet, crucially, suburbanization now dominates twenty-first-century urban development. This book series is the first to systematically take stock of worldwide developments in suburbanization and suburbanisms today. Drawing on methodological and analytical approaches from political economy, urban political ecology, and social and cultural geography, the series seeks to situate the complex processes of suburbanization as they pose challenges to policymakers, planners, and academics alike.

For a list of the books published in this series see page 375.

EDITED BY ILJA VAN DAMME,
RUTH McMANUS, AND
MICHIEL DEHAENE

Creativity from Suburban Nowheres

Rethinking Cultural and Creative Practices

UNIVERSITY OF TORONTO PRESS
Toronto Buffalo London

ISBN 978-1-4875-0829-6 (cloth)   ISBN 978-1-4875-3795-1 (EPUB)
ISBN 978-1-4875-2579-8 (paper)   ISBN 978-1-4875-3794-4 (PDF)

Global Suburbanisms

Library and Archives Canada Cataloguing in Publication

Title: Creativity from suburban nowheres : rethinking cultural and creative
    practices / edited by Ilja Van Damme, Ruth McManus, and Michiel Dehaene.
Names: Damme, Ilja Van, editor. | McManus, Ruth, editor. | Dehaene, Michiel,
    editor. Series: Global suburbanisms.
Description: Series statement: Global suburbanisms | Includes bibliographical
    references and index.
Identifiers: Canadiana (print) 20230186866 | Canadiana (ebook) 20230186912 |
    ISBN 9781487525798 (paper) | ISBN 9781487508296 (cloth) |
    ISBN 9781487537944 (PDF) | ISBN 9781487537951 (EPUB)
Subjects: LCSH: Suburban life. | LCSH: Suburbanites – Intellectual life. |
    LCSH: Creation (Literary, artistic, etc.) | LCSH: Suburbs – Social aspects. |
    LCSH: Sociology, Urban.
Classification: LCC HT351 .C74 2023 | DDC 307.74 – dc23

Cover design: Val Cooke
Cover images: (top) iStock.com/wacomka; (bottom) *Moscow I*, 1916 (oil on canvas),
1916, by Wassily Kandinsky (1866–1944) / Tretyakov Gallery, Moscow, Russia /
Bridgeman Photos

We wish to acknowledge the land on which the University of Toronto Press
operates. This land is the traditional territory of the Wendat, the Anishnaabeg, the
Haudenosaunee, the Métis, and the Mississaugas of the Credit First Nation.

This book is funded by the Flemish Research Council (FWO), through the program
Urban Agency II: The Historical Fabrication of the City as an Object of Study.

University of Toronto Press acknowledges the financial support of the Government
of Canada, the Canada Council for the Arts, and the Ontario Arts Council, an agency
of the Government of Ontario, for its publishing activities.

To our children: Romy, Aiko, Ciarán, Áine, Francis, Anna, and Andrea – an everlasting source of joy and creativity.

– Ilja Van Damme, Ruth McManus, and Michiel Dehaene

Contents

Figures and Tables

Tables

Preface

This book is the outcome of a large international research community, spearheaded by the Urban Studies Institute and the Centre for Urban History (both University of Antwerp) and funded by the Flemish Research Council. The scientific research network is focused on facilitating interdisciplinary collaboration on the study of the city as a historical and theoretical construct. Thus, the network aims to engender discussions on recent urban theory and findings in history, expanding the historical-geographical and methodological horizon of both urban studies and urban history (see https://www.uantwerpen.be /en/research-groups/urban-studies-institute/research/urban-agency /historical-fabrication-city/).

The current project started from recent observations arguing that processes of planetary sub/urbanization call into question classic definitions and causalities associated with a reified notion of the "city." An international research community "kick-off" event was organized in October 2016, with follow-up meetings and a workshop in the following years. Much of the content of this book was discussed during a small international workshop organized in April 2019 at the University of Antwerp, giving most participating authors the first chance to present and discuss their main findings and ideas. The next year, of course, the global COVID-19 pandemic brought the world to a standstill. Nevertheless, despite all professional and personal hardship endured, we are now very pleased to present the outcome of this conference in this book – expanded and enhanced with some additional chapters that were not part of the original set-up.

The discussions we had and the suggestions we received from direct colleagues, participants, and members of the broader research network when making this book all had a huge impact on the way this volume eventually materialized. We wish to thank all those who participated at these meetings, including presenters whose work for various reasons was eventually not included in the book. Their papers have nevertheless been very helpful in finding the right focus and balance of the volume. We would like to explicitly thank all participating authors for contributing their chapters; it was a delight working with them on this project. Last but not least, special thanks goes to Bert De Munck and Stijn Oosterlynck (University of Antwerp), direct colleagues and chairpersons of the Urban Studies Institute, and to Roger Keil (York University), who generously "adopted" our endeavours by linking them to his own "Global Suburbanisms" network. It seems fitting then to publish our results in his inspiring and ongoing book series on the matter.

Ilja Van Damme, Ruth McManus, and Michiel Dehaene

PART 1

Openings

1 Rethinking Creative and Cultural Practices from the Outside In: An Interdisciplinary Exploration

ILJA VAN DAMME, RUTH McMANUS, AND MICHIEL DEHAENE

Introduction

Unsympathetic Martians blasting their way through Woking, a typical late nineteenth-century suburb at the outskirts of London. A white church decapitated. Rustic, gardened villas trampled and destroyed. People scavenged like ants and, eventually, eaten. Oh dear, oh dear, oh dear …

There is much vicious irony to the Victorian invasion novel *The War of the Worlds* by H.G. Wells (2005 [1898]). Living and working at the time in Woking, the famous English novelist (1866–1946) is known to have cycled around the broader metropolitan area of London, marking out locations for the story that would make him world-famous (Aldiss, 2005). Clearly, Wells (1934, 1:542) felt entitled to laugh at suburbia, ridiculing in his autobiography the "small resolute semi-detached villa with a minute greenhouse" in which he lived and contriving in his work the type of horrific destructions and cruel endings that would have scared and shocked his fellow middle-class suburbanites (Cunningham, 2004: 432). At one point in the novel, "the sort of people" who could have been his neighbours are despised as lacking any spirit – "no proud dreams and no proud lusts" (Wells, 2005 [1898]: 155) – being too bored and too conservative to even make an impact in Hell. From the moment the Martians enter inner London, they are killed by the most common and despised domicile of its overpopulated centre: human bacteria (not a COVID-strain, we may safely presume; Wells probably had the dangerous cholera bacillus in mind, the "killer bacterium" of nineteenth-century cities).

At the end of the nineteenth century, the population in the counties of London's hinterland – comprising, among others, Middlesex, Hertfordshire, Surrey, and Kent – had risen to almost 1,500,000 people

(Georgiou, 2016: 96). Wells was but one of many contemporary commentators yammering about this rapid expansion of suburbia as a cultural and physical phenomenon (Hapgood, 2005). For a reader today, however, the trouble with suburban Woking is not so much it being linked to dullness and insipidness, to values and architectural forms that seem to justify – according to Wells – its pitiable, but inevitable downfall. Such literary, visual, or otherwise representations have been debunked by a wave of recent revisionist suburban research as worn-out, stereotypical tropes (Huq, 2013). According to historian Mark Clapson (2016: 336), this suburban "myth" has been concocted by "a self-consciously urbane commentariat who could never live somewhere so vacuous." And geographer Richard Harris (2018: 35) agrees that negative stereotypes about suburbia have been mainly products of class struggle, upheld by the self-righteous and condescending attitudes of both the aspiring intellectual urbanites and the more conservative and complacent exurbanites.

Thus, the trouble with Woking becomes the fact that the celebrated author H.G. Wells – an artistic and cultural icon of his generation – did live in such a despicable "geography of nowhere" (Kunstler, 1993), as suburbs have been disdainfully described. However, as this book will make clear, H.G. Wells was not alone in his class-based disdain but equally paradoxical preference for suburban living, work, and creation. Indeed, far from being geographies of nowhere – dull, materialistic, and monotone – suburbs throughout this volume appear as socially complex, geographically heterogeneous, and historically layered places. In short: places as interesting to study as inner-urban life. If this approach comes as a surprise, consider the following: in many ways, suburbs have been the most underacknowledged spaces in a wide and still growing academic literature on so-called "creative cities," which is at least partly because suburbs continue to be framed – even among academics! – through precisely the same sort of stereotypes that appear in the writing of Wells and many others.

In order to mitigate such oversight, this book dissects how creative and cultural practices have shaped suburban living in often unexpected and unforeseen ways, both in the process of their making as well as in the myriad ways in which suburbs are inhabited and experienced, and have become generative for shaping multiple identities throughout the nineteenth and twentieth centuries. In exploring and thematizing what we will denote and explore as "suburban creativity," we have followed an interdisciplinary trajectory, setting up a conversation with urban theory, history, geography, and a number of related urban disciplines. These complementary insights will shed light on crucial questions, asking how and in what forms creative and cultural practices should be understood in suburbs; why and when creativity can be found there; and how the notion of suburban creativity overthrows ingrained and dominant normative viewpoints on the issues at stake. Therefore, the chapters assembled in this volume should be taken as multiple threads, exploring and questioning the meaning and implications of suburban creativity from different academic angles and viewpoints. Taken as a whole, this volume aims to produce a rich interdisciplinary tapestry of the variety and diversity of creative and cultural interactions and processes that occur in suburban contexts.

We begin by delving deeper into the historiographical debates relating to the central notions of "creativity" and the "suburban." In so doing, we seek to align ourselves to a recent strand within urban studies aiming to reload urban theory from the outside in, thus foregrounding an ill-studied and badly understood urban outside: the urban periphery, margins, or disjunct fragments of non-central population (Keil, 2017, 2018). The following section gives content and meaning to the conceptual "outsides" that are key to this book – underexposed suburbia – and begins to question and analyse how suburban spaces can be connected to creative and cultural practices. Looking at the suburbs as places shaped by creative and cultural activities gives rise to novel and thought-provoking narratives, which succeed in rethinking the ordinary, mundane, and everyday in opposition to more widespread but restrictive definitions and narratives of creative cities shaped by economic and political-managerial bias. We end this introductory chapter by detailing the individual contributions that will make up the remainder of this volume, outlining the main themes, questions, and contradictions that arise when focusing on creativity from "suburban nowhere." Far from leading to a dead-end, a metaphorical cul-de-sac in urban theory, it will be argued that new interdisciplinary urban insights can arise from the discussions assembled here.

Hide and Seek: Theory on Creative Cities from the Inside Out to the Outside In

This book takes discussions on creative and cultural practices that have been most commonly associated with inner-urban life into the suburbs. In so doing, the participating authors seek to displace mainstream notions of creative cities as well as come to a more historically and geographically sensitive understanding of the suburban or non-central urban outside. Let us begin with the former.

For more than two decades now, a veritable avalanche of academic writing has delved into what is now commonly referred to as the "creative city debate" (Bianchini, 2018). This debate has entailed understanding the contemporary rise of so-called "cultural or creative economies" (Scott, 2000; Howkins, 2001) to develop and target "cultural and creative industries" (Caves, 2000; Hesmondhalgh, 2005; Hartley, 2005) and setting up urban policy models of "cultural and creative management," aimed at turning cities into creative and innovative "knowledge hubs" or "cultural crucibles," susceptible to accommodate the "creative class" (Landry, 2000; Miller and Yúdice, 2000; Florida, 2002b, 2005).

Much of this literature should be more properly historicized against the background of the crisis of the functional-industrial urban complex in Atlantic economies, gaining traction from the end of the 1960s and 1970s onwards due to processes of de-industrialization and economic globalization (Reckwitz, 2017: 173–200; Van Damme and De Munck, 2018: 5–10). The intellectual response to these historical changes was immediate. It engendered various Western epistemic communities and fragmentary voices making influential statements and urban-architectural messages commonly referred to as the "post-industrial" and "postmodern": the city that would take shape after the demise of Fordist modernity and the concomitant deficit in legitimacy of the welfare state. What needs to be stressed in the context of this introductory chapter was a growing emphasis on what Reckwitz (2017: 176, 179–80) has recently called "culturalization strategies." They entailed urban interventions aimed at turning cities into permanent sites of new creation, innovation, and aestheticization to the extent of making them attractive for cultural and creative workers/industries, new tourists, and the aimed-for creative class gentrifiers and new business investments. Within this growing concert of ideas and projects, "creativity" became a powerful and equally amorphous signifier for urban policymakers, city planners, and business circles alike. The outcome has been a gradual change in the character and outlook of Western inner cities from New York to Barcelona, and Glasgow to Athens, over the past thirty years.

Past academic focus on creative cities, however, has not been without fierce criticism, a tendency that has only grown more vehement in the last decade. Three bodies of critique can be discerned. First, and paralleling growing public discontent at the tail end of the global financial crisis of 2007–9, creative city debates have been criticized as being too one-sidedly reduced to a very specific neoliberal or utilitarian market-capitalistic logic. Thus, "creativity" – or "innovation" in its broader entrepreneurial meaning – has been made instrumental for (neo)liberal business use and an agenda of post-industrial urban revitalization and planning, which, moreover, has only served an unequal and unsustainable form of urban economic growth (Wilson and Keil, 2008; Pratt, 2011; Harris and Moreno, 2012; Scott, 2014). Second, and related, creativity has in many ways been abstracted and depoliticized from its concrete urban actors and historical contexts, naturalizing or neglecting how cultural and creative practices have always been intimately tied to power relations (of class, race, gender, age), strategies of governance, and urban political priority-making (Peck, 2005; Parker, 2008; McLean, 2014; Van Damme and De Munck, 2018). It has been remarked how urban-based power groups of the last twenty years have set largely elitist definitions of creativity, which frequently negate or invalidate its more diverse, subversive, non-economic, or day-to-day vernacular elements (Edensor et al., 2009; Mould, 2015). Third, and finally, creativity has almost always been appropriated and repackaged as something belonging solely to an inner-urban core or city centre, ignoring, or altogether casting away, the many ways in which culture and creativity is produced and engendered in non-central interurban or suburban places (Hutton, 2009; Shearmur, 2012).

All these fundamental critiques have had their bearings on the content and direction of this book, but the latter debates especially speak directly to the notion that is at stake here: suburban creativity. Only very recently has creativity been discussed in literature on suburbs, suburbanization, and suburbanism: the places, processes, and lifestyles connected to what is traditionally acknowledged as being "under," "close to," or directly "outside" the central city (referring to the etymological meaning of *sub-urbium* in Latin). This neglect of non-central populations and their spatial expansion has had deep epistemological roots in what Roger Keil (2017) calls a dominant mode of theorizing, concerned with looking at the urban from the "inside out" rather than from the "outside in." Crucial is an intellectual willingness to move beyond a dominant tendency in urban theory referred to as "methodological cityism" (Angelo and Wachsmuth, 2015). Here, suburbs become defined as the bleak mirror image or opposite of central cities, which receive all the attention, thus ignoring suburbs' more socially, geographically, and historically complex build-up and relationships with a wider region (Van Damme and Oosterlynck, 2021).

Inside-out urban theorizing can be seen as being heavily indebted to the sort of bounded city ideal type proposed by Weber (1978 [1920–1]) and, subsequently, the influential Chicago School mode of thinking (Park and Burgess, 1925). At the heart of these stylized urban models stands the Western urban germ or market centre – usually represented as a concentration of political, economic, and cultural power – from which all other urban processes are thought to blossom in radial extensions of diminishing function, complexity, and importance. Influential writings from Lewis Mumford (1961), through Jane Jacobs (1969), to Peter Hall's (1998) *Cities in Civilization* have stressed how political democracy, liberal economic development, culture, and creative innovation were essentially outgrowths of such an expanding Western city core, which is believed to be simultaneously cause and beneficiary of economic growth and urban-spatial expansion throughout history.

Thus, the spatial, temporal, and scalar complexities of cities are usually reduced in inside-out theory to the city centre as a distinctive, fixed ecology, whose properties and mechanics are theoretically knowable and empirically traceable (see also Brenner, 2019). Typically, the defining tenets making up an abstracted notion of "the City" *as a whole* are deduced from only *one part* of the urban fabric, namely its overrepresented and ideologically hegemonic core (Wachsmuth, 2014). The densely populated, diverse, and congested historic cores where urban theory was being written – Berlin, Paris, London, New York City, Chicago – became the defining models to generalize and theorize about cities in their worldwide and historical diversity (Edensor and Jayne, 2012). It is within such creative ecologies or creative milieus that Alfred Marshall (1920: 271) famously remarked how innovative ideas – the mysteries of industry – were something to be found "as it were in the air" (for a historic treatment, read Hessler and Zimmermann, 2008). Only in those spatially compact and crowded city centres – accentuated by a dense material build-up and maintenance of markets, institutions, and infrastructures – are agglomeration economies believed to operate. Basically – as inside-out theory goes – one needs city cores for the innate, natural advantages that geographical proximity and spatial clustering provides: it is in the emerging urban-land nexus of city cores that a spillover of ideas and diffusion of knowledge, face-to-face interactions, cooperation, and creative competition can arise from the many and diverse people and industries buzzing and huddling together (Scott and Storper, 2015).

No wonder our Martians from the introduction were only defeated in the City!

Indeed, in order to successfully operate as a theoretical construct, inside-out theory also needs a mirror-pole or spatially denigrated "other": an idea that beyond the central city lies a peripheral wasteland, a marginal nowhere that lacks the necessary structural traits to attract cultural and creative workers and to stimulate innovation, cultural and creative production as such. These are the less crowded and less diverse types of non-central, in-between, or intermediate places – basically our

stereotyped suburbs (Phelps, 2017a). Thus, whereas inner-urban interactions are glamorized as creating civilization, Shearmur and colleagues
humorously remark that interactions between non-central populations
are often stylized in the form of "potatoes in a sack": passively bump-
ing against each other without the necessary spark of energy or ideas
(Shearmur et al., 2016a: 171). Terry Flew (2013: 5) points out:

> The city is viewed as dynamic, edgy, diverse and spontaneous, and hence
> conducive to creativity, whereas suburbs are viewed as static, dull, homo-
> geneous and overly planned, whether by city authorities primarily con-
> cerned with a "safe" environment, or by the developers of master-planned
> estates seeking to realize land values over a medium-term time horizon.

The suburb becomes a blind spot between the dominant binaries of
city-countryside and culture-nature (Kaika, 2005; Angelo, 2017). Suburbia, defined as generic peripheral "nowhere" – a place between city
and countryside, or between culture and nature – appears as devoid of
precisely those qualitative characteristics (aesthetics, excitement, spillovers, and so on) that supposedly make both city and natural life inspiring for cultural and creative production (Webster, 2000: 2). The suburb
becomes a substandard place or peripheral limbo: a zone where rural
life has come to a halt, and the carousel of urban life has yet to begin
(Gibson, 2012: 1–10; Bain, 2013: 30). It is a transit zone between nature
and culture (Ameel et al., 2015: 5–8). Suburbs, and more broadly the
peripheral and interstitial nowhere places between city and countryside (Brighenti, 2013), lend their meaning from everything that negates
the centre: these are places that are thought to lack political authority,
economic dynamism, and the sort of cultural creativity that defines the
historic core.

Most of our spatial and historical understanding of suburbia is, in
one way or another, indebted to such flat, contrapositional framing of
inside-out theory. It is the academic equivalent of what recently has been
exposed as suburban myth, meaning a stereotypical historical depiction of these spaces as white, conformist, planned, residential, mainly
middle-class; sustained by transportation lines starting in the central
city; and populated by distinction-seekers who have grown tired of the
obnoxious smells, sights, and sounds of city life but are basically still
earning their money in the urban historical core (Ewen, 2016: 35–44). To
the extent that this model is based on concrete Anglo-Saxon empirical
trajectories, it is hugely incomplete, ignoring non-English peripheral
places (*faubourg, banlieu, Stadtrand, borgata, favela, jiaqu, buitenwijk*, and
so on), shaped by different historical evolutions and varying social and

cultural contexts (Harris and Vorms, 2017). And to the extent that it is only based on recent suburban studies, this model reduces non-central people and places to the role of passive onlookers, an unchanging ahistorical and a-spatial decor, internally not adjusting, and defined in opposition to an inner core that is believed to be constantly experiencing inner turmoil, interaction, and historical adaptation (McManus and Ethington, 2007; Van Damme and Oosterlynck, 2021).

For Marxist-inspired thinkers, the reasons for this supposed subordination and passivity of non-central zones are obvious: it is the same meaningful historic centre that is holding down the peripheral outsides and designating them as marginal according to hegemonic categories (Shields, 1991: 3–11; Merriman, 1991). Henri Lefebvre, for instance, sees the dualism of centre/periphery as one of rich versus poor. He links it to the same urban theoretical notion of the traditional city (clearly Paris), where socially and culturally deprived outsiders in the *banlieues* have to fight for the right to the city's centrality, meaning access to the concentration of political and economic power, culture, and education in the historical inner-city core (Marchal and Stébé, 2014: 117). Ironically enough, it is also Henri Lefebvre whose work is often referenced as being foundational in forging a paradigmatic shift in thinking the urban from the outside in (Keil, 2018). After all, it was Lefebvre (2003 [1970]: 14) who started to look more carefully at the neglected and denigrated outsides or "numerous disjunct fragments (peripheries, suburbs, vacation homes, satellite towns) into space." Lefebvre (2014 [1989]), by pointing out how a series of urban implosions/explosions – a veritable urban revolution – was transforming the traditional city as bounded and compact ecology into something much more amorphous and totalizing, shifted theoretical focus from the city to the process of planetary sub/urbanization of society – a notion that recently has been enthusiastically embraced by theorists such as Neil Brenner (2014) and Roger Keil (2017).

Currently, the multiple splintering of the urban reality into sprawling limbs and lobes makes it more and more difficult to retain the classic city/periphery binaries and to think of the suburb as "nowhere" (Graham and Marvin, 2001; Gandy, 2011; Keil, 2013). The increasing sub/urbanization of what was once seen as the periphery of a historic core makes urban theorists question the notion of centrality head on. The idea that city centres can be marginal, while the margins or peripheral "nowhere" hold essential characteristics of centrality, has gained broader acceptance (Keil, 2018; Brenner, 2016, 2019). The older idea that centre and periphery are relative in time (with suburbs turning into inner-city cores, for instance) is now also expressed in spatial terms.

Instead of a fixed vertical urban hierarchy, a notion of horizontal relatedness and "central flow theory" between unlocated and ever shifting centres and peripheries is gradually emerging (Taylor et al., 2010). Indeed, geographer Doreen Massey was at the vanguard in arguing for the complexity of cities through space and time. Rather than bounded entities, she emphasized the permeability of urban spaces (Massey et al., 1999).

To Roger Keil (2017, 2018), most prominent, contemporary urbanization in its extended, generalized form is mainly conceptualized as suburbanization. Suburbs have to be reappraised as "centers of work and industry, and not simply centers of lifestyle and consumption" (Flew et al., 2012: 200), as traditional reading determined. Discovering what role cultural and creative practices played in "infinite suburbia" (Gibson et al., 2017) is the logical next step to understand their continuous evolution, existence, and growing importance for the present. Thus, the suburban in-betweens from times past have become increasingly recognized as existing globally in all different shapes, constellations, names, and complexities (Clapson and Hutchison, 2010; Keil, 2013; Harris and Vorms, 2017). However, whereas this new revisionist suburban research has been especially influential in undermining established inside-out distinctions and turning away from mainly Anglo-Saxon homogenizing treatments of suburbia in the world, it still lacks proper historical footing. A concern with the empirical diversity of suburbs and processes of suburbanization on a worldwide scale – as well as with the growing internal and spatial differentiation of suburbs and suburban living – has been mainly applied to a present-day global research agenda. Despite the insightful and much-needed call by McManus and Ethington (2007) to compare suburbs, not only through space in the present day but also *through time* as ever-changing "interactive ecologies," such an empirical-historical, case-based approach has still to merge with urban theory (Van Damme and Oosterlynck, 2021). In order to understand suburbia, we cannot simply look at the present and adjust our theories accordingly; we also have to look back and realize how suburbs functioned differently from how they are commonly understood in urban histories.

The need to bring together a contemporary spatial perspective on the one hand, and a historical research approach on the other, has been precisely one of the goals of this book. The volume juxtaposes and compares cases on cultural and creative practices in suburban "nowhere" from different historical periods and spatial contexts, bringing together work that ranges from the nineteenth century to the present day, and from Europe to Australia. Only by doing so can we hope to break open rigid notions of what a suburb is or was and to juxtapose what their inhabitants ought to be doing (according to dominant suburban stereotypes and myths) with concrete historical realities. So, what is there to learn by looking at suburban creativity in diverse historical and regional suburban settings?

Lost and Found: Creativity and the Sub/urban

This book aims to re-evaluate suburban creativity in its historic and regional complexity. Such endeavour, logically, should start from previous research that consciously seeks to expose and tear down the often-unacknowledged methodological cityism in much of the dominant creative city script – the specific tradition of inside-out theory discussed earlier that negates creativity and innovation happening in its worldwide and historically extended sub/urban forms. Recent research on suburban creativity is taking place in different contexts and from different disciplines, although Australian and Canadian geographers have been particularly prominent in these discussions. It might be interesting to speculate how the relatively comparable geography of these countries – characterized by remoteness, rurality, and a mixture of extensive suburban living conditions, as well as their political history as British dominions – plays a role in this growing urgency to move beyond inner cities and shift attention to activities and processes happening in suburban, exurban, small, rural, remote, and other non-central, intermediate, or in-between places. More relevant to our goals, however, is to analyse how such an emerging body of literature can open up classic understandings of creativity and the city.

A crucial starting point in much of this recent research relates to a decoupling of the notion of creativity and innovation from its reductionist market-capitalistic meanings. This approach requires a conceptual shift in questioning why some forms of cultural and creative practice are perceived as being creative (and by whom?) and explaining why more mundane, vernacular, or unspectacular activities and processes are not. Australian geographer Louise Johnson (2012), for instance, in a special issue of the *International Journal of Cultural Studies* on "Creative Suburbia," explained why the notion of creativity and innovation did not figure prominently in twentieth-century discourses on suburban Australia. On the one hand, its existence was demonized and negatively stereotyped by a complex of male-urban political, cultural, intellectual, and other prominent voices, denying the suburb some of the "Promethean fire" that was believed to shine so brightly in Australia's major centres like Sydney or Melbourne. On the other hand, Johnson pointed to a more complex historical reality in which the perceived safe, dull, and materialistic suburban life appeared to be receptive for both cultural activity (artists' colonies), forms of vernacular creativity by mainly women (including gardening, mending, child educational activities, and the like), as well as, for instance, lesser-known architectural innovations and design experiments. Her insightful and historically attuned analysis shows how perceptions about suburbia, and widespread notions about what people are supposed to be doing there, always need to start from a deconstruction: one that uncovers our common understandings of innovation and creativity as coloured not only by inner-urban bias but often by a male gaze and restricting focus on the economic value of things – an important point to which we will return in the last section of this introductory chapter.

Understanding better the creative and innovative potential of suburban activities and processes is a start: suburbs are not solely the stereotyped places of consumption and lifestyles but are equally places of work and industry. But why would people decide to locate to suburban nowhere in the first place? Why would they value such living and working contexts above the much mythologized and talked about city centres? The most widely cited "push" explanations are given by economist Nicholas Phelps (2012). He argued that the suburban belt of London has always been susceptible to creativity throughout history, for instance, by making room for innovative industries and craftsmanship that cannot pay the rental prices or find the acquired space in the inner city – an argument that can be confirmed for a lot of other European and North American cities, both in the past and the present.

Explanations about agglomeration costs and benefits, however, are not sufficient, and one needs to dig deeper, especially when considering cultural and creative practices. Moving to, and working from, suburban nowhere is not only a necessity for creative workers; it can also be a choice, while simultaneously remaining connected to overarching political-economic geographies – depending on the type of activities one is considering. Whereas the spatial concentration of certain creative and innovative business in an inner city (or precisely a lack of it) might well be connected to political-economic decision-making to prioritize and subordinate certain places in the wider urban fabric, the suburban location might simultaneously also be conducive and stimulating to other types of activities. Work on artists' colonies, for instance, has made clear how groups of creatives moved out of the city to distance themselves from stiffening academic training and institutionalism in the main artistic centres of their time, such as Paris, Brussels, London, Dublin, Copenhagen, New York, and the like (Jacobs, 1985; Lübbren, 2001). On a more individual level, people might or might not have consciously chosen to live in suburbia, but their particular living environment can set them on the track of, for instance, do-it-yourself (DIY) forms of creativity, gardening, and so on, activities that are less called for – or even impractical or impossible – in crowded, inner cities (Hawkins, 2016: 241–67).

One dominant refrain within much of the creativity city literature relates to the notion that creative practitioners need the buzz of big inner cities, since it is in those densely built and populated city centres that Jane Jacobs–like knowledge spillovers and proximity effects tend to occur. Moreover, core cities are considered to be important, since they tend to cluster all sorts of cultural and creative amenities, from museums, to creative quarters, and to informal and much talked about meeting spots, like coffee bars, galleries, warehouses, factories, waterfronts, and so on. Finally, it is routinely assumed that inner cities are more inclusive and tolerant, and hence more susceptible to ethnic and sexual diversity and the sort of bohemian lifestyles that are unreflexively assigned to people active within cultural and creative production. Famous in this aspect are, for instance, the much quoted "gay," "bohemian," and "melting pot" indices of Richard Florida (2002a), which, according to him, prove his point about creativity and innovation correlating with inner-city diversity.

The problem with these stylized assumptions, however, is that they do not live up to both theoretical and empirical research on the spatiality of cultural and creative production that has been going on in the last ten years (primarily on present-day contexts). The most extensive discussions about the important notion of proximity can be found in the work of Canadian and French geographers Richard Shearmur, Christophe Carrincazeaux, and David Doloreux. In their *Handbook on the Geographies of Innovation* (Shearmur et al., 2016b), they basically kick away the scaffolding that has been holding up many of the theoretical constructs behind creative city/cluster thinking. What follows from their extensive discussions on the subject is that, clearly, geographies of innovation and creativity are by no means limited to the idea that agglomeration, information spillovers, localized knowledge bases, and urban diversity are the structuring factors underpinning them. Cultural and creative activities do not necessarily cluster together, and if they do, co-location and spatial concentration in (major) city centres is no requirement either. Proximities, then, can exist in many forms and formats, and can perfectly, and paradoxically enough, exist at a distance (Shearmur, 2012). Creative outposts have, for instance, been detected and analysed for such diverse contexts as the Canadian North and Arctic regions (Petrov, 2007, 2008), rural and coastal England (Bell and

Jayne, 2010; Brouder, 2012; Harvey et al., 2012), and remote and periph-eral outback Australia (Gibson, 2010, 2012).[1] Alison Bain (2016, 2018) and research projects led by Terry Flew (Flew et al., 2012) and Mark Gibson (2012) have specifically focused on cultural and creative production in, respectively, outer suburban Canada and Australia.

Some of these recent empirical findings on present-day suburban contexts have strong parallels with each other, which is probably also because they share a similar ethnographically inspired methodology and research set-up based on in-depth interviews with suburban work-ers. They give a good insight into why people are pulled (rather then pushed) to all sorts of marginal or peripheral suburban "out there's," understood both in a physical sense – as geographical reality – and as a state of mind or imaginary (Hawkins, 2016: 241). At the same time, they show how the city centre precisely matters, not in the stylized way of much creative city literature but more as a centripetal force, which simultaneously controls and shapes suburban narratives and sets out political-economic policies and processes that have a direct influence on cultural and creative activities in the wider metropolitan area.

The relative dislocation of creative workers in suburban Canadian contexts, according to Bain (2013), posed both challenges and oppor-tunities in comparison to the more visible and studied creative classes of inner cities. Disadvantages were related, among other factors, to the lack of cultural institutions and infrastructures, and/or the absence of government-led investments, structurally supporting creative initia-tives and processes taking place in suburbs. Forms of peer network-ing, face-to-face feedback, interconnectivity, and public visibility and acknowledgement common to inner-urban cores were mostly absent, while suburban locations in Canada – similar to Australia – often had to overcome stigma and the entangled centripetal force and attraction of nearby central cities. Yet, Bain also made apparent why non-central, in-between, and intermediate landscapes could equally be particu-larly valued by cultural and creative practitioners. These assets were connected to the same sort of advantages that the traditional aspir-ing middle classes have always assigned to suburbia: "homes that are affordable and spacious in family-oriented neighbourhoods that value closeness to nature, domestic convenience, freedom, and privacy" (Bain, 2013: 215). More directly connected to the process of creativ-ity and innovation, however, was the labelling of suburban spaces as "emancipatory." Paradoxically, remoteness from cut-throat competi-tion and financial and institutional support in inner cities, and relative invisibility or disrespect of the critical eye of the urban commentariat, made suburban contexts ideal for under-the-radar experimentation and piecemeal, often DIY, adjustments and retrofitting of suburban living environments. In this sense, Bain has made a plea for reappraising cultural and creative workers as important "place-makers" of suburbia, essential for the vitality and sustainability of the broader metropolitan regions and in dire need of more recognition and government support.

The research of Gibson and colleagues brought similar results and policy recommendations to the fore (Gibson et al., 2017: 356). They turn the classic Florida "cookbook" almost on its head by starting from a description of the sort of "negative externalities" connected to central cities, rather than repeating again their usual trump cards for creative and cultural production. Thus, whereas generations of scholars have believed the crowded inner city to be a simulating environment, such sensory bombardment was seen by suburban creatives mainly as a distraction: time and energy needed to be spent on following all the latest cultural events, social networks, and places of interconnectivity, which could have been more wisely used for developing one's own creative practice. Similarly, whereas the inner city has been frequently heralded as being crucial to stay in tune, suburban creatives also thought the hipster qualities of such inner-urban cliques to be constraining in their own right, imposing pressures to conform and clipping individual freedom and originality – essential assets for cultural and creative production. Moreover, whereas the role of diversity as such could not be excluded as being beneficial to foster creativity and innovation, Gibson and colleagues indicated that Australia's inner cities could hardly still be labelled as diverse and inclusive. Rather, "the inner city has become characterized by high-end consumption and retail outlets, and a predominantly professional population, making it relatively culturally homogeneous" (Gibson et al., 2017: 357). Thus, the charges of a stiffening, overly gentrified, and materialist environment that have been so widely made against the suburbs are now being connected to the inner city. What a reversal of positions!

Instead, suburban homesteads in Australia were embraced for their frontier mentality, which expressed not only a romantic yearning for nature and countryside living but also one that captures a notion of exploration and day-by-day inventiveness and vernacular creativity. Similar to the research in Canada, however, central cities were seen as being successful in binding amenities, support, promotion, and other policies and strategies to their own activities and practices, eventually excluding the wider metropolitan region from making any significant contribution to cultural and creative production. Local suburban politicians and economic stakeholders often failed to recognize the specific needs and demands of suburban creative workers, for instance, in setting up alternative forms of interconnectivity and networking or in dealing with problems related to stigma and bad image, often connected to these localities.

While much of this present-day research has effectively turned urban theory outside in and has shifted empirical research from the centre to the Lefebvrian "disjunct fragments" – being essentially much more complex than previously assumed – the question remains as to what a historical and more broadly comparative perspective can add to these insights. A response to this question is precisely where this volume wants to make an original contribution to recent and ongoing urban theoretical and urban historical debates.

Cultural and Creative Practices from Suburban Nowhere: New Building Blocks for an Interdisciplinary Exploration

As the previous discussion emphasizes, this volume brings together interdisciplinary approaches to an exploration of creativity beyond the city: suburban creativity. In these explorations of creativity, we aim to broaden, first, our understanding of the nature of creativity and the work of creatives. Rejecting the narrow focus of the creative city literature – or more broadly speaking, inside-out approaches – we embrace forms of creativity ranging from the small-scale, often unseen, vernacular or unrecognized work of individuals to better-known larger scale creativity in the form of craftsmanship, artistic expressions, and (sub)cultural endeavour. By deliberately casting a wider disciplinary net, we aim to consider and draw into the discussion those aspects of creativity that have been typically edited out of the mainstream, whether because they are seen as lacking in value (monetary or otherwise) or because they do not fit with the rigid notions of creativity as defined by elite, generally male, urban commentators. These broader notions of creativity add a depth and richness to our understanding of creative lives and experiences both in urban areas and beyond.

Second, we deliberately utilize a very loose definition of suburbs. As alluded to earlier, debates around defining suburbs continue to rage, sometimes to the point where they become meaningless (Vaughan et al., 2009; Forsyth, 2019). We accept suburbs as being far more heterogeneous than outmoded theorizations would have us believe. Rather than confining contributions to rigid suburban definitions, however, we have allowed each chapter to explore the concept within their own regional and historical context. In so doing, each chapter will interrogate the meaning and implications of suburban creativity, engendering in the overall volume a rich exploration of the complex interactions that occur in a range of past and contemporary contexts.

Finally, our volume aims to historicize the discussions around creativity in suburban settings. Through dual lenses – both historical and spatial – this book will make a new and original contribution to ongoing debates on cultural and creative practices in suburban settings. With chapters ranging across time and space, allowing for the entanglements and configurations of multiple trajectories and multiple histories, we begin to uncover the manifold ways of seeing and understanding suburban creativity. While recent discussions about suburban creativity have mainly prioritized a contemporary focus on Australia and Canada, this book brings these debates to the nineteenth century and to under-acknowledged suburban contexts in Europe, such as those in Ireland, Belgium, Finland, and Hungary (on suburbanization in Europe, read Phelps, 2017b). Our case-based, in-depth approach in the following chapters facilitates a nuanced dialogue between different disciplines, thereby unpacking complex and historically and spatially diverse experiences to reveal deeper understandings of the theme at hand.

In order to structure such interdisciplinary dialogue, the book is divided into four parts. Part 1 of the book, "Openings," follows with a short essay by geographer David Gilbert, who elaborates on the content of this introductory chapter in several ways. Most importantly, he connects our introduction to the COVID-19 pandemic, insisting that the virus has brought into focus what has always been hiding in plain sight: the creative qualities of suburbia.

Each of the three following parts of this book works across a different but overlapping and complementary scale to approach and discuss the notion of suburban creativity in its full complexity: the "suburban home" (part 2), the "suburban creative milieu" (part 3), and the "suburb" itself, as a newly created spatial-material fragment of urban extension (part 4). Part 2, "The Suburban Home as Locus of Creativity," begins by looking into the most basic micro unit within which creative and cultural practices in suburban areas operate: the home as locus of suburban living, work, and creation. The home, indeed, takes on a central role in any discussion of suburban creativity, since suburbs are always – irrespective of differences across time and space – places that, often initially, become appropriated for residential purposes. Many types of suburban constellations and home dwelling types, of course, exist throughout history, leading Jauhiainen (2013), for instance, to discern between "terraced suburbs," "villa suburbs" with spacious gardens, "industrial and working-class suburbs" – often characterized by tenements, high-rises, and a much denser street layout – "suburban squatting and shanty towns," and so on. Since suburban homes, in general, were created *ex nihilo* on open tracts of land surrounding older historical centres or, in any case, stood in a peripheral relationship to centrally provided goods and services, it follows that these places offered new opportunities and challenges but equally elicited new activities and responses that were impractical, impossible, or uncalled for in the centre.

Thus, literary historian Sarah Bilston argues in chapter 3 that the actual process of physically mastering and appropriating the suburban home in Victorian England became in itself a source of new, genuinely female, vernacular creativity. By unearthing underrepresented and long-dismissed books on late nineteenth-century home design, DIY home decoration, and craftsmanship, she succeeds in reinterpreting suburban homes as spaces of new invention and creative possibilities. While most mainstream, inside-out narratives have focused on the creative flourishes of the male-dominated Arts and Crafts movement, many more suburban women took a keen interest in home decoration as a form of artistic fulfilment – publishing influential but now grossly neglected manuals and advice literature on the subject. By entering the world and practice of one such female author, Jane Ellen Panton, we see how she cautiously navigated contemporary gender norms, while seeking to license women's artistic and creative labour in the suburban home. Going into dialogue with an older body of research aimed at reassessing the role of Victorian women in public life, Sarah Bilston re-evaluates the private life of suburban women "hiding in plain sight." Theirs is a world focused on the privacy of the material home – true – but one that is far less dull, meaningless, and solipsistic than dominant suburban stereotypes have led us to believe.

Home decoration as emerging, creative practice was in itself not new, nor confined to the Victorian suburban home (Hamlett, 2021). However, its suburban articulation was intimately tied to more openly expressed middle-class values by growing groups of middle-class suburbanites, who sought the freedom and emancipatory potential these new spaces seemed to promise. Other rapidly urbanizing parts of continental Europe saw similar activities, as discussed in chapter 4, which shifts focus to the emerging nineteenth- and early twentieth-century suburban peripheries around the old artistic urban cores of the new nation-state and industrial powerhouse, Belgium. Historians Müller and Van Damme bring back to life the creative practice of "home collecting," the assemblage of art, curiosities, and antiquities by private individuals. Initially an aristocratic, male-elitist endeavour, reaching all the way back to humanist models from Renaissance Europe, Belgian suburban homes around 1900 became the *terra nova* or creative battleground from which new generations of private collectors could start to communicate different societal roles, aesthetic and material interests, and cultural values. Once again, women can be found among these new generations of private collectors, eager to develop new ways to creatively stage and interact with their home collections in order to shape alternative subjectivities and identities. The chapter juxtaposes two turn-of-the-century art and antique collectors: one male, Édouard Van den Corput, and one female, Anna Boch, both of whom were living in the immediate suburban surroundings of the capital and industrial city of Brussels. Boch in particular seems to have enjoyed the new possibilities and freedoms afforded by her suburban home to stage a more individual, modern, and democratized form of art collecting, breaking away from established aesthetics and dominant modes of collecting behaviour. Boch's new, more spacious, and convenient suburban home both facilitated and reinvigorated private enjoyment of art, while simultaneously becoming a new private salon of sociability, frequented by befriended artists and like-minded social circles.

Historian Susan Reidy takes the thematic focus on the home as locus of creativity across the oceans to suburban Melbourne, Australia, in chapter 5. Whereas the previous chapters explored vernacular creativity and shifting modes of cultural consumption as creative expressions of a rising middle-class sensibility, Reidy focuses on the suburban home as a potential site of creative and cultural production. Following the lives and work of successive generations of Australian artists taking up residence in Fairy Hills, a suburb on the northeastern fringe of metropolitan Melbourne, she revisits their homesteads as spaces of professional artistic practice and home-based studios. What binds these stories of "artists in suburban residence" together is an urge to break free from societal and professional constraints and produce art according to their own "tastes and preferences of practice." Like writer Jane Ellen Panton and artist-collector Anna Boch, the Melbourne artists sought to make their home studios the meeting point of like-minded friends and art lovers, thus fostering bottom-up artistic community ties, clustering, and newly emerging suburban art networks.

Central to these first chapters focused on the suburban home is a methodological eagerness to lift the veil on hegemonic suburban stereotypes and a concern to show what is actually hiding behind the broadly shared suburban disdain and generally dismissive opinions of suburban life and its activities. Moreover, as literary and visual art critic Simon Workman makes clear in chapter 6, widespread pejorative conceptions about suburban culture do not just belong to the past. Through a detailed critical analysis of contemporary Irish art installations, photography, and literary representations, present-day suburban homes are deconstructed as eliciting particularly strong reactions from artists working on universally produced "global suburbanisms" (Keil, 2018). Through their warped artistic lenses, suburban homes of the here and now become the opposite of sedate middle-class ideals, instead being transformed into genuine "ghost estates" – a material expression of conquering "zombie capitalism" (Harman, 2010). According to Irish artists – Workman seems to suggest – a suburban "winter is coming," draining all lifeblood from contemporary societies. Yet, paradoxically enough, these artworks, inspired by and often made within Irish suburban homes, also reveal a wider contemporary reappreciation of present-day, worldwide suburbia as a crucial site conducive to cultural creativity and radical artistic expressions.

Part 3 of this book, "The Suburban Creative Milieu," groups the writings of academics working within related social disciplines including geography, sociology, and spatial ethnography. Whereas the chapters in part 2 are heavily indebted to the more empirically inductive, case-based approaches of urban history and art/literary theory, the chapters assembled in part 3 ask bigger questions, moving beyond the suburban home to analyse how the suburb as urban-spatial complex or milieu can engender creativity and cultural production. Questions in this part of the book relate to three major subthemes, probing an outside-in approach to suburban creativity. These interlocking sets of questions relate, first, to *conditions and location*: When, why, and under what conditions does a suburb or suburban milieu become favoured as a location of cultural and creative activity? Why, for instance, do cultural and creative workers decide to locate and stay in suburban places, and what are the specific benefits/deficiencies of such location? Second, they tackle issues related to *expressions and organization*: What sort of cultural and creative activities are happening outside major city centres, and in what sense are such activities and processes different in expression, organization, and so on? How and in what forms does cultural and creative activity become linked to suburban identities and processes of place-making and why? Third, and finally, they focus on *power and relationships*: What type of power relations are embedded in cultural and creative praxis taking place in suburbs? How can a particular in-between or out-there location be seen as both curse and blessing – often at the same time – influencing expressions and the way these are perceived and commented upon?

In chapter 7, historical geographer Tatiana Debroux unearths the social dynamics behind a relatively unknown artist colony that took up residence in interbellum Uccle, a suburb of Brussels. On the basis of detailed archival and geographical research, she asks why this artist group specifically clustered in this suburb and what they were hoping to find there. In answering these questions, classic economic-geographic reasoning – cultural and creative workers in search of cheaper real-estate in urban peripheries – still holds sway. But one needs to dig deeper than push factors alone for explanations. Without a doubt, suburban creativity was facilitated in the course of the nineteenth and twentieth centuries due to changes in transport technologies and mobility infrastructures (such as tram and rail lines), making a locational preference for disjunct outsides more likely. Moreover, as Debroux convincingly demonstrates, social and cultural explanations also played a role. Uccle rapidly became a creative cluster in itself, with artists clustering close to their potential, wealthy suburban patrons, thus creating new networks of sociability and artistic intercourse very similar to the ones we encountered in part 1. By producing rather sedate landscapes of nostalgia, members of this artist colony benefitted from their less densely urbanized and greener suburban surroundings in Uccle, while pandering to the sort of conservative style and aesthetics in demand with suburban patrons. Within suburban Uccle, these artists were able to free themselves from the avant-garde reputation of Brussels's centre and its powerful metropolitan art critics (the latter reproducing eventually the sort of suburban disdain that led to the neglect of the Uccle group in the first place).

Chapter 8, written by Ruth McManus, perfectly complements the larger findings of Debroux. Both academics share a similar background in historical geography, and McManus shows a similar eagerness to unearth how the socio-economic and physical make-up of suburbia impacts on the nature of creative activity there and the degree to which such forms of cultural and creative practice are accepted (or not) by mainstream metropolitan arbiters of taste. Just as Debroux's chapter can be spatially linked to that of Müller and Van Damme, so the chapter by McManus historicizes the suburban Ireland analysed by Simon Workman. On the basis of several case studies drawn from the suburban hinterland of twentieth-century Dublin, the author of this chapter both diversifies our classic understanding of creativity – comparing practices of vernacular creativity, craftsmanship, and "proper" artistic (literary) production – while simultaneously speculating on the role of the "suburban life-cycle." Is a suburban creative milieu more likely to arise during the initial, emergent phases of a suburban community? The question stands open but is tantalizing enough to trigger further research.

Chapters 9 to 12 by János Kocsis, Johanna Lilius, Margaret Crawford, and Alison Bain, respectively, return the discussion to the present day, but do so for understudied suburban geographies and/or suburban communities. Chapter 9 by urban sociologist János Kocsis is a case in point, analysing the historically complex and multilayered experience of post-socialist Budapest, Hungary, a spatial context and suburban "community-in-the-making," which hardly figures in mainstream Anglo-Saxon research. Macro-statistical evidence is combined with micro-findings, which draw on in-depth interviews conducted over the last twenty years, showing the rapid and disruptive changes taking place in the broader metropolitan region of Budapest. The emergence of aspects of suburban creativity is only one element in this complex of change, but certainly one of the most interesting evolutions. The suburb, here, seems to arise as a creative milieu due to the out-of-centre migration of wealthy creative classes, which the author dubs "gentrification of the suburbs." However, Kocsis makes clear how stylized, Richard Florida–like explanations (relating to the appearance of creative classes, proximity of businesses, diversity in population and activities, openness in values, and the like) and local policies acting upon these assumptions through concrete, physical interventions in the spatial fabric can be seen as hiding deeper and more structural historical changes in explaining the rise of suburban creativity. Equally important has been a growing willingness among local suburban inhabitants and new suburbanites (including a growing percentage of women) to reconnect with local and pre-socialist activities, such as wine production, the organization of farmers' markets, and diverse forms of authentic, home-made craftsmanship. Thus, a grassroots willingness and enthusiasm of existing and new inhabitants to recreate and reinvest in previously neglected and ill-perceived peripheral communities has been crucial to suburban creativity.

In chapter 10, Johanna Lilius goes deeper into the role of urban policies in engendering forms of suburban creativity. Like Kocsis, she does so by examining an understudied spatial context and community, namely the high-rise suburbs of Helsinki, Finland. The current remaking of these formerly despised suburban peripheries has much in common with the Budapest case, with planning policies in the last decade aimed at attracting creative classes and gentrifying this suburban belt. However, as Lilius cogently reminds us, the use of top-down regeneration and government-enforced planning strategies to institutionalize creativity also, paradoxically, runs the risk of destroying endemic, bottom-up forms of suburban creativity (creativity from within). More open-ended and less economically reductionist cultural policy strategies might, eventually, be more conducive to improving a suburb's reputation and creative activities while simultaneously enhancing the self-respect, liveability, and social chances of local inhabitants. On the basis of ethnographically inspired observational research and in-depth interviews, the author of this chapter explores one such project set up by the National Theater in Kontula, a particular problematic suburb of Helsinki, plagued by crime, drug abuse, and social tensions between aging inhabitants and a new, rapidly growing multicultural population.

Architectural theorist and spatial ethnographer Margaret Crawford picks up exactly where Lilius ends. In chapter 11, she delves deeper into one such suburban ethnic community lifting themselves – and the reputation and visibility of their "hood" in broader society – through home-grown forms of creativity and cultural inventiveness. In getting under the skin of a so-called "ethnoburb" in the wider Los Angeles area, Crawford discovers young Asian American creatives using YouTube to – both proudly and humorously – establish their newly arising group identity and the particular qualities of their suburban neighbourhood (centred around Asian food and classic middle-class suburban values). Ironically, with growing media and public attention, emergent forms of local creativity became encapsulated in the tried and tested creative city – strategies of place promotion by officials and business leaders. Feeling trapped in their own suburban milieu or bubble, at least some of these Asian American creatives have located elsewhere, signalling again how strategies of economic-political appropriation of creativity have the potential to start working counter-productively (Gibson and Klocker, 2005; Mould, 2015).

In the last chapter of part 3, Alison Bain continues her pioneering research on suburban creativity in the broader Toronto area. This time she assesses the LGBTQ+ community, a group which has traditionally – and rather stereotypically – been associated with inner-city creativity (Florida, 2002a, 2005). One of the persistent suburban clichés, of course, is thinking of suburbia in terms of heteronormative reproductive futurity. But, as Bain convincingly argues, non-central suburban regions are not solely bastions of male-dominated nuclear families; in terms of sexuality and gender identification, much more is happening behind our overly mediatized white picket fences and crabgrass frontiers. Through socio-geographic analysis and ethnographic fieldwork, Bain uncovers socially significant LGBTQ+ suburbanites and community organizations, adjusting in creative fashion to a formal lack of civic recognition and support. In analysing the strategies of vernacular, non-market-based forms of creativity practised by often informal LGBTQ+ organizations, Bain unearths nothing less than a suburban "counterculture queerness" – different from inner-city gay quarters – which is in dire need of more scholarly attention.

Part 4, "Creating Suburbia," brings the volume to a close, shifting emphasis – innovatively enough – to the planning and making of suburbia as a creative endeavour in itself. Chapter 13, by architectural historians Tom Broes and Michiel Dehaene, delivers a case study that questions the central architects and actors behind the rapid suburbanization of Antwerp, Belgium, at the start of the twentieth century. Antwerp, at the time one of the three most important port cities of Europe, had been struggling to keep pace with rapid population growth around the turn of the century, leading to congestion, unsanitary conditions, and a rapidly deteriorating inner-city housing stock. However, the first decades of the twentieth century became characterized by a wave of massive suburbanization and radical socio-economic change. This shifting historical context gave rise to a new generation of architects and planners, who consciously turned away from the classic, nineteenth-century bourgeois city to rethink and refashion Antwerp's new suburban landscapes. As an innovative breed of spatial agents, working in close cooperation with, among others, landowners and developers, building contractors, and financial and public interests, these architects took a key role in creatively shaping the suburban environments in which they, eventually, ended up living themselves. Thus the broader socio-economic necessity of suburbanization once again triggered new creative practices and solutions that were unneeded or uncalled for in the city centre. And rather than seeing suburbanization as a degenerate or unwanted outgrowth of the urban process, Broes and Dehaene call for a better understanding and appreciation of their making, from creative idea to envisioned social outcome.

Renowned suburban geographer Richard Harris ends on a similar note in his concluding chapter. His is a passionate plea to reinterpret the materialization of the suburbanization process itself as a major historical feat of creativity and ingenuity. In reinterpreting the genesis and historical transformation of North American suburbs, Harris discovers unacknowledged and unsung creativity and inventiveness in the design, production, and adaptation of the built suburban environment. These historical changes have depended on a variety of actors, finding solutions to unprecedented challenges, and genuine inventiveness in the conquering of adverse terrain and landscape. It has not just been the work of engineers, architects, and planners but also that of suburbanites themselves actively trying to reshape in consecutive cycles their own homes and suburban spaces into sought-after (ideal?) living environments – practical and ongoing DIY creativity, an "ongoing example of human adaptability and innovation."

And with these theoretical insights, we invite the reader to explore in the next pages and chapters for themselves what is hidden in suburbia. Go and peek behind those much-maligned suburban façades, houses, and landscapes, which, however, warrant much more academic attention and serious scrutiny. Pondering upon the creativity of cities is not complete without taking its suburbs into account as well. If only the Martians of Woking had known …

NOTE

1 For a related but specific focus on "small cities," read Bell and Jayne (2006), Lorentzen and van Heur (2012), and Waitt and Gibson (2009: 1223–46). The latter article is part of a special issue on processes of regeneration and dislocation in the inner city.

REFERENCES

Aldiss, B. 2005. "Introduction." In H.G. Wells, *The War of the Worlds*. London: Penguin Books, xiii–xxix.
Ameel, L., J. Finch, and M. Salmela. 2015. "Introduction: Peripherality and Literary Urban Studies." In L. Ameel, J. Finch, and M. Salmela, eds., *Literature and the Peripheral City*. Basingstoke, UK: Palgrave Macmillan, 1–17.
Angelo, H. 2017. "From the City Lens toward Urbanization as a Way of Seeing: Country/City Binaries on an Urbanizing Planet." *Urban Studies*, 54(1): 158–78. https://doi.org/10.1177/0042098016629312
Angelo, H., and D. Wachsmuth. 2015. "Urbanizing Urban Political Ecology: A Critique of 'Methodological Cityism.'" *International Journal of Urban and Regional Research*, 39(1): 16–27. https://doi.org/10.1111/1468-2427.12105
Bain, A. 2013. *Creative Margins: Cultural Production in Canadian Suburbs.* Toronto: University of Toronto Press.
Bain, A. 2016. "Suburban Creativity and Innovation." In R. Shearmur, C. Carrincazeaux, and D. Doloreux, eds., *Handbook on the Geographies of Innovation*. Cheltenham, UK: Edward Elgar Publishing, 266–76.
Bain, A. 2018. "Cultural Production in the Suburban Context." In B. Hanlon and T.J. Vicino, eds., *The Routledge Companion to the Suburbs*. New York: Routledge, 23–331.
Bell, D., and M. Jayne, eds. 2006. *Small Cities: Urban Experience beyond the Metropolis*. New York: Routledge.
Bell, D., and M. Jayne. 2010. "The Creative Countryside: Policy and Practice in the UK Rural Cultural Economy." *Journal of Rural Studies*, 26(3): 209–18. https://doi.org/10.1016/j.jrurstud.2010.01.001

Bianchini, F. 2018. "Reflections on the Origins, Interpretations and Development of the Creative City Idea." In I. Van Damme, B. De Munck, and A. Miles, eds., *Cities and Creativity from the Renaissance to the Present*. London: Routledge, 23–42.

Brenner, N. 2014. *Implosions/Explosions: Towards a Study of Planetary Urbanization*. Berlin: Jovis Verlag.

Brenner, N. 2016. *Critique or Urbanization: Selected Essays*. Berlin: Bauverlag.

Brenner, N. 2019. *New Urban Spaces: Urban Theory and the Scale Question*. Oxford: Oxford University Press.

Brighenti, A.M., ed. 2013. *Urban Interstices: The Aesthetics and the Politics of the In-Between*. Abingdon, UK: Ashgate.

Brouder, P. 2012. "Creative Outposts: Tourism's Place in Rural Innovation." *Tourism Planning & Development*, 9(4): 383–96. https://doi.org/10.1080/2156 8316.2012.726254

Caves, R.A. 2000. *Creative Industries: Contracts between Art and Commerce*. Cambridge, MA: Harvard University Press.

Clapson, M. 2016. "The New Suburban History, New Urbanism and the Spaces In-Between." *Urban History*, 43(2): 336–41. https://doi.org/10.1017/S0963926816000067

Clapson, M., and R. Hutchison, eds. 2010. *Suburbanization in Global Society*. Bingley, UK: Emerald Publishing.

Cunningham, G. 2004. "Houses In Between: Navigating Suburbia in Late Victorian Writing." *Victorian Literature and Culture*, 32(2): 421–34. https://doi.org/10.1017/S1060150304000579

Edensor, T., and M. Jayne, eds. 2012. *Urban Theory beyond the West: A World of Cities*. London: Routledge.

Edensor, T., D. Leslie., S. Millington, and N. Rantisi, eds. 2009. *Spaces of Vernacular Creativity: Rethinking the Cultural Economy*. London: Routledge.

Ewen, S. 2016. *What Is Urban History ?* Cambridge: Polity Press.

Flew, T. 2013. *Creative Industries and Urban Development: Creative Cities in the 21st Century*. London: Routledge.

Flew, T., M. Gibson, C. Collis, and E. Felton. 2012. "Creative Suburbia: Cultural Research and Suburban Geographies." *International Journal of Cultural Studies*, 15(3): 199–203. https://doi.org/10.1177/1367877911433755

Florida, R. 2002a. "Bohemia and Economic Geography." *Journal of Economic Geography*, 2(1): 55–71. https://doi.org/10.1093/jeg/2.1.55

Florida, R. 2002b. *The Rise of the Creative Class: And How It's Transforming Work, Leisure, Community and Everyday Life*. New York: Basic Books.

Florida, R. 2005. *Cities and the Creative Class*. New York: Routledge.

Forsyth, A. 2019. "Defining Suburbs." In B. Hanlon and T.J. Vicino, eds., *The Routledge Companion to the Suburbs*. New York: Routledge, 13–28.

Gandy, M. 2011. *Urban Constellations*. Berlin: Jovis Verlag.

Georgiou, D. 2016. "Weaving Patterns in the Suburban Fabric: Carnival Procession Routes, Mapping Place and Experiencing Space on London's Changing Periphery, 1890–1914." In S. Griffiths and A. von Lünen, eds., *Spatial Cultures: Towards a New Social Morphology of Cities Past and Present*. London: Routledge, 95–113.

Gibson, C. 2010. "Guest Editorial – Creative Geographies: Tales from the 'Margins.'" *Australian Geographer*, 41(1): 1–10. https://doi.org/10.1080/00049180903535527

Gibson, C. 2012. "Introduction – Creative Geographies: Tales from the 'Margins.'" In C. Gibson, ed., *Creativity in Peripheral Places: Redefining the Creative Industries*. London: Routledge, 1–10.

Gibson, C., T. Flew, and C. Collins. 2017. "Creative Suburbia: Cultural Innovation in Outer Suburban Australia." In A. Berger, J. Kotkin, and C.B. Guzman, eds., *Infinite Suburbia*. Princeton, NJ: Princeton Architectural Press, 350–8.

Gibson, C., and N. Klocker. 2005. "The 'Cultural Turn' in Australian Regional Economic Development Discourse: Neoliberalising Creativity?" *Geographical Research*, 43(1): 93–102. https://doi.org/10.1111/j.1745-5871.2005.00300.x

Graham, S., and S. Marvin. 2001. *Splintering Urbanism: Networked Infrastructures, Technological Mobilities and the Urban Condition*. London: Routledge.

Hall, P. 1998. *Cities in Civilization: Culture, Innovation and Urban Order*. London: Weidenfeld & Nicholson.

Hamlett, J., ed. 2021. *A Cultural History of the Home in the Age of Empire*. London: Bloomsbury.

Hapgood, L. 2005. *Margins of Desire: The Suburbs in Fiction and Culture 1880–1925*. Manchester: Manchester University Press.

Harman, C. 2010. *Zombie Capitalism: Global Crisis and the Relevance of Marx*. Chicago: Haymarket Books.

Harris, A., and L. Moreno. 2012. *Creative City Limits: Urban Cultural Economy in a New Era of Austerity*. London: University College London.

Harris, R. 2018. "Suburban Stereotypes." In B. Hanlon and T.J. Vicino, eds., *The Routledge Companion to the Suburbs*. New York: Routledge, 29–38.

Harris, R., and C. Vorms, eds. 2017. *What's in a Name? Talking about Urban Peripheries*. Toronto: University of Toronto Press.

Hartley, J., ed. 2005. *Creative Industries*. Oxford: Blackwell.

Harvey, D.C., H. Hawkins, and N.J. Thomas. 2012. "Thinking Creative Clusters beyond the City: People, Places and Networks." *Geoforum*, 43(3): 529–39. https://doi.org/10.1016/j.geoforum.2011.11.010

Hawkins, H. 2016. *Creativity*. London: Routledge.

Hesmondhalgh, D. 2005. *The Cultural Industries*. London: Sage.

Hessler, M., and C. Zimmermann, eds. 2008. *Creative Urban Milieus: Historical Perspectives on Culture, Economy, and the City*. Frankfurt: Campus Verlag.

Howkins, J. 2001. *The Creative Economy: How People Make Money from Ideas.* London: Allen Lane.

Huq, R. 2013. *Making Sense of Suburbia through Popular Culture.* London: Bloomsbury.

Hutton, T. 2009. "The Inner City as Site of Cultural Production Sui Generis: A Review Essay." *Geography Compass*, 3(2): 600–29. https://doi.org/10.1111/j.1749-8198.2008.00201.x

Jacobs, J. 1969. *The Economy of Cities.* Harmondsworth, UK: Penguin.

Jacobs, M. 1985. *The Good and Simple Life: Artist Colonies in Europe and America.* London: Phaedon.

Jauhiainen, H.S. 2013. "Suburbs." In P. Clark, ed., *The Oxford Handbook of Cities in World History.* Oxford: Oxford University Press, 791–808.

Johnson, L.C. 2012. "Creative Suburbs? How Women, Design and Technology Renew Australian Suburbs." *International Journal of Cultural Studies*, 15(3): 217–29. https://doi.org/10.1177/1367877911433744

Kaika, M. 2005. *City of Flows: Modernity, Nature and the City.* New York: Routledge.

Keil, R., ed. 2013. *Suburban Constellations: Governance, Land and Infrastructure in the 21st Century.* Berlin: Jovis Verlag.

Keil, R. 2017. *Suburban Planet: Making the World Urban from the Outside In.* Cambridge: Polity Press.

Keil, R. 2018. "Extended Urbanization, 'Disjunct Fragments' and Global Suburbanisms." *Environment and Planning D: Society and Space*, 36(3): 494–511. https://doi.org/10.1177/0263775817749594

Kunstler, J.H. 1993. *The Geography of Nowhere: The Rise and Decline of America's Man-Made Landscape.* New York: Simon and Schuster.

Landry, C. 2000. *The Creative City: A Toolkit for Urban Innovators.* London: Earthscan.

Lefebvre, H. 2003 [1970]. *The Urban Revolution.* Minneapolis: University of Minnesota Press.

Lefebvre, H. 2014 [1989]. "Dissolving City, Planetary Metamorphosis." *Environment and Planning D: Society and Space*, 32(2): 203–5. https://doi.org/10.1068/d3202tra

Lorentzen, A., and B. van Heur, eds. 2012. *Cultural Political Economy of Small Cities.* New York: Routledge.

Lübbren, N. 2001. *Artists' Colonies in Europe 1870–1910.* Manchester: Manchester University Press.

Marchal, H., and J.-M. Stébé. 2014. "From the City to Crumbling Urbanism: Beyond Centre/Periphery Dualism. A Re-Examination of Henri Lefebvre's Concept of Centrality." In G. Erdi-Lelandais, ed., *Understanding the City: Henri Lefebvre and Urban Studies.* Cambridge: Cambridge University Press, 117–38.

Marshall, A. 1920. *Principles of Economics.* 8th ed. London: Macmillan.

Massey, D., J. Allen, and S. Pile, eds. 1999. *City Worlds*. London: Routledge/ Open University.

McLean, H. 2014. "Digging into the Creative City: A Feminist Critique." *Antipode*, 46(3): 669–90. https://doi.org/10.1111/anti.12078

McManus, R., and P.J. Ethington. 2007. "Suburbs in Transition: New Approaches to Suburban History." *Urban History*, 34(2): 317–37. https://doi.org/10.1017/S096392680700466X

Merriman, J.M. 1991. *The Margins of City Life: Explorations on the French Urban Frontier, 1815–1851*. New York: Oxford University Press.

Miller, T., and G. Yúdice. 2000. *Cultural Policy*. London: Sage.

Mould, O. 2015. *Urban Subversion and the Creative City*. London: Routledge.

Mumford, L. 1961. *The City in History: Its Origins, Its Transformations, and Its Prospects*. Harmondsworth, UK: Penguin.

Park, R.E., and E.W. Burgess. 1925. *The City: Suggestions for Investigation of Human Behavior in the Urban Environment*. Chicago: Chicago University Press.

Parker, B. 2008. "Beyond the Class Act: Gender and Race in the 'Creative City' Discourse." In J.N. DeSena, ed., *Gender in an Urban World*. Vol. 9 of *Research in Urban Sociology*. Bingley, UK: Emerald Publishing, 201–32.

Peck, J. 2005. "Struggling with the Creative Class." *International Journal of Urban and Regional Research*, 29(4): 740–70. https://doi.org/10.1111/j.1468-2427.2005.00620.x

Petrov, A.N. 2007. "A Look beyond Metropolis: Exploring Creative Class in the Canadian Periphery." *Canadian Journal of Regional Science*, 30(3): 451–74. https://idjs.ca/images/rcsr/archives/V30N3-PETROV.pdf

Petrov, A.N. 2008. "Talent in the Cold? Creative Capital and the Economic Future of the Canadian North." *Arctic*, 61(2): 162–76. https://doi.org/10.14430/arctic15

Phelps, N.A. 2012. "The Sub-Creative Economy of the Suburbs in Question." *International Journal of Cultural Studies*, 15(3): 259–71. https://doi.org/10.1177/1367877911433748

Phelps, N.A. 2017a. *Interplaces: An Economic Geography of the Inter-Urban and International Economies*. Oxford: Oxford University Press.

Phelps, N.A. 2017b. *Old Europe, New Suburbanization ? Governance, Land, and Infrastructure in European Suburbanization*. Toronto: University of Toronto Press.

Pratt, A. 2011. "The Cultural Contradictions of the Creative City." *City, Culture and Society*, 2(3): 123–30. https://doi.org/10.1016/j.ccs.2011.08.002

Reckwitz, A. 2017. *The Invention of Creativity: Modern Society and the Culture of the New*. Translated by Steven Black. Cambridge: Polity Press.

Scott, A.J. 2000. *The Cultural Economy of Cities: Essays on the Geography of Image-Producing Industries*. London: Sage.

Scott, A.J. 2014. "Beyond the Creative City: Cognitive-Cultural Capitalism and the New Urbanism." *Regional Studies*, 48(4): 565–78. https://doi.org/10.1080/00343404.2014.891010

Scott, A.J., and M. Storper. 2015. "The Nature of Cities: The Scope and Limits of Urban Theory." *International Journal of Urban and Regional Research*, 39(1): 1–15. https://doi.org/10.1111/1468-2427.12134

Shearmur, R. 2012. "Are Cities the Font of Innovation? A Critical Review of the Literature on Cities and Innovation." *Cities*, 29(S2): S9–S18. https://doi.org/10.1016/j.cities.2012.06.008

Shearmur, R., C. Carrincazeaux, and D. Doloreux. 2016a. "Cities, Innovation and Creativity." In R. Shearmur, C. Carrincazeaux, and D. Doloreux, eds., *Handbook on the Geographies of Innovation*. Cheltenham, UK: Edward Elgar Publishing, 171–3.

Shearmur, R., C. Carrincazeaux, and D. Doloreux, eds. 2016b. *Handbook on the Geographies of Innovation*. Cheltenham, UK: Edward Elgar Publishing.

Shields, R. 1991. *Places on the Margin: Alternative Geographies of Modernity*. London: Routledge.

Taylor, P.J., M. Hoyler, and R. Verbruggen. 2010. "External Urban Relational Process: Introducing Central Flow Theory to Complement Central Place Theory." *Urban Studies*, 47(13): 2803–18. https://doi.org/10.1177/0042098010377367

Van Damme, I., and B. De Munck. 2018. "Cities of a Lesser God: Opening the Black Box of Creative Cities and Their Agency." In I. Van Damme, B. De Munck, and A. Miles, eds., *Cities and Creativity from the Renaissance to the Present*. London: Routledge, 3–22.

Van Damme, I., and S. Oosterlynck. 2021. "Seeing through the Darkness of Future Past: "After Suburbia" from a Historical Perspective." In R. Keil and F. Wu, eds., *After Suburbia: Urbanization on the Planet's Periphery*. Toronto: University of Toronto Press.

Vaughan, L., S. Griffiths, M. Haklay, and C.E. Jones. 2009. "Do the Suburbs Exist? Discovering Complexity and Specificity in the Suburban Built Form." *Transactions of the Institute of British Geographers*, 34(4): 475–88. https://doi.org/10.1111/j.1475-5661.2009.00358.x

Wachsmuth, D. 2014. "City as Ideology: Reconciling the Explosion of the City Form with the Tenacity of the City Concept." *Environment and Planning D: Society and Space*, 32(1): 75–90. https://doi.org/10.1068/d21911

Waitt, G., and C. Gibson. 2009. "Creative Small Cities: Rethinking the Creative Economy in Place." *Urban Studies*, 46(5–6): 1223–46. https://doi.org/10.1177/0042098009103862

Weber, M. 1978 [1920–1]. *Economy and Society: An Outline of Interpretive Sociology*. Edited by G. Roth and C. Wittich. Berkeley: University of California Press, 1212–1372.

Webster, R., ed. 2000. *Expanding Suburbia: Reviewing Suburban Narratives*. New York: Berghahn.

Wells, H.G. 1934. *Experiment in Autobiography*. 2 vols. London: Gollancz.

Wells, H.G. 2005 [1898]. *The War of the Worlds*. London: Penguin Books.

Wilson, D., and Keil, R. 2008. "The Real Creative Class." *Social & Cultural Geography*, 9(8): 841–7. https://doi.org/10.1080/14649360802441473

2 The Uncool Hunt: Searching for the Creative Suburb

DAVID GILBERT

Perhaps it took a global pandemic for the popular imagination to catch up with what some of us had been saying about suburbs for years. As cities across the world ground to a halt in early 2020, there was a boom in writing about "the death of the city." The pandemic was being seen as a turning point for an "inside-out" model of economies and cultures that had put cities, and particularly the seemingly unstoppable great global cities, as the centres of dynamism and innovation. In a paean to those great cities, and for the urgency of their recovery, a leader article in *The Economist* couldn't resist adding yet another entry to a catalogue of suburban disparagement as old as the modern city: vibrant and successful cities that "cram together talented people who are fizzing with ideas" were contrasted with "the joy of suburban life" that "derives from the houses and gardens that are more affordable there" (*Economist*, 2020). But elsewhere commentators were rapidly discovering that the world beyond the metropolitan might have more to offer than the supposed passivity of domestic life and horticulture. Buried in endless columns and blog posts targeted at urban professionals making sense of their new lockdown lives – on working from home, on Zoom dress codes, on balancing work and home schooling – has been a broader and more profound re-evaluation of certain kinds of suburban spaces and their creative potential.

As this collection demonstrates, the creative qualities of suburbia have been hiding in plain sight across time and cultures, and have a history that is far richer and longer than the 2020 "discovery" of the potential of the home office or the inspiring qualities of a jog around a local park (and guilty as charged – I am indeed writing this chapter during the third UK lockdown in a home office in a west London suburb after my legally permitted local morning run). But before leaving the obsessions of a relatively privileged group in particular kinds of cities and suburbs, I want to make one further dip into that response to the pandemic. The pattern is not completely consistent, but in many major cities affected by the pandemic there has been a financial as well as a cultural re-evaluation of the suburbs. It, of course, has been primarily about safety and space in the face of a potentially deadly infection and restrictions on movement. But it might also reflect that some of the most stubborn cultural associations with suburbia are changing. The property pages of *The Times* in August 2020 identified a surge in suburban house prices around London but also commented on suburbia's new "coolness," for as "city centres lie silent, it is now where the young aspire to be" (Lewis, 2020).

This remark might have been just a throwaway line in a throwaway property supplement, but the mention of coolness and young aspirations hits a nerve for those of us interested in the creativity of suburbia. The idea of coolness, seemingly hard to define, is a key cultural element of inside-out thinking about cities and suburbs, a kind of central place theory of the imagination (see chapter 1). That sense of the cultural superiority of the urban has a particularly blunt expression in the celebration of those crammed-together talented people fizzing with ideas, the so-called creative classes in their creative cities lauded by a certain school of urban boosterism (Florida, 2002; Landry, 2008 [2000]). But this creative superiority of the city is much more long-standing and engrained. Its apotheosis comes perhaps in the veneration of Paris as, in Walter Benjamin's term, "the capital of the nineteenth century" (Benjamin, 1999). The very invention of modernity becomes an urban phenomenon bound up with concentrated, central changes in power, consumption, and culture. "Cool" before the term, a special place is given to the Parisian poets of urban modernity; for Benjamin, it is Baudelaire's creativity that unlocks the paradoxes and contradictions of modern life, a creativity that is intimately bound to the practices of being in a city of crowds, spectacle, and commodity capitalism (Benjamin, 2006).

Writing in *The New Yorker* in 1997, cultural critic Malcolm Gladwell coined the term "coolhunt." He tracked the ways that multinational sports shoe companies sent "coolhunters" out to journey into Harlem and the Bronx, "giving chase to the elusive prey of street cool" (Gladwell, 1997: 78). This practice was a particularly late twentieth-century version of that long-running connection of urban authenticity and commodity capitalism, with a view that determining "cool" could only come from "certain kids in certain places" – certain places that mass-market consumers would understand as the streets and "hoods" of the great cities. Gladwell discusses the circularity of the process, particularly the way Nike and Reebok started to seed new experimental limited-run versions of shoes to just those "certain kids in certain places." Gladwell's essay points to the double-sided and racialized relationship between fashion culture (and popular culture more widely) and inner-city life in the United States; built into the coolhunt are both fears and romanticization of African American and Latinx urban cultures as somehow more real, more dangerous, more cutting-edge than mundane lives in places beyond the city.

We can adopt and expand the idea of the coolhunt to think about other kinds of journeys in search of urban cool. One of the most common tropes in literature and biography is the journey from village, small town, or suburb to the city. Such journeys are also coolhunts, not just a search for the bright lights and excitement of the big city but also a creative remaking of self in an environment of possibilities, happenstance, and liberating anonymity. In Hanif Kureishi's classic semi-autobiographical account of the move from suburb to city, *The Buddha of Suburbia*, the central character, Karim, expresses the change to his sense of self and his creative potential brought about by a move from suburban Bromley to central London:

> So this was London at last, and nothing gave me more pleasure than strolling around my new possession all day. London seemed like a house with five thousand rooms, all different; the kick was to work out how they connected, and eventually walk through all of them. (Kureishi, 1990: 126)

However, if we look more closely at these coolhunting narratives, we often start to see something different that destabilizes these familiar tropes. This perspective thinks about the spatiality of creativity differently, undercutting the supposed cultural superiority of the city. One of my favourite examples comes in the 2008 documentary *Beautiful Losers*. At first sight, this film is an archetypal coolhunt journey. A disparate group of artists working in graffiti, punk, and skateboard subcultures is picked up and pulled together by the Alleged Gallery in New York's Lower East Side. The film contrasts the group's suburban roots with the hipness of the city, and a key scene in the documentary is of their arrival in the streetscapes of Manhattan. Aaron Rose, curator of the Alleged Gallery's *Beautiful Losers* exhibition, was one of the directors of the film, and the story it tells is one of discovery, fame, and, for some of the artists, significant commercial success and uptake by corporate marketing. Yet what the film unintentionally also does is draw attention to the intimate and creative relationship between suburban landscapes and the art of Margaret Kilgallen, Barry McGee, Ed Templeton, and the other beautiful losers. Read differently, the story becomes one about the creative possibilities of everyday marginal spaces of skateboard parks, scrubland plots, and domestic garages – and also a story of the perils of co-option into a metropolitan scene.

There is a version to these stories, which involves a return to the place supposedly escaped from and recognition of its previously unacknowledged importance, that we might also think of as an "uncoolhunt." The musician Tracey Thorn (Everything but the Girl, Massive Attack) in her memoir *Another Planet: A Teenager in Suburbia* tells of growing up in Brookman's Park, an outer commuter suburb sitting "in a sea of green just off the coast of London" (Thorn, 2019: 9). In some ways, it's a conventional coolhunt journey: "I'm not the only person to have grown up stifled and bored in suburbia; it's almost the law" (96). She writes of the growing distance between her parents and her teenage self in terms of competing cultural geographies of city and suburb: "Did they sense something in me breaking away, turning my back on them? Youth culture, tribalism, music, creativity, all of this was a kind of modern, urban misbehaviour, and more alarming to them than pubs, snogging older boys, or cars on country lanes" (140). And yet, reflecting later while walking around the landscape of those teenage days, she writes of her "suburban bones" and the complexities of her relationship with that past: "I feel terribly at home here, and terribly out of place" (204).

This sense of an ambivalent relationship with an edge place and a life of creativity is beautifully expressed in a recent essay by poet and novelist Lavinia Greenlaw. She talks of her ongoing refusal to engage with "the landscape that has been most formative to me … one that I don't remember looking at" (Greenlaw, 2021: 32): the landscape of 1970s Essex, a place that "was blurring into the city while refusing to become a suburb." Again, the supposed constrictions on creativity are expressed geographically, emphasizing the physical and creative flatness of the edgelands of Essex: "I wanted to be in a place that met me with drama. This meant the city and if not the city then at least a steep hill, high cliff or ruined cathedral. Essex didn't offer such easy charms" (32). But this between-scape leaves its creative mark, its plainness demanding attention to detail, reflected in the poet's craft. Greenlaw's ambivalence becomes bound up with the creative potential of resistance and refusal: "The Essex I am describing is one I remember and invented. It's also one I absorbed far more than I knew. It went in deeply because of my resistance to it – as well as its resistance to me" (36–7).

The narratives of escape, but the reminder of the suburban (and edgeland) conditions of early creative development, have become one of the themes of the approach of what has sometimes been described as the "new suburban studies" (although now not particularly new) (Gilbert, 2004: 444). These approaches seek to challenge and complicate deep-seated tropes about the suburbs as dull, conservative, inward-looking, and distinctly uncreative. In English new suburban studies, the London suburb of Bromley often stands for a much wider phenomenon, the "most significant suburb in British pop history" (Frith, 1996: 271). Bromley produced not only David Bowie in the late 1960s but also Souxsie Sioux, Billy Idol, and the wider "Bromley contingent" in the British punk wave. As Rupa Huq suggests, this connection between suburbia and popular music is found across genres, periods, different national contexts, and different forms of suburbia (Huq, 2013). Of course, escape, rebellion, and rejection are key elements of much of this music, but so too is a fine-grained attention to detail and the nuances of class, identity, and geography, a sensibility often transplanted into a suburban gaze on the city.

These forms of creativity can be seen as a particular form of uncool-hunt, a recognition of the creative importance of the journey from suburb to city, of the lasting echoes of sensibilities acquired in suburban landscapes, but also of the creative significance of the kind of continuing double refusal suggested by Greenlaw. However, this collection, with its historical and geographical range, can be taken as a broader kind of uncoolhunt that seeks to destabilize the geographies of cool more radically. It is not just about recognizing the hidden suburban histories in the creative careers of those who left to fizz in the big city. Instead, I think we can suggest a number of ways in which this uncoolhunt proceeds.

First, as the introduction suggests, this uncoolhunt broadens and reshapes our understanding of the nature of creativity. Much recent work has challenged the co-option of creativity as the primary driver of economic value and profit in post-industrial neoliberalism. In a recent polemic *Against Creativity*, the geographer Oli Mould (2018) suggests that the term "creativity" is now so comprehensively debased by this discourse that we should abandon its use. By contrast, Harriet Hawkins points to multiple forms of creativity, suggesting a whole range of understandings of the term beyond the narrow confines of the creative economy, indicating its significance as a "psychological trait and philosophical concept" and suggesting that it is also "an embodied, material and social practice that produces both highly specialist cultural goods and is a part of everyday life, and it offers myriad possibilities for making alternative worlds" (Hawkins, 2017: 7). This shift towards the creativities of everyday life is important in consideration of distinctively suburban creativity. This thinking, of course, is not to deny one of the key themes of the collection: that the economic signifi-cance of suburban creativity has been overlooked. However, it draws attention to the importance of forms of creativity that have particular significance for suburban geographies, notably those associated with the domestic, the vernacular, and the amateur (Edensor et al., 2009; Gilbert et al., 2020).

Second, the uncoolhunt changes the geographies of creativity. One of the key themes running through this collection concerns the per-vasiveness of inside-out thinking, which gives priority to the creative experience and stimulus of the modern city. When we break with that thinking, we can think about creativity in different ways. One such example thinks about suburban spaces as potentially creative sites. The current pandemic has highlighted the flexibility of suburban domestic space, but the role of bedrooms, garden sheds, and garages as sites of invention and creativity has a history as old as suburbia itself. Suburban built forms are often more flexible, extendable, and adaptable than those found in city centres. We can also think about the creative potential of suburban landscapes. The importance of green space has been reinforced in the pandemic, but more often than not suburban landscapes have been treated as inferior, less than either the urban or the wild. The undramatic "flatness" that Greenlaw attributes to Essex is a characteristic that, in various ways, is often applied to suburbs and edgelands (even those with rather hillier topography). But if we treat suburban landscapes not as a contamination of both the urban and the rural but as a hybrid and a particular meeting of cre-ative interactions between humans and nature, then we might think differently. Consider archetypal figures in the engagement with land-scape – the explorer in the wilderness, the flaneur in the city, and, maybe, the gardener in the suburb. The first two figures have feet of clay, long exposed as self-heroic, masculinist, and colonialist; the tedious clichés of the "urban explorer" often reveal little more creativ-ity than limitless self-obsession. If we depart from inside-out thinking (and maybe also from an outside-in thinking that starts from the wil-derness or the rural), we see a model for engaged, embodied creativity in the gardener, one that has too often been disparaged through its his-tory, yet with its human scale, its sensitivity to climate and season, to growth and decay, and to the aesthetic and the material has never been so significant. Beyond the garden too, there has been a cultural revalu-ing of in-between spaces, the landscapes of edgelands, which can be characterized as a kind of unruly associate of more ordered suburban landscapes. In scrublands, abandoned industrial land, the remnants of older farmland, and woodlands is a distinctive imaginative world that can also reward creative exploration and attention to detail (Keiller, 1999; Roberts and Farley, 2012).

Finally, our uncoolhunt draws attention to how we can reconfigure our understanding of creative networks and flows. This collection across its chapters counteracts the expected centripetal flow of creativity associated with those coolhunting life stories discussed earlier. A counterbalance to the pervasive trope of the escape from the stultifying suburb is another kind creative escape *from* the city. This flight is often imagined as a retreat or as an artistic colony, often to the sea or deep country; yet, as some of the contributions to this collection show, very often such moves were to sites on the margins of cities that provided a change but not disconnection. Such creative flows and fluidity may work over a creative life course, but may also be built into the rhythms of everyday life. The twentieth-century English visionary artist Stanley Spencer is most closely associated with his home village of Cookham in Berkshire and is often thought of as a rural artist. Yet Spencer commuted into London for a significant part of his career and was certainly not isolated from the London art scene. Spencer's Cookham with its rail connection into London (more direct and faster than the current service) was part of an outer commuter belt, a place in the process of becoming more suburban in the mid-twentieth century. One of the key messages of this collection is to think about suburbs not as a passive ring around vibrant core cities but as part of dynamic spatial systems with changing interconnections and flows. This concept has certainly been the case for the creative industries, but we can also track the changing patterns of other dimensions of creativity.

The pandemic has revealed the limits of urban cool, but that re-evaluation was perhaps already well underway. While some moves from the cities may be temporary, there are longer duration pressures to centralized creative activity, particularly in some of those global centres eulogized by *The Economist*. To take one example, the fashion industry is increasingly squeezed by the hypercapitalized property markets of its global centres. The kinds of cheaper inner-city districts that once provided urban interstices for young designers and independent boutiques have all but disappeared (McRobbie, 2016; Gilbert and Casadei, 2020). The digital world has been shaking up the geographies of cool; Gladwell's urban coolhunt came just before the main impact of digital media on style and influence. The rise of the digital influencer means that often it is a home space rather than the "street" that is the site of style. As fashion design businesses have increasingly moved online, they have also often moved physically out of the metropolitan cores.

What this collection shows is that, distinctive as the effects of digital media and pandemic are, such moves are neither without historical precedent nor are they to places without complex, often hidden histories of creativity.

REFERENCES

Benjamin, W. 1999. "Paris, the Capital of the Nineteenth Century." In W. Benjamin, *The Arcades Project*. Cambridge, MA: Belknap Press, 3–13

Benjamin, W. 2006. *The Writer of Modern Life: Essays on Charles Baudelaire*. Cambridge, MA: Belknap.

Economist, The. 2020. "Great Cities after the Pandemic." *The Economist*, 11 June 2020. https://www.economist.com/leaders/2020/06/11/great-cities-after-the-pandemic

Edensor, T., D. Leslie, S. Millington, and N. Rantisi. 2009. *Spaces of Vernacular Creativity: Rethinking the Cultural Economy*. London: Routledge.

Florida, R. 2002. *The Rise of the Creative Class: And How It's Transforming Work, Leisure, Community and Everyday Life*. New York: Basic Books.

Frith, S. 1996. "The Suburban Sensibility in British Rock and Pop." In R. Silverstone, ed., *Visions of Suburbia*. London: Routledge, 269–79.

Gilbert, D. 2004. "Review of Mark Clapson, *Suburban Century: Social Change and Urban Growth in England and the USA*, Berg, Oxford, 2003." *Journal of Historical Geography*, 30(2): 444–6. https://doi.org/10.1016/j.jhg.2004.03.018

Gilbert, D., and P. Casadei. 2020. "The Hunting of the Fashion City: Rethinking the Relationship Between Fashion and the Urban in the Twenty-First Century." *Fashion Theory*, 24(3): 393–408. https://doi.org/10.1080/1362704X.2020.1732023

Gilbert, D., J. Hawley, H. Nicholson, and L. Worth. 2020. "On Amateurs: An Introduction and a Manifesto." *Performance Research*, 25(1): 2–9. https://doi.org/10.1080/13528165.2020.1736754

Gladwell, M. 1997. "The Coolhunt: Who Decides What's Cool?" *The New Yorker*, 17 March 1997. https://www.newyorker.com/magazine/1997/03/17/the-coolhunt

Greenlaw, L. 2021. "The Refusal of Place." In L. Greenlaw, *Some Answers Without Questions*. London: Faber and Faber, 32–7.

Hawkins, H. 2017. *Creativity*. London: Routledge.

Huq, R. 2013. *Making Sense of Suburbia through Popular Culture*. London: Bloomsbury.

Keiller, P. 1999. *Robinson in Space: And a Conversation with Patrick Wright. Topographics*. London: Reaktion.

Kureishi, H. 1990. *The Buddha of Suburbia*. London: Faber and Faber.

Landry, C. 2008 [2000]. *The Creative City: A Toolkit for Urban Innovators*. 2nd ed. London: Earthscan.

Lewis, C. 2020. "The Top 20 Suburbs to Move to in the UK." *The Times*, 7 August 2020. https://www.thetimes.co.uk/article/the-top-20-suburbs -to-move-to-in-the-uk-bm7fq8bqj

McRobbie, A. 2016. *Be Creative: Making a Living in the New Culture Industries*. Cambridge: Polity.

Mould, O. 2018. *Against Creativity*. London: Verso.

Roberts, M.S., and P. Farley. 2012. *Edgelands: Journeys into England's True Wilderness*. London: Vintage.

Thorn, T. 2019. *Another Planet: A Teenager in Suburbia*. Edinburgh: Canongate Books.

PART 2

The Suburban Home as Locus of Creativity

3 "Pictures, Plants, and Ornaments": Jane Ellen Panton and Creative Practice in the British Victorian Suburbs

SARAH BILSTON

Pictures, plants and ornaments are the hallmarks of a home, which cannot exist without them, for these things judiciously chosen and arranged, and not overdone, are after all what turn the worst suburban villa that ever was designed into an artistic abode.

– Jane Ellen Panton (1896: 126)[1]

Introduction

The origin of the modern suburb is often traced to Britain and the nineteenth century: as the architectural historian Andrew Saint has put it, "the elusive relationship between London and its suburbs set off a revolution that still resonates the world over ... [H]ere the positive conception of the suburbs first saw the light of day" (Saint, 1999: 9). The Eyre estate of "villa" housing in London's St. John's Wood, first planned in 1794, was successfully developed in the early decades of the nineteenth century, and suburban growth picked up pace rapidly thereafter (Thompson, 1982a; Barratt, 2012; on St. John's Wood, see Galinou, 2010). From the beginning, the new homes were pitched not just as attractive places to live but also as symbols of material and social success: the *Times*'s infamous "Houses to Let" section bristled with "desirable" and "elegant" semi-detached properties that were hopefully puffed, in grand terms, as "gentlemen's residences" (Bilston, 2019: 39).

Literary critics have long argued that, in spite of the rapid pace of growth, in spite of suburbia's remarkable takeover of the British architectural and cultural landscape, there was surprisingly little engagement with the suburbs by writers until the end of the nineteenth century (Cunningham, 2000: 51; Hapgood, 2005: 1). Certainly, many of the authors we know best (Charlotte Bronte, George Eliot, Thomas Hardy, even Charles Dickens) did not have a great deal to say about them. However, as I've shown in *The Promise of the Suburbs*, plenty of less well-known writers were actively debating the suburbs' complex, evolving meanings and possibilities (Bilston, 2019). Some presented the suburbs in ways that may sound familiar: the "positive conception" of the suburbs – as places of bourgeois success – was loudly rejected as early as the 1820s. Popular author Edward Bulwer-Lytton exposed the "gentleman's residence" as nothing but a show, a façade, in an 1829 novel that evokes the new suburbs as places of strange, almost unearthly absence – of taste, kindness, integrity, honesty, and even room (the house the hero inhabits is built, for cost-saving purposes, on a miniature scale). The association of the new suburbs with absences and deadening dullness was to become a literary commonplace, reaching a zenith in novels by writers like George Gissing and H.G. Wells at the century's end. George and Weedon Grossmith's (1892) infamous comic tale *The Diary of a Nobody* picks up on what was already a well-established literary tradition of associating the suburbs with nobodies and nothingness.

The association of the suburbs with dullness and absence is entirely familiar to us, of course; the stereotype endures (Harris, 2018: 29–38). But not every nineteenth-century writer presented the suburbs in these terms. The new suburbs – still unfolding, undefined, perennially undefinable – were treated by a number of authors as places of creativity, interest, and possibility. The texts in which these unexpected, less familiar readings of the suburbs emerge are ones that are rarely found on university syllabi today; they are not, in any sense, canonical, produced by "great" writers known for their literary centrality. Rather, they are produced by women writers who lived, and worked, in the suburbs; by writers on the margins, geographically and culturally.

Recent scholars, in human geography especially, have turned our attention to the creative lives unfolding in our own contemporary margins, still so often dismissed and stereotyped as "sterile and uninteresting," as Alison Bain puts it (Bain, 2013: 3). To do so has required picking apart the "unacknowledged urban bias" in scholarship that has expected to find the most vibrant creativity in cities, and so has found it there; that did *not* expect to find creativity in the suburbs, and so has comfortably discovered its absence (Gibson, 2012: 3; Van Damme and De Munck, 2018: 3–10). A new scholarly direction, yielding fresh readings of the worlds we inhabit today, requires renewed, refocused attention on the suburbs and other apparently peripheral spaces and a reassessment of what a creative life *means*. It requires recognizing the degree to which lingering stereotypes of the suburbs have shaped our interpretations of suburban experiences and practices; it requires recognizing the rich potential of "vernacular" creativities, for so long dismissed as not creative at all. And it demands evolving a new analytical approach, one that is not founded on and shaped by an urban-based model but rather takes the suburbs as their own kind of Ground Zero (Bain, 2013: 3–28).

The nineteenth-century British suburbs reward a similar approach. Turning our attention away from the more familiar texts of the literary canon, we find in less familiar, woman-authored literature, long disdained as "merely" ephemeral – non-fiction manuals, conduct literature, family advice texts – very different interpretations of the suburbs and of their opportunities for creative experience and practice. In interior design manuals especially, a new genre of literature becoming wildly popular from the 1860s, positive articulations of the suburbs as places of creative opportunity are commonplace. Interior design texts did not suggest that the suburbs were perfect – far from it; indeed, they began from the premise that suburban homes were architecturally flawed. Recycling the already-familiar arguments about suburban nothingness and absence, writers proceeded to examine and explain the rich possibilities of creative practice to address problems and thereby enrich individuals, families, communities, and the nation. Indeed, suburban interior design was presented in this new tidal wave of design literature as a means of bringing beauty to Britain and fostering social cohesion in a time of rapid population growth and increasing geographical mobility (Cohen, 2006: 64). Moreover, middle-class women – with increasing amounts of time on their hands at home, denied access to professional work, yet with increasing access to literacy and print media – were represented and celebrated as the best-placed individuals to learn about art and bring beauty to suburban family dwellings. Design works repeatedly argued that women were the ones who should and could bring taste and art into the suburban landscape and interior by learning about and putting into practice vernacular creativities that would counter the "mistakes" made by male architects and developers of the fast-growing suburbs.

Design manual texts offer a fascinating insight into how the suburbs were imagined, discussed, and staged by the first generations of residents, female occupants especially. They give us access to an intriguing new vision of suburban possibility, one that has received comparatively little scholarly attention. The design figures we know best from the period – those typically credited with seeding the rich flourishing of the Arts and Crafts movement at the turn of the twentieth century – are mostly men, like William Morris. But women played a key part in encouraging artistic interest from around the middle of the century, and many staked out a claim for the importance of suburban art, presenting vernacular creativities as no mere "woman's pastime," limited and small scale, but rather as vital labour that fulfilled urgent social and practical needs.

Given the constraints of space, this chapter will focus on the literature and practice of just one such writer, the celebrated and best-selling Jane Ellen Panton (1847–1923). Panton, a resident of Shortlands, a new suburb southeast of London, expressly identified suburban-set vernacular creativities as activities that offered women powerful forms of artistic self-expression; though she was no suffragette, Panton advocated for women's entrance into the architecture profession. Yet, as suspicion of women's employment and even self-fulfilment lingered in an era that advocated "separate spheres," Panton cautiously navigated contemporary gender norms as she sought to license women's artistic and creative labour. To that end, the common fear of the dull, identikit "suburban nowhere" is deployed by Panton as a way to explain and justify female creativity to readers and, implicitly, society writ large. Women's design work is carefully depicted as a practical solution to what so many contemporaries decried – the supposedly deadening suburban nowhere. Yet Panton's writing and her own lived experiences in fast-growing Shortlands reveal just how rich the creative possibilities of the suburb could be.

"A Country Wherein Nothing Is"?

The idea of the creative, enabling Victorian suburb sounds, on the face of it, ridiculous; it flies in the face of so much we think we know. At the turn of the twentieth century, British satirical journalist T.W.H. Crosland described the new suburbs as "a country wherein nothing is; save villas ... where the principal objects of cultivation are the stunted cabbage and the bedraggled geranium" (Crosland, 1905: 15–16). Crosland's references are pointed: people, he implies, are *not* "cultivated" in the suburbs, and the reference to geraniums is freighted, for these cheap-and-cheerful annuals – easy to pot but not winter-hardy – had become a much-derided alter ego of supposedly rootless suburban residents themselves. The geranium's ubiquitous presence along suburban streets was itself taken as a sign of a lack of aesthetic appreciation, a signal of the suburbanite's preference for unoriginal show over rarified good taste; there was little enough "real" art or taste, Crosland sneered, in the suburbs. A suburbanite's walls were apparently covered by cheap prints surrounded by gilt, a preference Crosland drolly summarized as *"art for frame's sake"* (179).

Few artists lived in suburbia, the journalist concluded, and those who did painted only trite, sentimental scenes, such as rose buds and poor children on crutches: "You can no more help painting for Suburbia, consciously or unconsciously, than you can help breathing the circumambient air" (183).

Women's suburban creativity is, for Crosland, virtually unimaginable. Indeed, he suggests that women do nothing at all once "Mr. Suburb" heads to the "railway-station, or to [the] bus with four shining morning horses," leaving "what is already suburban in essence more suburban still. Practically, it gives over the household and all that dwell therein to the unquestioned rule of woman" (Crosland, 1905: 21). Without men around, he explained, continuing his misogynistic daydream, the suburbs are voids, lacking any economic, social, or cultural production whatsoever: the wife and servants lie around vacantly until the hour before Mr. Suburb's return. Jumping temporally in his satirical sketch from the moment the man leaves to the hour when "Mrs. Suburb" finally springs into action, Crosland seems literally unable to put into words what filled up the days of women in the suburbs.

Middle-class suburban women did not, needless to say, lie around all day waiting for their husbands' return. As historian Erika Diane Rappaport has persuasively shown in *Shopping for Pleasure*, "one of the key features of the late-Victorian suburban woman was her refusal to stay at home or even remain in her local high street" (Rappaport, 2000: 23). Indeed, as Rappaport explains, "as early as the 1850s, she seemed to be in constant motion, ever traveling to and through the city center" (23). The suburbs were by definition connected: the energy with which the Victorians developed new omnibus routes, carved train tracks, and expanded tram lines assured that women of means could move as easily towards shopping, acquaintances, and friends as commuting men moved towards work. The suburb may have offered, in theory at least, a world separated from the public sphere, but it did not mean that the new zones required remaining "isolated from the world of business, commerce, or public amusement" – far from it (23). Transportation routes from the new suburbs into the entertainments of London's West End, into Tottenham Court Road's new department stores, to the markets and bazaars of multicultural suburbs like Bayswater, and to bustling neighbouring high streets allowed women to purchase textiles, foods, plants, spades, furniture, wool, crewel hooks, tools, decorative objects, and more with ease. Transportation brought friends on visits; transportation offered opportunities for chance meetings and acquaintanceship with strangers.

This depiction is not the vision of the Victorian woman we know best, of course. We still tend to think of such women as trapped in stifling homes, bored, isolated, and struggling to live up to the unattainable ideal of the "Angel in the House." And certainly the geographical separation of home life from work life in the suburbs may seem to render that ideal in concrete, architectural form, as if the entire suburban movement strove through its very bricks and mortar to keep women away from the marketplace (Hapgood, 2005: 8). Yet community emerged and was shaped in new ways: train tracks and omnibus routes connected the suburbs with each other, with the provinces, and with urban centres, while burgeoning genres of print text facilitated new forms of community based on shared interests (Barratt, 2012: 207–12). The suburbs, then – so often described as dead-ends, cul-de-sacs, places "wherein nothing is; save villas" – were in practice dynamically networked environments whose opportunities for movement and interaction were available to women from the first (Rappaport, 2000: 25).

And indeed, as I've argued in *The Promise of the Suburbs*, memoirs and fictions of the period set in suburban spaces insistently narrativize not isolation, nothingness, and dullness but chance meetings and startling encounters. They turn on what happens when an alien boarder arrives, when a new lodger answers an advertisement, when people from different communities meet and strive to coexist, when an intriguing stranger is spied on an omnibus. What anchors meetings between people from different backgrounds, people who have moved towards the city in search of new opportunities, is the suburban house itself, which – building on precisely the architectural sameness so many critics denounced – is treated as something that offers people across classes, faiths, and even nationalities something in common, a shared point of contact from which new relationships develop. (On Victorian migration and mobility, see Feldman, 2000: 189–97; on the "flight from the land," see Armstrong, 2000 [1981]: 118–30.) A shared garden in the midst of a block, for instance, offers immigrants and migrants alike literal common ground. A shared chimney design forms a talking point for women of different classes (these are the plots of Bertha Buxton's *Great Grenfell Gardens* [1877] and Emily Eden's *The Semi-Detached House* [1858], respectively; see Bilston, 2019: 62–72). And in the literature of interior design, as we shall see, the suburban home layout is similarly evoked as a connection, for good and for ill; it is a shared spatial arrangement that often exasperates, whose architecture often annoys but whose very common features offer a starting point for community, connection, engagement with the marketplace, and creative practice.

Vernacular Creativities in the Victorian Suburbs

Crafting flourished in women's circles from around the middle of the nineteenth century. As literary and cultural critic Talia Schaffer explains, interests in activities like needlepoint and Berlin wool work, in shell sculptures, cut-paper flowers, fish-scale embroidery, and wax coral construction burgeoned partly as ways to fill the time gap that opened as middle-class women stopped working the land, yet were unable for ideological reasons to engage in professional labour. "Deprived of the occupations that had made them important managers of their farms," Schaffer summarizes, "forbidden to engage in virtually any kind of remunerative employment, middle-class women were searching for activities to help pass the time in their new domiciles" (Schaffer, 2011: 5).

Crafts were not exclusively suburban pastimes, but the craze for interior design, emerging in the late 1860s, regularly referenced the suburban villa as the "new domicile" in need of artistic attention through the purchase of small objects – "pictures, plants and ornaments," as Panton put it. Advocating the layering of textiles over doors, bookcases, mantelpieces, and pianos; the scattering of potted plants and decorative objects across tables, shelves, and even musical instruments; the application of new wallpapers and paints, of oilcloths, carpets, and dados, design writers presented the homes in which so many lived as both the everyday setting and the focus for a new generation of creative women. The end-product was not intended to be extraordinary, unusual, or one of a kind: on the contrary, the appeal of the design displays advocated in the new interior design books was precisely that they could be produced by ordinary women in an ordinary home.

"Creativity is commonly conceived as the property of particular artists, thinkers and philosophers, exceptional, individual geniuses who are able to transcend the bounds of what is presumed to be ordinary thought and practice," Tim Edensor and his co-writers observe, before working to unsettle the long-standing association of creativity with radical change or startling innovation (Edensor et al., 2009: 8). They argue that creativity "should rather be conceived as an improvisation quality that, across all forms of cultural activity, requires people to adapt to particular circumstances." Creativity may be found not only in the small scale and homely, the hidden and the domestic, but also in the drive towards continuation, even in the maintenance of traditions. In this analysis, the everyday is not merely jogtrot and banal, and ongoing daily lives are not failing to capture creativity's pioneer thrill. Rather, the everyday is part of an ongoing scene and cycle of creativity – the creativity of the home-dweller, say, who works hard to maintain the rhythms of daily life in the face of constant challenges to their shape.

Accepting this vision of the creative does not mean rejecting creativity's transformative power, however – quite the contrary. Indeed, scholars of vernacular creativity argue that such practices help render the ordinary sustaining, offering the individual a sense of purpose and identity (Hawkins, 2016: 228). Vernacular creativities allow people to connect and bond with others and build a sense of belonging and place: scrapbooking, quilt-making, model train building, allotment gardening, and more help foster collective identity, particularly in situations where difference is felt, "forming modes of conviviality," as Harriet Hawkins puts it, that allow people to feel likeness, relationship (Hawkins, 2016: 115). In situations where people are dispersed, and at difficult life moments, creative practices may bring especially valuable meaning to life and connect individuals with networks that support them. Fostering community, vernacular creative practices are facilitated in turn by the communities that support, guide, nurture, and educate. Resisting the celebration of the isolated genius, the disruptive pioneer, may help us more fully recognize how groups – families, spouses, friends, colleagues, neighbours – enable and facilitate a wide range of artistic forms and how creative practice in turn helps generate and build an abiding sense of community and place.

Matthew Arnold, the great Victorian poet and cultural critic, defined culture as "the best which has been thought and said in the world"; its opposite is "our stock notions and habits, which we now follow staunchly but mechanically, vainly imagining that there is a virtue in following them staunchly which makes up for the mischief of following them mechanically" (Arnold, 1869: viii). The suburbs were already evoked by the middle of the nineteenth century as the place of the undifferentiated, the many, the ordinary; of dull nobodies dressed alike, acting alike, "following ... mechanically" the mores, tastes, and norms of the day.[2] But just as twentieth-century scholars questioned Arnoldian distinctions between "high" and "low" culture, so we increasingly question the distance between the creative city and the mechanical suburbs. The kinds of creativities that flourish in the suburbs may be different from those that arise in the cities, but, as Roger Silverstone insists, "suburbia is creative": "Suburban streets are complex and subtle signifiers, offering, for those who can read the signs, delicate statements of style and styles ... The modern suburb is a social as well as a cultural hybrid" (Silverstone, 2007: 6–7). Suburban creativities in the Victorian suburb may also, just as today, take work to discover and decode; "delicate statements" of difference push subtly at apparent homogeneity and sameness.

In an era of regular house-moving, the Victorian suburban home was, on the one hand, a site of difference, a signal of a new architectural and spatial arrangement catering to mobile and dispersed populations (Panton is explicit that her readers are unlikely to be buying their houses but rather renting on "the usual three-year tentative lease" – which, she remarks, "is so rarely renewed" [Panton, 1896: 58]). Meanwhile the suburban house was, precisely because the floorplans, aesthetics, and even problems were so similar across geographical areas, a powerful means of connection for diverse populations (see Harold Dyos's foundational study of Camberwell, Dyos, 1961: 70; Bilston, 2019: 8–9). Creative solutions to common design issues further helped bond mobile people, fostering new communities oriented around shared interests, difficulties, experiences, and practices. Produced by and in the suburban house, women's vernacular creativities ultimately helped establish the suburb *as* home, a site where strangeness and difference were managed by reference to commonalities and sameness.

Jane Ellen Panton and Shortlands

The new genre of the interior design text catered to a rapidly growing market of suburban dwellers, particularly women with a little spending money and leisure time. Jane Ellen Panton is little read today, but she was a familiar name at the end of the nineteenth century: when she joined the staff of the popular weekly magazine *The Gentlewoman* in 1891, the editor Joseph Snell Wood crowed: "Mrs. PANTON is the best-known and admittedly the highest authority on matters of the Home. She was the first lady to make it a speciality" ("Other Folks' Homes," 1891: 504). Panton's breakout work, *From Kitchen to Garret: Hints for Young Householders*, was first published in 1887 and earned seven reprints over the course of the next decade. She was still advising householders after the First World War in the significantly retitled *In Garret and Kitchen: Hints for the Lean Years* (1920).

From Kitchen to Garret opens with a chapter on "Choosing a Home" that explicitly recommends the southern suburbs of London – "Bromley, Beckenham, Shortlands, and all the Crystal Palace district are to be spoken well of," the author explains (Panton, 1898 [1887]: 1–16, 5). Panton herself moved to Shortlands in 1882, as a wife of some thirteen years and mother of five children. James Albert Panton was a brewer, but after feuding with his brother and sister-in-law, the couple departed the family company and fled rural Dorset. They selected Shortlands,

Figure 3.1. Shortlands, a suburb of Bromley, ca. 1900. A nanny with a child in a perambulator is captured walking along the sidewalk. Such quiet scenes may appear to confirm biases about suburban dullness, but Panton's work suggests creative lives unfolded behind doors. Source: Bromley Historical Society, https://www.bblhs.org.uk/shortlands#ShortlandsImages.

on the Kent border, for reasons both practical and familiar: the couple had friends living nearby, and at just ten miles from the city, the region was affordable yet accessible. The property they rented, while James attempted to retrain as a chemist in the London University Chemical Laboratory, was commodious, with a generous garden, eight bedrooms, and stabling.

Shortlands was in the midst of a boom, and Panton's first-hand experience of the suburb's rapid growth was to drive her career (Figure 3.1).[3] The London *Times* from the 1860s onwards regularly reported sales of land parcels, for development, in and around Beckenham Road, in easy reach of the Shortlands Railway Station (which opened in 1858).[4] Beckenham, the once-village from which Shortlands was carved as a separate suburb in 1870, grew from 2,391 residents in 1861 to 13,011 in 1881 (*Beckenham Directory*, 1885: 28).[5] Residents were shocked at the pace of development, yet proud (in Panton's time, at least) of the area's still-rural feel. The 1885 *Beckenham Directory* shook its head at "the loss of so much that was beautiful," yet reflected: "We cannot forget that we are living in a practical age, in which there seems to be no standing still, and with which few suburbs are going more rapidly or successfully forward than favoured Beckenham" (28).[6] Shortlands itself was little more than a dozen streets when Jane and James first moved there, and the couple seems to have been the home's first tenants; Panton later described how "we made the garden out of a field" (1909: 355).

This kind of creative transformation – field into garden, nowhere into home – hovers as the alluring promise in all Panton's design writing. In Panton's work and life in suburban Shortlands, creativity blossoms in the small scale, the ordinary. Nothingness meets presence: through "pictures, plants and ornaments," the suburban "nowhere" becomes an "artistic abode." Yet the property in which she lived sounds, on the face of it, hardly in need of rescuing or rehabilitation. "Gable-end, Short-lands" [sic] stood approximately where 23–27 Scott's Lane, Bromley, stands today and, like so much suburban architecture of the late century, was partially timbered in an "Old English" style.[7] It is described in a 1909 bill of sale as "of a handsome and imposing elevation of stucco and half-timber work, with tiled roof. It is approached by carriage drive to bold entrance with massive pitch-pine door, then to a spacious and lofty tiled Entrance Hall (having high anaglypta dado)" – this last, perhaps, was introduced by Panton herself.[8] The bill of sale trumpets the same kind of amenities that drew the Pantons to the house in the first place: "excellent shops" and the proximity of the station, a mere seven minutes away, strongly signal its suitability for a commuter ("there is an excellent service to trains to the City and the West End").[9] Gable End was well-positioned, then, spacious and modish – yet still, in Panton's estimation, far from perfect: "The suburbs, take them how you will, are not Paradise and can never now be made so," she sighed (Panton, 1896: 1). The clay soil of Shortlands, the frustrations of neighbours, the layout of suburban houses (in particular, the separation of nurseries from parent bedrooms), the sounds of the railway, and above all the poor quality of suburban building were problems she returned to repeatedly. The suburbs – even at their very best – were, according to Panton, seriously in need of intervention.

Panton was no New Woman; she never tied herself to railings or threw herself under a horse. Yet her works are full of addresses to the reader about the need for more female architects, whom she regarded as far more likely to design houses well. Panton was unabashedly angry that male architects dominated a profession whose creations were run and managed by women, and she framed her advice and practice repeatedly with reference to gender. Until women entered the profession, women's daily suburban creativities were, in Panton's view, absolutely required in order to make up for men's many design errors. Her expressiveness on the matter is perhaps surprising; Panton was not raised to speak out. Daughter of a mid-Victorian painter of royalty, William

Powell Frith, Panton was to recall with bitterness that her own artistic ambitions were roundly and repeatedly scotched by her parents, especially her father.

The move to Shortlands changed Panton's life. In the new, fast-growing suburb Panton encountered a network of women earning a living and building a name for themselves as writers, artists, and activists. Dinah Mulock Craik, author of *John Halifax Gentleman* (a highly successful novel that later spawned two films and a TV series), occupied a large property, Corner House – now 114 Shortlands Road – built with money she'd earned from her writing. The property was under half a mile from Panton's front door, and the two women soon met: Panton describes a suburb of easy sociability, "where we could go out to dinner walking, with shawls over our shoulders and snow-boots against the mud over our evening shoes; where we could 'run in and out' to beloved neighbours; and where, if Mrs. Craik ruled us all somewhat with a rod of iron, we all bowed beneath her scepter and much enjoyed the evenings we spent and the people we met at her lovely home, the 'Corner House'" (Panton, 1909: 352). Panton, who had encountered plenty of the rich and famous as a child in her father's home, found herself moving in a new kind of community on the edge of the city, including abolitionist James Russell Lowell, American actress Mary Anderson, actor-manager Wilson Barrett, writer F. Anstey-Guthrie, children's authors Juliana Horatia Ewing and Jean Ingelow, and suffragist Clementina Taylor. The women modelled for Panton a professional literary life, while a number of the men actively helped advance her career: Panton had already begun a design business advising neighbours on their home furnishings, and a meeting with the editor of the *Lady's Pictorial*, Alfred Gibbons, proved crucial. He invited her to try writing a column, and it quickly took off; by the time she joined the staff of *The Gentlewoman*, she was, as editor Joseph Snell Wood made plain in his triumphant celebration of her arrival, a modern celebrity. If Shortlands was a creative inspiration for Panton, then, it was also a practical means of advance: "Our proximity to London and the cheap return fare made the Shortlands years delightful to me, and if we could have been settled there I should have never wanted to move again," she later explained (Panton, 1909: 355). Indeed, Panton attributed a subsequent mental breakdown to the loss of this supportive, enabling community when she was forced once again to move house.[10]

On the face of it, though, Shortlands was not an obvious creative hub. Indeed, it would be easy to dismiss its streets of brand-new houses as the kind of "suburban nowhere" Crosland so emphatically disdained. Carved out from a larger suburb, Shortlands was typically characterized in relation to its more distinctive neighbours – as, for instance, "situate [sic] midway between Bromley and Beckenham" (*Strong's Bromley Directory*, 1891: 217). Shortlands, with no high street, lacked obvious spurs to creativity: it had no museums, no restaurants, and no public meeting places (other than the Shortlands Tavern, which stood – and still stands – by the railway, serving commuters). Shortlands was also resolutely non-commercial: the 1891 *Strong's Bromley Directory* designates it "a purely residential neighbourhood" (217). It possessed a lone church, St. Mary's, "built of Kentish rag, in the Decorated style"; its only other unusual feature, as far as contemporary guide books and directories were concerned, was the Kent Waterworks (Walford, 1895: 104).[11] The number of writers and artists living there was not unusual or distinctively large; guide books and local histories tend to mention George Grote, born in 1794, and his wife and biographer Harriet Lewin as the suburb's most illustrious residents.[12] Photographs from Panton's time suggest quiet streets of terraced, lookalike middle-class homes.

Yet the suburb was not as homogeneous as it may sound or appear. Census data reveals that middle- and working-class families alike lived in Shortlands in Panton's time, on the same streets and even next door, while listed birth places indicate that heads of household, employed residents, and visitors came from all across Britain and even around the globe (see Tables 3.1 and 3.2 in the appendix to this chapter). Meanwhile, excellent postal and telegraph services facilitated connections with the metropolis and beyond: the Shortlands Grove pillar box alone was collected eight times a day. And Shortlands's very relationship to its neighbours gave residents access to further amenities: Shortlands residents benefitted from Bromley and Beckenham's rapid development. The public recreation ground of Martin's Hill, Bromley, for instance, dotted with benches, was just a mile away to the north, over the footbridge at the station. Bromley Road connected with Panton's Scott's Lane, and it offered access in the mid-1880s to a greengrocer, a bootmaker, a hairdresser, a stationer, and a chemist, as well as a mission and the Victoria Nursery (which regularly advertised mass sales of bedding plants by auction in the press, touting its proximity to Beckenham Junction).[13]

Bustling Beckenham High Street, meanwhile, was a mile and a half to the west, a short walk for a Victorian or a trip on the number 71 bus: "Luggage free, parcels moderate charge" (*London Guide*, 1878: 76). Amenities in Beckenham included a lending library, two bakeries, a bank, an "art needlework repository," a bootmaker, a builder-decorator, a cabinet maker, a coffee tavern, a cheesemaker and poulterer, a confectioner, a chemist, a coach-builder, a coal merchant, a hotel, a hairdresser, two greengrocers, a piano tuner, a printer and stationer, a saddler, a silversmith, a tailor, and

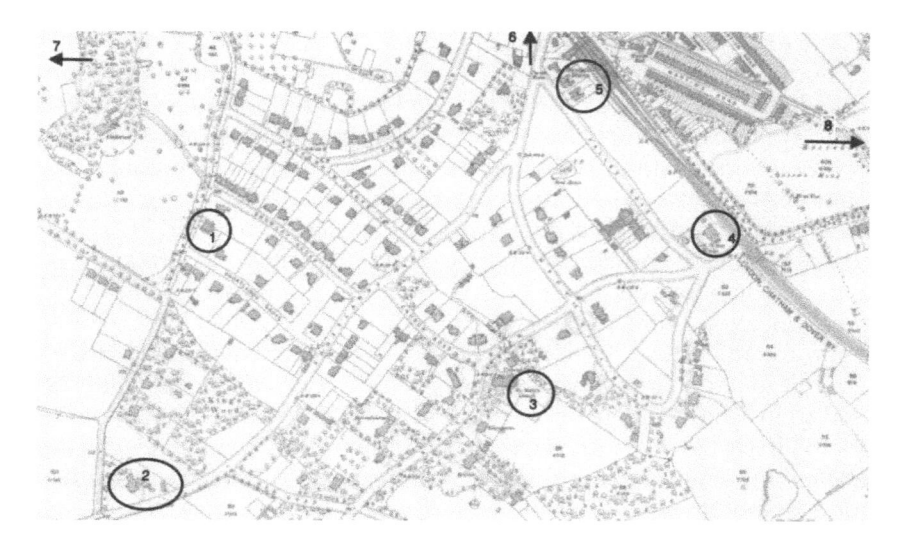

Figure 3.2. Map of Shortlands, a suburb of Bromley. Indicated are (1) Panton's house, (2) Craik's house, (3) St. Mary's church, (4) the Water Works, (5) the Station, (6) Bromley Road (to the north), (7) Beckenham High Street (to the west), and (8) Bromley (to the east). Explore this area in more detail at www.promiseofthesuburbs .com. Source: Reproduced with the permission of the National Library of Scotland, CC-BY (NLS), https://maps.nls.uk/view/101202912.

an upholsterer. Even more opportunities awaited on cross streets – in, for instance, two fancy toy warehouses, one selling Berlin wool, and a china shop. With no fewer than twelve churches and missions, an orchestral society (formed in 1883), two coffee houses (the second opened in 1884), a large public hall seating 550 and a smaller one seating 200, as well as multiple charitable boards, Beckenham's amenities catered to men and women alike. Even more facilities were available in Bromley, a bare mile and a quarter from Panton's front door – including, apart from its numerous shops (and a hairdresser who intriguingly offered "hair brushing by machinery"), many taverns, hotels, a public bath, a reading room, a meeting house, and four train stops (two on the London, Chatham, and Dover line; two on the South Eastern line). Bromley was also served by three newspapers, the *Bromley Journal*, the *Bromley Record*, and the *Bromley Telegraph*.[14]

The point of detailing these amenities is to remind us that even a seeming "suburban nowhere" was a short walk or bus ride from bustling high streets, communal meeting places, and rail hubs into the London centre (Figure 3.2).[15] For Panton, the net effect was obviously enabling, and the design aesthetic and practice she advocated encouraged and expected participation in exactly the kinds of networks and commercial premises she encountered at home. For even as Panton spoke, practically, to a mobile readership renting the properties they needed to "fix," to women hardly likely to plaster walls or tear out fireplaces themselves, she also expected shopping, mailing, travelling. As we shall see, Panton's home design aesthetic was one that specifically expected women's movement in and through the wider suburban and urban environment as a precursor to their practical, creative work in the house.

Panton's Design Aesthetic: Covering a Staircase

Panton complained that builders thought more about curb appeal than the practicalities of family life. Far too often the suburban house was little more than the sum of parts that could be puffed in the *Times*'s "Houses to Let" section, she protested, charging that builders and developers set their own profits above family need. The same woman whose own house was "approached by carriage drive to bold entrance" later complained loudly about the construction of "ridiculous 'carriage approaches' (see house agents' advertisements)" that were damaged by hasty tradesmen and postmen attempting to get through (Panton, 1896: 6). But these problems could be ameliorated through creative interventions. Room by room – "from kitchen to garret" – Panton suggested how looming mantelpieces could be managed, how draughty windows could be draped, how awkward bedrooms could be rearranged for children and the sick.

Staircases were especially serious challenges for the suburban dweller, Panton claimed. Dominating the house, they could hardly be diverted or rebuilt, at least not without the permission of a landlord and a significant financial investment. "The ordinary suburban staircase" is thus to be approached "with fear and trembling," Panton observed wryly, because treads are often too steep yet too narrow, while ill-formed edges wear through the carpeting (Panton, 1896: 53). Worn carpets are expensive to fix, and stair treads can't be easily rebuilt. Panton advises the purchase of textiles, and more textiles, in response:

> As a rule, Jacob's ladder has suggested [suburban stair] design, and it is so proud of its appearance that it thrusts itself on our notice the instant we enter the house. In this case we can only grin and bear it, the while [*sic*] we make its long expanse of open balustrade and wooden understructure as bearable as we can by covering in the first with Eastern dhurries or Khelim

curtains, and filling in any panels in the latter with Japanese leather paper, which is invaluable for this purpose. (Panton, 1896: 51–2)

Theories of vernacular creativity help us unpack this remarkable spectacle, for vernacular creativities, as Edensor and colleagues point out, often take on "interstitial" spaces, those typically seen as outside, or beyond, creativity (Edensor et al., 2009: 13). Moreover, Panton's approach is emphatically not one of radical revision: she does not consider demolishing or redirecting the steps. Rather she proposes a navigation, a creative making do – or, as she puts it, "we make its long expanse … bearable." Panton's vernacular creativity is one of managing, holding on (until the next move).

Panton's is also a design that directs women to traverse the city and enter commercial, trade, and transportation networks. The "Eastern dhurries," "Khelims," and "Japanese leather paper" encourage readers to feel the thrill of buying and bringing back the "exotic," the modern, the modish at a time of fascination with all things pan-Asian (see Cohen, 2006: 127–8, 243–4n25). While it may be tempting to read the shopper as seduced by the marketplace, Rappaport argues that this notion misses the ways in which the Victorian woman challenged gender norms as she moved about the city. As a shopper, she pursued pleasure through travel, occupied the streets, participated in the marketplace, and chose to be a part of modernity:

> For women with few public activities and limited employment and educational options, shopping allowed them to occupy and construct urban space … When Victorian and Edwardian women shopped, they were central actors in the English economy; they altered the city and ideals of bourgeois femininity; they inhabited and built the public spaces of the modern metropolis. (Rappaport, 2000: 221–2)

Panton is explicit that design required adventurous shopping, and her books overflow with references to a wide range of stores, products, and specialty manufacturers. In the case of the staircase, for instance, she recommends a visit to William Wallace & Co., at 155 Curtain Road, Hackney, to purchase a fretted arch to hold up the curtain (Wallace, on the cheaper end of the new department stores, offered "artistic" furnishings to the new middle classes).[16] Readers are also advised to visit, in this and other works, Smee & Cobay ("close to Moorgate Street Railway Station"), Strode & Co. on Portland Road (Regent's Park), Godfrey Giles's on Old Cavendish Street, Verity on Regent Street, Oetzmann on Hampstead Road, Maple's and Shoolbred's on Tottenham Court Road,

Burnett & Co. in Covent Garden, and many more purveyors of lights, furniture, and textiles.[17] While Panton's references are surely de facto advertisements (it can't be accidental that many of the same companies hawk their wares in the front and end pages of her books), the long lists of shops also offer readers access, on the page and in reality, to a wide network of vendors, acting as a guide to the teeming multiplicity of fast-growing London. Products named after Panton, available from multiple distributors – Aspinall's paint came in "Panton Blue," for instance, while William Wallace made a "Panton" corner wardrobe, and J. Wilson made a cloth after Panton's design called "The Gentlewoman" – signal the vitality of the Panton aesthetic and further serve to bond and anchor a new design community, to identify them with one another and with Panton.

The very obviousness of the staircase as an end-product – the swathes of cloth draped across it from all over the world, hanging from an industrially produced curtain rod, with leather wallpaper to boot – is, then, finally the point: the whole remarkable production demonstrates the woman's active engagement with the staircase, her creative reworking of its form, and the many stages of design labour and commercial engagement she has undertaken to produce the transformation. This kind of visual staging has its roots in earlier craft crazes. Talia Schaffer has asked, of a mid-century handicraft in which women covered watch-hooks in fabric: "Why did the women … feel a need to swathe [a] cheap disposable product in layers of decoration? The answer is that the craftswoman was literally surrounding and wrapping the bought object in the fabric that testified to her labor, skill, taste, and affection" (Schaffer, 2011: 46). Covering a staircase surely performs a similar purpose, revealing care and attention, creative engagement, commitment to the home. And, like the watch-hook, the staircase – a feature built by male developers and speculators – is rendered new and visibly transformed through the creative process.

Vernacular creativities, as Hawkins observes, typically "stand in opposition to the time-saving, alienating promises of consumer culture"; the handmade and managed suggest the personal, the individual, over the mass-produced (Price and Hawkins, 2018: 15). Panton's advice simultaneously requires and expects engagement in the marketplace, the city, with consumer culture. Still, for Panton this engagement offers hopeful compromise, not contradiction. Her aesthetic embodies an attempt to reconcile the forces of modern consumerism with those of domestic life, of public with private, commercial with home. Bringing a mass-produced table or roll of leather wallpaper into the suburb creatively connects the different with the familiar, the "foreign" and the domestic: the "Old English" Gable End filled with "Eastern dhurries" embodied in material form, a creative intermingling of styles that staged and performed their connectedness. The creativity of Panton's approach lies fundamentally in this aesthetic: the very ways in which her design makes harmonious and connected so much that often seemed distanced and in conflict.

Conclusions: Locating Suburban Creativities

Women's suburban creativity has been hiding in plain sight for a number of reasons, beginning with an ongoing willingness to accept a lingering opposition of urban vibrancy and suburban tedium. We have perhaps, too, inherited textual and aesthetic biases – about what constitutes art and "legitimate" creativity – from late nineteenth-century art critics who worked hard to distinguish their own aesthetic practice from that undertaken by "average" or "amateur" women. When fin-de-siècle decadents and aesthetes developed keen interests in areas long associated with women's labour, such as home decoration and fashion, they were at pains to characterize their design interests as professional, skilled, and artistic, to distinguish their collections of decorative objects and textiles from a "mere" housewife's dabbling (Schaffer, 2000: 85). When an amateur woman experimented with cloths from Liberty in the 1890s, she was to be mocked or disdained, her practice treated as emphatically *not art*; when an aesthete purchased the same textiles, he was deploying his cultural capital as a connoisseur. Aesthetes were forced into this rhetorical practice precisely because the number of women "dabbling" in art and home decoration was so very large. In fact, women had been responding creatively to the suburban house for decades, its layouts and designs centring and fostering a thriving creative community. Perhaps aesthetes and decadents celebrated the hushed Jacobin House precisely to distance themselves from the creative lives unfolding inside the rows of ordinary modern terraces.

Women's rich vernacular creativities in the suburbs have also been obscured because woman-authored literatures remain under-read. Attention to writers like Jane Ellen Panton offers access to long-lost narratives of the new suburbs as networked spaces that facilitated new kinds of communities formed around shared experiences, rooted in creative engagement. And, as Panton's story makes clear, some women even developed creative lives into thriving careers, finding in suburban-set practices not only an opportunity for personal fulfilment but a portal into paid work, a new identity as a professional writer and designer. Far from being places of nothingness and tedium, then, the fast-growing Victorian suburbs were places of rich, creative possibility. Filling *nowhere* with something – with pictures, plants, and ornaments – Victorian designers like Panton created new interiors, new communities, and new lives.

Appendix

Table 3.1. Shortlands Road, Shortlands, Bromley: Selected residents in 1891, showing occupations and places of birth

Address	Name(s) of heads of household, employed residents, and visitors	Occupation of heads of household, employed residents, and visitors	Places of birth of heads of household, employed residents, and visitors	Number of and places of birth of servants (if any)
2 Shortlands Road	Joseph P. Ashton	Merchant, textile goods	Surrey	3: Middlesex, Norfolk, [illegible]
	Florence Ashton	Occupation not given	London	
4 Shortlands Road	Hellen Wilkinson (head)	Living on own means	London	2: Kent, London
	John J. Shenlove (visitor)	Artist, painter, sculptor	[illegible]	
8 Shortlands Road	George Swain	Engraver on wood / sculptor	Oxford	1: Kent
	Katherine Swain [daughter]	Student of music	London	
Shortlands Road [number not given]	Thomas Basson	Railway servant	Bristol	n/a
	Sarah Basson	Caretaker	Whitechapel	
Shortlands Road [number not given]	John G. Carswell	Living on own means	Notting Hill	4: Surrey, Northants, Lee, Kent
	Sarah C. Carswell	Occupation not given	Lancashire	
	James Paxton (visitor)	Architect	Scotland	

Address	Name(s) of heads of household, employed residents, and visitors	Occupation of heads of household, employed residents, and visitors	Places of birth of heads of household, employed residents, and visitors	Number of and places of birth of servants (if any)
Shortlands Road [number not given]	Samuel Fry	South American merchant	Rio de Janeiro	4: London, Essex, Kent, Oxfordshire
	Mary Fry	Occupation not given	Birmingham	
171 Shortlands Road	William Copping	Brick maker / Field labourer	Kent	n/a
	Maria Copping	Occupation not given	Kent	
189 Shortlands Road	Thomas Nuttall	Bookkeeper to brickmaking company	Cheshire	n/a
	Sarah Nuttall	Occupation not given	Yorkshire	
Corner House, Shortlands Road [former home of Dinah Craik]	John Charles Taite	Assistant to civil engineer	London	4: Essex, Plymouth, Cambridgeshire, [illegible]
	Jane Taite	Occupation not given	Highbury	

Source: *Census Returns of England and Wales, 1891*. Kew, Surrey, England: The National Archives of the UK (TNA): Public Record Office (PRO), 1891.

Table 3.2. Scotts Lane, Shortlands, Bromley: Residents in 1891, showing occupations and places of birth

Address	Name(s) of heads of household, employed residents, and visitors	Occupation of heads of household, employed residents, and visitors	Places of birth of heads of household, employed residents, and visitors	Number of and places of birth of servants (if any)
Westfield, Scotts Lane [House listed in *1895 Beckenham Directory* as 1]	Philip Benton

Julia Benton | Accountant, General Post Office

Occupation not given | Essex

Northumberland | 2: Norfolk, Norfolk |
| Carelon, Scotts Lane [House listed in *1895 Beckenham Directory* as 3] | Thomas Leaton Leadam

Isabel Constance | Timber merchant

Occupation not given | London

Middlesex | 4: Kent, Cornwall, Buckinghamshire, Somersetshire |
| Scotts Lane: Clincart [House listed in *1895 Beckenham Directory* as 5] | John Carswell

Cristina Crawford

John Carswell

Samuel Carswell | Insurance manager

Occupation not given

Brewer and brewers' chemist

Insurance surveyor | Ireland

Scotland

Scotland

Scotland | 1: Dorsetshire |
| Scotts Lane: Gable End [House listed in *1895 Beckenham Directory* as 7 Scotts Lane. Note: former home of Pantons] | Arthur Sturt

Constance Mary Sturt | Hosiery manufacturer and merchant

Occupation not given | London

London | 6: Norfolk, Oxfordshire, Aldershot, East Mersea, Kent, Essex |

Address	Name(s) of heads of household, employed residents, and visitors	Occupation of heads of household, employed residents, and visitors	Places of birth of heads of household, employed residents, and visitors	Number of and places of birth of servants (if any)
Stables: Scotts Lane	Charles Wellstead	Coachman, groom	Dorset	n/a
	Susan Wellstead	Occupation not given	Dorset (Wareham)	
	May Wellstead	Teacher	Dorset (Wareham)	
9 Scotts Lane	Maria Bulwer	Living on own means	Norfolk	2: Kent, Surrey
	Rosina Bulwer (visitor)	Living on own means	Norfolk	
11 Scotts Lane	Henry Gretton	Gardener and caretaker	Derbyshire	n/a
	Emma Litchfield Gretton	Occupation not given	Westminster	
13 Scotts Lane	Catherine A. Lee	Living on own means	Middlesex	5: Buckinghamshire, Herefordshire, Essex, Kent, Montgomeryshire (Wales)
	Henry Lee	Solicitor	Middlesex	
	George Lee [sons of Catherine]	Clerk in London Parcels Delivery Company	Middlesex	
Scotts Lane: Heathfields [no number listed on census or in directory]	George Cowen	Wine merchant clerk	Kent	n/a
	Rebecca Cowen	Occupation not given	Kent	

Source: *Census Returns of England and Wales, 1891*. Kew, Surrey, England: The National Archives of the UK (TNA): Public Record Office (PRO), 1891.

NOTES

1 I would like to thank the indefatigable Simon Finch in the Central Library, Bromley, for his assistance, together with the generous Max Batten in the Bromley Borough Local History Society.

2 Lara Baker Whelan first argued that Arnold "had a significant influence on the way late-century commentators framed the discussion surrounding Culture and the suburbs" (Whelan, 2010: 151).

3 For a detailed account of Bromley's evolution from a market town in Kent to a suburb of London, see Rawcliffe, 1982: passim. Bromley was left behind in the great suburban expansion of the mid-century because it did not have a railway. When rail stations finally opened, high fares initially precluded all but the richest commuters. The Shortlands train did not even depart until 8:35 a.m., which impeded its development as a commutable suburb. Even after the development of a second branch in Bromley, the local community's resistance to workmen's return fares meant that neighbouring areas like Penge attracted more lower-middle-class families (35, 40).

4 The first departure, at 8:40 a.m. on 3 May of that year, is described as a rather anticlimactic event, though people lined up to watch from nearby Martin's Hill. See "Summary of Local News," 1858: 1.

5 The directory, which is both detailed and full of mistakes, gets the population sizes of 1861 and 1871 the wrong way around; in the version I read, a patient Victorian hand has corrected this error and many others.

6 Panton's husband James listed the property for rent in the *Times* in 1885: it is described as having five sitting rooms and eight bedrooms, not to mention a garden with a lawn suitable for tennis. The listed price was 5 pounds 8 shillings, though perhaps he was unable to find someone to take over the lease at such a large sum; the couple did not leave until 1887.

7 The street was later renumbered. Panton's next-door neighbours in the 1880s included A. Paget Wade, a West Indian merchant born in St. Kitts. By 1891 the Pantons had moved, and Gable End was occupied by Arthur Sturt, a hosiery manufacturer and merchant. A carriage house on the property was listed as a separate home. Census data from the street in 1891, listed in Table 3.1, shows that residents came from all around Britain and were employed in a range of different professions and trades.

8 *Freehold Family Residence*, 1909: 2. Panton explicitly recommended the anaglypta dado (1896: 179, 243). For more on the use of tile and timber in the later Victorian suburbs, see Long, 1993: 38–40; on the fin-de-siècle preference for stucco over stone, see Long, 1993: 81–3.

9 Victorian and Edwardian auctioneers and developers were notorious for making elevated claims about suburban developments; jokes about the

"puffing" of the suburban villa were circulating as early as the 1850s. The sellers of Gable End nimbly move in *Freehold Family Residence*, true to form, from pitching the house as a practical and middle-class choice to framing it as an aristocratic gentleman's pleasure ground: it offers wonderful opportunities, they insist, for golf, cricket, tennis – and hunting; it is "an ideal spot for a gentleman seeking rural surroundings … within easy reach of the Metropolis" (*Freehold Family Residence*, 1909: 1).

10 "When we moved to Watford, in Hertfordshire, that move, and not my work, was at the bottom of my first breakdown" (Panton, 1909: 365).

11 The church was destroyed in 1944. The directory repurposes its description of Shortlands from Walford's 1895 work.

12 See, for instance, Walford, 1895: 104.

13 See "Victoria Nursery," 1879: 614.

14 I have compiled these lists of shops and services available in Panton's time from *Beckenham Directory for 1885*, *Beckenham Directory for 1887*, and *Strong's Bromley Directory for 1887*. Doyle's Bromley Hair Cutting and Shampooing Saloons stood at 135 High Street, Bromley (see *Strong's Bromley Directory*, 1887: 25).

15 Users can explore and uncover more of the sights and sounds, the homes, amenities, and transportation networks of the region at www .promiseoftheubsurbs.com.

16 An advertisement by Wm. Wallace & Co. entreats readers to send off "for our New Catalogue, entitled 'Beauty, Skill and Economy,'" if they are furnishing a "Flat or House." See Panton, 1896: xv. On Wallace as a department store, see Cohen, 2006: 52.

17 The services and shops in the text are recommended in *Homes of Taste* (Panton, 1890). For more on Maple's, Shoolbred's, and Oetzmann's, see Cohen, 2006: 50–2. Smee and Cobay trumpeted its handy proximity to the station in its advertisements; see Panton, 1890: facing 152.

REFERENCES

Armstrong, W.A. 2000 [1981]. "The Flight from the Land." In G.E. Mingay, ed., *The Victorian Countryside*. London: Routledge, 1:118–30.

Arnold, M. 1869. *Culture and Anarchy*. London: Smith, Elder.

Bain, A. 2013. *Creative Margins: Cultural Production in Canadian Suburbs*. Toronto: University of Toronto Press.

Barratt, N. 2012. *Greater London: The Story of the Suburbs*. London: Random House.

Beckenham Directory for 1885. 1885. Beckenham: T. Thornton. https://irp-cdn .multiscreensite.com/c3844fd3/files/uploaded/1885_Beckenham.pdf

Beckenham Directory for 1887. 1887. Beckenham: T. Thornton. https://irp-cdn
.multiscreensite.com/c3844fd3/files/uploaded/1887_Beckenham.pdf

Bilston, S. 2019. *The Promise of the Suburbs: A Victorian History in Literature and Culture*. New Haven, CT: Yale University Press.

Cohen, D. 2006. *Household Gods: The British and Their Possessions*. New Haven, CT: Yale University Press.

Crosland, T.W.H. 1905. *The Suburbans*. London: John Long.

Cunningham, G. 2000. "The Riddle of Suburbia: Suburban Fictions at the Victorian Fin de Siècle." In R. Webster, ed., *Expanding Suburbia: Reviewing Suburban Narratives*. New York: Berghahn, 51–70.

Dyos, H.J. 1961. *Victorian Suburb: A Study of the Growth of Camberwell*. Leicester, UK: Leicester University Press.

Edensor, T., D. Leslie., S. Millington, and N. Rantisi, eds. 2009. *Spaces of Vernacular Creativity: Rethinking the Cultural Economy*. London: Routledge.

Feldman, D. 2000. "Migration." In M.J. Daunton, ed., *The Cambridge Urban History of Britain: Vol. 3. 1840–1950*. Cambridge: Cambridge University Press, 185–206.

Freehold Family Residence: Gable End, Scott's Lane. 1909. Pamphlet. Bromley Historic Collections. Bromley: Strong & Sons.

Galinou, M. 2010. *Cottages and Villas: The Birth of the Garden Suburb*. New Haven, CT: Yale University Press.

Gibson, C., ed. 2012. *Creativity in Peripheral Places: Redefining the Creative Industries*. London: Routledge.

Grossmith, G., and W. Grossmith. 1905 [1892]. *The Diary of a Nobody*. Bristol: Arrowsmith.

Hapgood, L. 2005. *Margins of Desire: The Suburbs in Fiction and Culture 1880–1925*. Manchester: Manchester University Press.

Harris, R. 2018. "Suburban Stereotypes." In B. Hanlon and T.J. Vicino, eds., *The Routledge Companion to the Suburbs*. New York: Routledge, 29–38.

Hawkins, H. 2016. *Creativity*. London: Routledge.

London Guide: How to Get From or To Any Part of London, Or Its Suburbs. 1878. London: Edward Stanford.

Long, H.C. 1993. *The Edwardian House: The Middle-Class Home in Britain, 1880–1914*. Manchester: Manchester University Press.

"Other Folks' Homes." 1891. *Gentlewoman*, 11 April 1891, 504.

Panton, J.E. 1890. *Homes of Taste: Economical Hints*. London: Samson, Low.

Panton, J.E. 1896. *Suburban Residences and How to Circumvent Them*. London: Ward & Downey.

Panton, J.E. 1898 [1887]. *From Kitchen to Garret: Hints for Young Householders*. London: Ward & Downey.

Panton, J.E. 1909. *Fresh Leaves and Green Pastures*. London: Eveleigh Nash.

Price, L., and H. Hawkins. 2018. "Introduction: Towards the Geographies of Making." In L. Price and H. Hawkins, eds., *Geographies of Making, Craft and Creativity*. New York: Routledge, 1–30.

Rappaport, E.D. 2000. *Shopping for Pleasure: Women in the Making of London's West End*. Princeton, NJ: Princeton University Press.

Rawcliffe, J.M. 1982. "Bromley: Kentish Market Town to London Suburb, 1841–81." In F.M.L. Thompson, ed., *The Rise of Suburbia*. Leicester, UK: Leicester University Press, 27–91.

Saint, A., ed. 1999. *London Suburbs*. London: Merrell Publishers.

Schaffer, T. 2000. *The Forgotten Female Aesthetes: Literary Culture in Late-Victorian England*. London: University Press of Virginia.

Schaffer, T. 2011. *Novel Craft: Victorian Domestic Handicraft and Nineteenth-Century Fiction*. Oxford: Oxford University Press.

Silverstone, R., ed. 2007. *Visions of Suburbia*. New York: Routledge.

Strong's Bromley Directory, Revised and Enlarged for 1887. 1887. Bromley: E. Strong & Sons. https://irp-cdn.multiscreensite.com/c3844fd3/files/uploaded/1887%20Strongs%20Directory.pdf

Strong's Bromley Directory, Revised and Enlarged for 1891. 1891. Bromley: E. Strong & Sons. https://irp-cdn.multiscreensite.com/c3844fd3/files/uploaded/1891%20Strongs%20Directory.pdf

"Summary of Local News." 1858. *Bromley Record*, 1 June 1858, 1.

Thompson, F.M.L. 1982a. "Introduction: The Rise of Suburbia." In F.M.L. Thompson, ed., *The Rise of Suburbia*. Leicester, UK: Leicester University Press, 2–25.

Thompson, F.M.L., ed. 1982b. *The Rise of Suburbia*. Leicester, UK: Leicester University Press.

Van Damme, I., and B. De Munck. 2018. "Cities of a Lesser God: Opening the Black Box of Creative Cities and Their Agency." In I. Van Damme, B. De Munck, and A. Miles, eds., *Cities and Creativity from the Renaissance to the Present*. London: Routledge, 3–22.

"Victoria Nursery." 1879. Advertisement. *Gardeners Chronicle*, vol. 11, 17 May 1879, 614. https://onlinebooks.library.upenn.edu/webbin/serial?id=gardenerchron

Walford, E. 1895. *Greater London: A Narrative of Its History, Its People, and its Places*. Vol. 2. London: Cassell.

Whelan, L.B. 2010. *Class, Culture, and Suburban Anxieties in the Victorian Era*. New York: Routledge.

4 Battlegrounds of Taste and Distinction: Art and Antique Collectors in the Suburban Hinterland of Nineteenth- and Early Twentieth-Century Belgium

ULRIKE MÜLLER AND ILJA VAN DAMME

He who has stood in the ways of a suburb, and has seen them stretch before him all shining, void, and desolate at noonday, has not lived in vain. Such a sight is in reality more wonderful than any perspective of Bagdad or Grand Cairo.
— Arthur Machen (2018 [1895]: 154)

Introduction

In his episodic novel *The Three Imposters*, late-Victorian Welsh writer Arthur Machen (1863–1947) introduces a character getting lost in one of London's turn-of-the-century sprawling suburbs. Having missed his late-night train, the main protagonist decides to walk a good nine miles through a newly built suburban "waste," with streets as "deserted as those of Pompeii" (Machen, 2018 [1895]: 155). However, before finishing his walk in this landscape of "unutterable monotony" – a "world's end" and "outer void of the universe," a region as unknown as the "darkest recess of Africa" – our hero, Frank Burton, stumbles into a vaguely familiar suburban acquaintance. The latter, a certain Mr. Mathias, invites Mr. Burton to sleep over in his suburban homestead. But this experience turns out to be as horrific for the narrator as its broader, suburban surroundings. After all, far from being a quiet place of domesticity, the suburban home of Mr. Mathias appears to be the abode of a Gothic private collection of torture instruments. Before the obviously strange suburban collector Mr. Mathias has had a chance to show off his peculiar antiquarian interests to the full, he makes a lasting impression on his *invité* by getting killed by one of his newly added collectibles, a so-called iron maiden from Germany.

Clearly, Machen's little conte cruel, the *Novel of the Iron Maid*, is not to be taken too seriously; the decadent author is known to have written mainly to shock and scandalize his fin-de-siècle audiences. Yet, his short stories frequently tap into suburban tropes and anxieties that were well understood and frequently shared and commented upon by late nineteenth-century contemporaries in England, and more broadly northwestern Europe (Whelan, 2010; Bilston, 2019; see also Bilston, chapter 3, this volume). To be noted are the ways in which suburban collecting and collectors around 1900 were ridiculed and represented as being something out of the ordinary. Location and collection – the outré collector's interests as well as his chosen place of abode, a nightmare-world suburb – all tend to merge into a well-narrated piece of chillingly dark humour. However, the humour and irony are of a snobbish kind. The author, Arthur Machen, obviously sympathizes with the metropolitan gaze of the London artistic connoisseurs, looking down on the odd and "disrespectful" cultural activities performed by their suburban counterparts, taking place in empty-headed suburban "wastelands." What "bad taste" to collect "toys of death," the author implies; good riddance to kill off such a collector!

In Machen's story, an often-analysed suburban disdain – the so-called "suburban stereotype" (Harris, 2018) – mingles with equally growing late nineteenth-century anxieties about the nature of, and cultural values associated with, collecting itself. Properly understood as the assembling of meaningful material objects by private individuals for often (semi-)public display, nineteenth-century collecting practices still carried strong overtones from their more noble and genteel origins. Tied to early modern humanism and heavily indebted to an enlightened quest for knowledge and aesthetic taste formation, collecting practices were originally regarded with almost the same symbolic esteem as the processes of scientific discovery and artistic creation itself (Evans and Marr, 2006; MacGregor, 2007). The process of assembling, designing, and setting up a collector's cabinet for (semi-)public display boiled down to a matter of skills and knowledge, choice and distinction, experimentation and creative audacity, and, eventually, a justification of aesthetic and even moral claims (on beauty, authenticity, nationality, and so on). Art and antique collections were believed not only to embody tactile stimulation but also to serve as a veritable social and intellectual template to be discussed among like-minded aficionados. Collectors aspired to distinguish themselves from the inquisitive crowd (the *curieux*), those mere "consumers" of ancient, luxurious, or exotic goods that could be found or bought on the market. Consciously shaping their own identity, collectors presented themselves as discerning, learned, true connoisseurs, or so-called virtuosi of the material culture (Cowan, 2004). As such, collectors assumed the role of important cultural and intellectual brokers, often playing a pioneering role in the institutionalization of the very first public art and archaeological museums in emerging nation-states like France, England, Germany, and Belgium (Crane, 2000; Sweet, 2004; Guichard, 2008; Nys, 2012; Van Damme, 2015).

Collecting, especially art and antiques but also books, antique coins, and other marvellous and exotic curiosities, was initially something reserved for the noble and cosmopolitan milieus in Europe, often spatially centralized in such important art cities as Paris and London but equally present in much smaller art-loving circles from Venice to Antwerp. Moreover, collecting was seen as a quintessentially male activity, something to be taken up in close connection to such "male" topics of inquiry as connoisseurship, science, technology, politics, and the progress of the nations. However, by the end of the nineteenth century, northern European collecting cultures had expanded both across different social strata and across an increasingly diversified and commercialized range of material interests (Sachko Macleod, 1996; Kuhrau, 2005; Silverman, 2008; Denis, 2016). Much of this evolution was tied to the rise of industrial, mass-consumer societies and global expansion, and became spatially fixed, among others, to the expanding suburbs – clustered around rapidly growing cities. At the same time, in the course of the nineteenth century, private collecting also became more explicitly linked to creative practices and artists' activities, an evolution that was, among other things, related to the artist's growing social status and increasing function as a model of taste, as well as to the emerging ideals of individualism and aestheticism that turned the private collection into an ideal means to demonstrate one's personality and preferences (Müller and Sterckx, 2022). Private collectors' choices, as it were, became increasingly susceptible to individual interpretations and one's own creative adaptations to an aesthetic and artistic norm. To the disdain of self-proclaimed arbiters of taste and art around 1900, suburban homes became literally the new locus of an increasingly heterogeneous and "democratized" form of collecting, one that was more privatized, individual, middle-class, and closely linked to other, more everyday pastimes, such as home decoration, ephemeral collecting, and so on. Last but not least, these were collecting activities in which women collectors, for many reasons, started to appear more frequently (Morowitz, 2006; Sachko Macleod, 2008; Verlaine, 2014; Brogniez and Debroux, 2017; Iskin and Salsbury, 2020).

Surprisingly, however, the history of such expanding collecting cultures from around the 1850s onwards has hardly been thematized in existing literature, nor has its geographical locus – the expanding suburb – been seriously problematized and questioned. No doubt, such oversight can be linked to powerful male, elitist narratives in the historical sources themselves, silencing or ignoring the creative activities of (often female) individuals that took place outside a classic, city-centred gaze. This disregard happened to the point of even radically downplaying the creative nature of suburban activities as such (from dull or uninteresting, to tasteless, mundane, uncreative, and so on). Such neglect, contempt, or negation of difference for what happens outside a narrowly defined "artistic centre" still continues to haunt, distort, and subvert present-day historiographies on collecting cultures.

In order to provide a much-needed historical corrective to such suburban stereotyping, the following chapter will trace the emergence of private collecting practices in Belgian suburbs from the early nineteenth century through the early years of the twentieth century, as well as the changing discourses related to suburban collecting. During this period, travellers from all over Europe characterized Belgium as a "country of private collectors," indicating that art and antique collectors played an important role in the public sphere (Müller, 2017). In the first section of what follows, the major spatial, temporal, and social shifts with regard to urban and suburban collecting practices during this period will be traced, based on the analysis of a database of 121 art and antique collectors active in the largest Belgian cities and dominant artistic centres of Brussels, Antwerp, and Ghent, including their surrounding suburbs. The analysis will show that, for much of the nineteenth century, the leading Belgian private collections represented an inner-urban and elitist endeavour – strongly connected to noble city palaces and upper-class bourgeois sociability. On the other hand, private collections found in the still mostly rural hinterland of these Belgian cities were – just as in other countries in western Europe – typically tied to a waning noble identity and conservative lifestyle (Jackson-Stops et al., 1989; MacGregor, 2014; Avery-Quash, 2018; for Belgium, see, for example, Duverger, 1974; Tollebeek, 1999). But by the turn of the twentieth century, the material, social, and spatial characteristics of private art and antique collecting in Belgium had undergone a considerable transformation comparable to other parts of Europe. Our analysis demonstrates how social profiles of Belgian art and antique collectors, as well as the range of their aesthetic interests, had diversified significantly, and at the same time, the suburb had become a regular place of residence for such collectors. Again, similar to other European contexts, Belgian women started to increasingly come to the fore as patrons, collectors, and hosts of cultural salons in suburban contexts.

In the second section of this chapter, we delve deeper into the practices, motivations, and aims of two of these suburban collectors in Brussels: Édouard Van den Corput (1821–1908) and Anna Boch (1848–1936). Both were prominent figures in the Brussels collectors' scene around 1900 and are therefore well-known in Belgian art and urban history. They resided not far from one another in the southeastern suburbs of the Belgian capital. While the case of Anna Boch has recently been discussed in the context of suburbanization processes in late nineteenth-century Belgium, and especially with regard to social and artistic practices in suburban areas (Brogniez and Debroux, 2013, 2017), Van den Corput and Boch have not previously been considered together in academic research. The two figures can, however, be considered illustrative of the ways in which, from the last quarter of the nineteenth century onwards, male and female Belgian art collectors used the expanding (sub)urban space to creatively stage their artistic preferences, give a personal expression of their taste and sensibilities, and position themselves in the art and cultural world more broadly. It is hypothesized that the creative practice of collecting outside of traditional urban centres in Belgium can be linked to the emancipatory freedom these spaces provided for certain social groups. Far from being dull or stifling under creeping conformity, Belgian suburbs gave rise to an increased splintering of identities typical for the Belgian art world in the later decades of the nineteenth century. Suburban collecting as a creative practice thus overlapped with changing social and political standpoints and a broadening range of positions on how to interact with modern urban life.

Private Collecting in an Age of (Sub)urbanization: Spatial Aspects of a Cultural Activity, ca. 1830–ca. 1914

The following examination of the spatial and social trends in the Belgian collectors' scene in the nineteenth and early twentieth century is based on the systematic analysis of a database consisting of 121 private collectors active in this period in three of the most important art cities in Belgium, namely Brussels, Antwerp, and Ghent, and their surrounding area. This database includes collectors' names, biographical data, addresses, social status, profession, and main areas of collecting interest.[1] Based on the collectors' period of activity and documented addresses, their collections were systematically grouped per decade and according to their geographical location, that is, whether they were situated in the city centres of Brussels, Antwerp, or Ghent, in a suburb and/ or in the rest of the province (of Brabant, Antwerp, and East Flanders, respectively).[2] As a research instrument, the database was created on the basis of explicit historical references to private collectors and their collections in a broad and diverse range of published and unpublished source material (travel literature, journals, compendia, archival material). Nevertheless, the database does not systematically map all new trends in amateurship and taste that emerged during the (second half of the) nineteenth century, such as the collection of photographs, posters, lithographs, or other ephemeral items. Since collecting became characterized by a broadening social base (as described earlier), and new generations of collectors were no longer necessarily labelled (or openly and self-confidently labelled themselves) in traditional terms as "amateurs," "collectionneurs," or "connoisseurs," our database makes no pretensions of exhaustivity. Rather than being an overview of all sorts of material collecting activities going on in the three studied cities and their surroundings, our empirical material mainly captures the most frequently and most extensively described collectors active in the Belgian cities and their suburbs. These were, for the most part, still heavily focused on the more traditional, canonic art objects such as paintings, sculptures, and antiques. As such, we mainly consider a representative sample of especially the top layers of collectors in their time and place. Their creative collecting practices and aesthetic goals served, in many ways, as models for the collecting practices of other social strata and, as such, mirror general trends in collectors' tastes, choices, and location quite well (see, for example, Sachko Macleod, 1996; Emery and Morowitz, 2003; Kuhrau, 2005).

During the first half of the nineteenth century, private collecting in Belgium was still very much an inner-urban phenomenon (Figure 4.1). For the 1830s, the database records sixty-two private collectors. Of these collectors, the majority (79 per cent) were located in the urban centres: 23 per cent were situated in Brussels; 19 per cent in Antwerp; and 37 per cent in Ghent. In the same period, a small but consistent number of collections (21 per cent) were located in the rural area. In the 1830s, just over 8 per cent of all collections could be found in the rural hinterland of Brussels; 3 per cent in the wider province of Antwerp; and 10 per cent in the county of East Flanders near Ghent. These rural collections were typically housed in the castles or country houses of members of the traditional nobility, most of whom possessed a city palace as well as one or more country estates – a habit that dated back to the early modern period (Baetens, 2013). The rural estates and collections of the established, often Catholic, noble families such as the Arenberg (in Heverlee, Flemish Brabant and Enghien/Edingen, Hainaut), de Ligne (Belœil, Hainaut), and Baut de Rasmon (Wannegem-Lede, East Flanders) usually reflected the rather conservative social and political ideas of their owners and an artistic taste inspired by the aesthetic ideals of the Enlightenment. As such, they form a contrast to most of the collections in urban centres such as those of Jacques André Coghen (Brussels), Florent van Ertborn (Antwerp), and Louis Minard (Ghent), which functioned more strongly in the context of typically early nineteenth-century trends of nation building and the formation of local and national taste and consciousness (Müller, 2019).

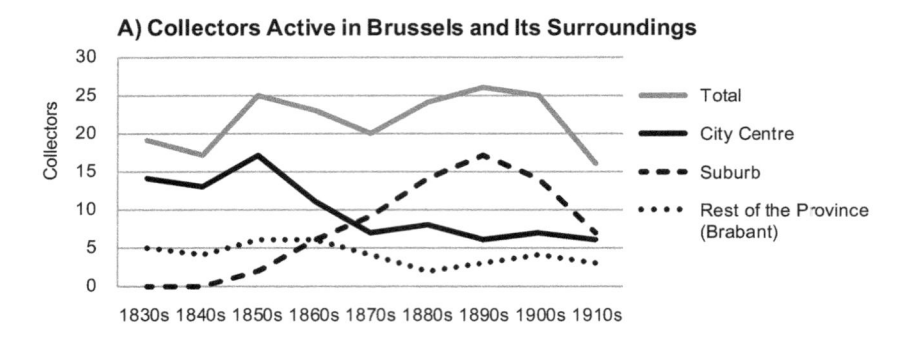

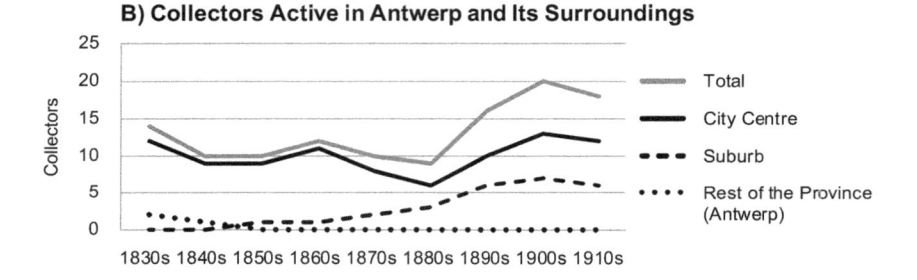

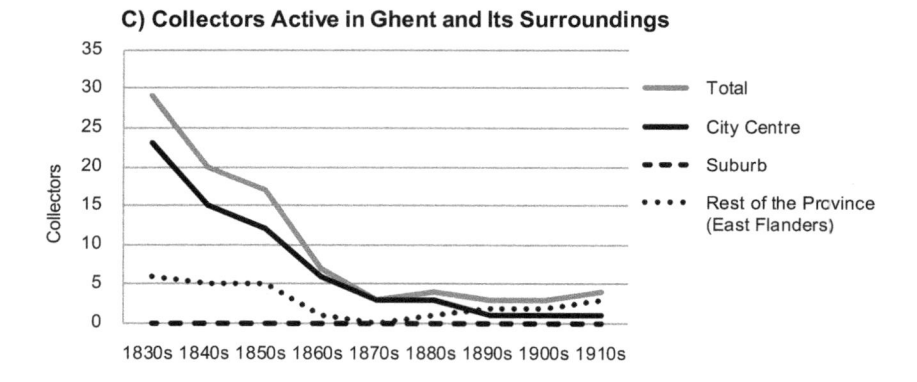

Figure 4.1.   Graphs of private art and antique collectors in A) Brussels,
B) Antwerp, and C) Ghent between ca. 1830 and ca. 1914. Source: Figures
based on data in Ulrike Müller, 2019: vol. 2, database.

From the middle of the nineteenth century onwards, the impact of suburbanization processes became increasingly visible, and suburbs started to gain prominence as places for private collections (Figure 4.1). However, the situation in and around the three studied cities developed quite differently. Of all Belgian cities, Brussels, without a doubt, experienced the most feverish suburban growth in the second half of the nineteenth century. The Belgian capital started to expand after 1830, and by 1900, Brussels's suburbs like Schaerbeek, Ixelles, Molenbeek-Saint-Jean, and Saint-Gilles figured in the top ten of most populated municipalities in the whole of Belgium (Greefs et al., 2005: 222).

From the 1850s onwards, the number of private collections in the Brussels suburbs grew continuously. In the 1860s, when a total of twenty-three collectors were active in the Province of Brabant, one in three collections (35 per cent) was located in the suburban area; a decade later the collections in the capital's suburbs had outnumbered those in the city centre of the capital, reaching 56 per cent in the 1870s and a high point of 74 per cent in the 1890s. This development can be explained by the strong population and industrial-economic growth in Brussels after Belgian independence and the resulting investment in (sub)urban expansions. The new inhabitants of the inner-city core of Brussels were, for a considerable part, workmen and women, but also clerks and artists, who came to Brussels from smaller towns and the countryside to improve their economic situation. Many of them settled initially in the city centre where they found employment and affordable housing. On the other hand, Brussels's wealthy urban elites – which included many collectors – increasingly moved out of the city centre to the newly established suburbs, where they found not only the proximity to green areas for leisure activities but also the space to build new bourgeois villas that provided improved possibilities for interior decoration and for showcasing and enjoying art and other collections at home. Here, they were joined by elite collectors from other parts of the country, attracted by and conscious to associate themselves with a national upper class in the vicinity of the Belgian monarchy. As the century progressed, and especially from the third quarter of the nineteenth century onwards, the Brussels suburbs became increasingly complex, attracting groups and professions (among which were also artists) from different social as well as geographical backgrounds (Debroux, 2013).

In the 1850s, the first art collections emerged in the Brussels suburbs of Saint-Josse-ten-Noode (in the northeast of the city centre) and Saint-Gilles (in the south). A decade later, Ixelles (in the southeast of the city centre) also established itself as a place for private collectors, and by the 1880s, it counted the highest number of collections in Brussels's suburbs. During the following decades, Ixelles not only maintained this prominent position as a centre for collecting in the suburban area, but by the 1890s, it even outnumbered the collections then located in the centre of Brussels, with 30 per cent of the collectors located in Ixelles as opposed to 26 per cent in the centre (see Figure 4.2). Furthermore, the collectors active in Ixelles distinguished themselves from those in the other suburbs by their dominant taste for modern and contemporary art. By contrast, collectors in Saint-Josse-ten-Noode focused on ancient art until the 1890s and directed their attention to modern and contemporary art only around the turn of the twentieth century, while in Saint-Gilles a preference for the Old Masters and historical objects remained dominant throughout the period under consideration. Admittingly, these socio-geographic patterns are based on relatively small numbers and warrant further research. While clearly having distinct characteristics, the social composition of Brussels's suburbs also shared many similarities. Yet, the preference for modern and contemporary art among collectors in Ixelles does seem to have been closely related to the specific identity of the suburb as an artists' neighbourhood, where artists lived in close vicinity to each other and to their patrons (Goldman, 2012; Brogniez and Debroux, 2013, 2017). Around the turn of the century, Ixelles was known as a rather unconventional place in political as well as cultural and artistic terms. It was in this suburb where foremost members of a cultural elite, like the influential jurist and writer Edmond Picard (1836–1924) and the renowned art critic Octave Maus (1856–1919), nourished socialist sympathies. Here, artists could experiment with novel modes of creative expression as well as new forms of sociability (Vanderpelen, 2011; Aron and Vanderpelen-Diagre, 2013). For collectors too, Ixelles was an ideal place to test the emancipatory potential of the suburb: far enough from the city centre to create a mental distance from the traditional artistic establishment but nevertheless connected to the centre so as to benefit from its cultural logistics such as exhibitions, galleries, and art market. The specific situation in Ixelles thereby provided an especially fertile ground for collectors to interpret their activity in a new way, for example by focusing more on their own creative freedom regarding what to collect and how to assemble or present their collections. We will delve deeper into this remarkable turn-of-the-century suburb of Brussels in the next section of the chapter.

The city of Antwerp also saw a continuous suburban expansion in the course of the second half of the nineteenth century, in line with the economic boom of its port and related industries. Working-class suburbs to the north and south of the city, close to the port activities near the Scheldt River, as well as more middle-class extensions and villa parks to the (south)east, were rapidly transforming the surrounding countryside (May, 2020). Accordingly, the number of private collections in the suburban area around Antwerp also grew from the 1850s onwards, but the phenomenon was less dramatic than in Brussels. A new generation of collectors mainly sought the newly built bourgeois neighbourhoods around the green zones in the southeast of Antwerp, such as the area around the city park, and later Zurenborg and the suburb of Wilrijk near the newly developed Nachtegalenpark. However, unlike Brussels, the number of private collections in the suburbs never overtook those in the inner city. Between 1900 and 1910, when private collecting reached its zenith, the majority of Antwerp collectors remained in the historic city centre.

Despite fluctuations, private collecting in Antwerp thus generally remained a rather traditional, urban matter. This development ran parallel to a dominant preference for Old Master paintings and antiques that characterized the Antwerp collections throughout the nineteenth century. In the early 1900s, 65 per cent of the Antwerp collectors preferred the Old Masters to more recent art, while in Brussels 63 per cent of the collectors were lovers of modern and contemporary paintings. The tastes of the Antwerp collectors thereby mirrored the rather conservative cultural and political climate of the city and the ideal to revive its traditional sixteenth- and early seventeenth-century reputation as an international capital of arts and trade. Such nostalgic ideals were equally defended by the local Antwerp academy, its public museums, art enthusiasts, and private collectors alike. In this orientation towards a glorious past, Antwerp considerably differed from the liberal and socialist belief in the future that characterized the cultural scene in Brussels (Buyck, 1994; Goddard, 1998). It was only from the early 1900s that Antwerp collectors such as Emma Lambotte, François Franck, Robert Murdoch, and Charles Good shifted their attention to contemporary paintings by Flemish artists, among whom James Ensor took a prominent position (Müller, 2020). These collectors, however, did not reside in the suburbs. Instead, they took up residence in Antwerp's newly developed South district near the new Museum of Fine Arts (opened in 1890), a modern urban area characterized by circular, star-shaped squares, straight streets, and an eclectic bourgeois architecture that combined historicist styles with influences from the Art Nouveau movement. While not a suburb and situated in the old historic core of Antwerp, the urban renewal of the South district did cater to similar urban ideals, including bigger and more modern housing, which were attractive to more forward-looking collectors.

Finally, Ghent and its surroundings counted a very high number of private collections during the first half of the nineteenth century, but numbers radically dropped during the second half of the century: only five well-known collections were mentioned by our source material at the turn of the twentieth century. The high number of art and antique collections in the early nineteenth century equalled the great pride in the local tradition of collecting, amateurship, and antiquarian study in the city. This tradition went back to the seventeenth and eighteenth centuries and was again reinforced at the beginning of the nineteenth century, among others, with the creation of Ghent University by King William I of the United Netherlands in 1817, the foundation of the Royal Commission of Monuments in 1818, and the influential journal *Messager des Sciences et des Arts* in 1823 (Tollebeek, 1998). After Belgian independence, the first archaeological museum of Belgium was founded in Ghent in 1833, an initiative in which local collectors, including Jean d'Huyvetter, played a decisive role (Müller, 2015).

The very early and thorough institutionalization of collecting and scholarship in Ghent likely had a strong impact on the local private collectors' scene. At a moment when the collection, documentation, and study of art and heritage were taken over by well-organized institutions such as the archaeological museum, city archive, university library, and several historical societies during the second half of the century, there was a radical decrease in the number of private collectors active in Ghent. Private collecting in Ghent also remained mainly linked to the inner-urban elite circles, rather than expanding into the newly established suburban areas, as it did in Brussels and Antwerp. In fact, thanks to the early industrialization of Ghent, which was initiated during the Austrian (1715–95) and French (1795–1815) regimes, and continued under the Dutch rule of William I (1815–30), the city had experienced a steady urban growth during the first half of the nineteenth century. As a result, however, the suburbs of Ghent were mainly industrial areas and home to the working classes (Boone and Deneckere, 2010). In turn, private collectors, as members of the social and cultural elite, remained concentrated in the historic city centre. Alternatively, they moved further outside to the rural area in an attempt to reconnect to the noble tradition of the country estate, as did, for example, the wealthy artist Fernand Scribe (1851–1913), who turned an old monastery in Bottelare into a summer residence, or the architect and politician Arthur Verhaegen (1847–1917), who inhabited a neo-Gothic castle in the village of Merelbeke. Around 1900, Sint-Martens-Latem, a village in the southwest of Ghent, became home to the "Latemse School" of painters who sought out the rural abode for its inspiring natural surroundings. Shortly after, the village became increasingly popular as a place of residence for the wealthy Ghent bourgeois elites who were particularly attracted by the place's aura of creativity and unconventional artistic spirit – a process particularly lamented by the painter Gustave van de Woestyne (1881–1947; see Van de Woestyne, 2020).

From this comparison, Brussels and its suburbs emerge as a particularly interesting case with regard to the spatial distribution of its private collections in relation to the preferences and practices of their owners. The case demonstrates that suburbs could become alternative locations for collectors to creatively stage their collections, while the examples of Antwerp

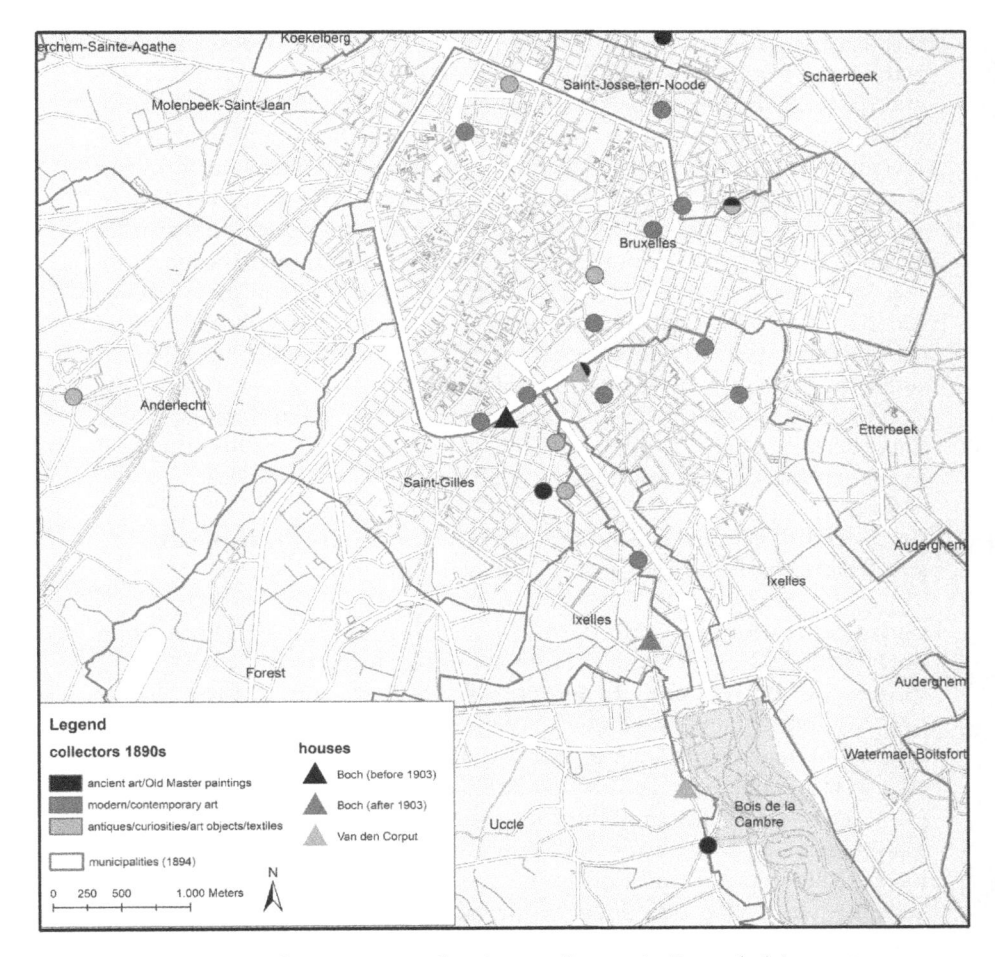

Figure 4.2. Map of private art and antique collectors in Brussels (city centre and suburbs) in the 1890s. Source: Based on Müller, 2019: vol. 2, database. Map by Margo Buelens-Terryn, EOS, on the basis of BhiGIS, 2020. Brussels historical Geographic Information System. IGEAT/ULB. https://bhigis.ulb.be.

and Ghent likewise show us that this trend was not inevitable. The city of Brussels distinguished itself by a remarkable presence of private collectors in the suburban areas, among which the southeast – and particularly Ixelles – comes to the fore with the largest number of suburban private collections in Belgium and a considerable density of new collections focused on modern and contemporary art (Figure 4.2). On the one hand, this development was strongly influenced by the rapid process of urbanization and population growth that the flourishing Belgian capital experienced during the second half of the nineteenth century. On the other hand, the concentration of private collectors in one specific suburban area also reveals the very specific social geography of Ixelles, a suburb that attracted a particular stratum of the art-interested audience. The following section will therefore take a closer look at two collectors active in this geographical area, the physician Édouard Van den Corput and the artist Anna Boch. We will further consider to what extent and how the suburb of Ixelles functioned as a particular place of emancipation, where collectors could develop new ways to creatively stage and interact with their collections in order to shape specific identities or alternative subjectivities. The discussion of the two case studies furthermore aims to unveil different layers of the collectors' social, cultural, and artistic practices in the suburbs, as well as the continuities and tensions that characterize their activities.

Brussels and Its Suburbs as Creative Hotspots: The Preferences and Practices of Édouard Van den Corput and Anna Boch

Born a generation apart, Édouard Van den Corput (1821–1908) and Anna Boch (1848–1936) were both prominent figures in the world of collecting in Brussels around 1900. In terms of social background, professional profile, gender, and aesthetic preferences, they represent opposing poles. Van den Corput – a member of the conservative, upper-bourgeois elite in late nineteenth-century Brussels – was a renowned pharmacist, physician, and professor at the Université Libre de Bruxelles, as well as a politician and senator for the Catholic Party (1894–1900). As a collector, he followed the tradition of the typically male connoisseur with a taste for Old Master paintings, historical furniture, and antique objects, emphasizing an attachment to local and national history. Anna Boch also descended from a wealthy and privileged bourgeois milieu. Her ancestors were the founders of an important manufactory of fine china and ceramics, still operational today under the name Villeroy & Boch. Additionally, she was bound to the art world through family ties: she was the elder sister of the painter Eugène Boch (1855–1941) and a cousin of the previously mentioned lawyer, art critic, and patron Octave Maus. Anna Boch became an important painter in her own right and was an active member of the modernist circles Les XX (1884–93) and La Libre Esthétique (1894–1914). While Van den Corput represented the "old," inner-urban political and cultural establishment, Anna Boch, by contrast, was a fervent supporter of the artistic avant-garde that blossomed in late nineteenth-century Brussels and its suburbs.

It is in the suburb of Ixelles that both played out their different, at times conflicting, cultural visions on collecting, taste, and societal roles. During the 1890s, Édouard Van den Corput and Anna Boch resided just a few houses away from one another on the Avenue de la Toison d'Or, at the southeastern edge of the urban centre of Brussels: Van den Corput at no. 21 (Deschaumes, 2007–9) and Boch at no. 73 (Thomas et al., 2005: 137). Already in the 1870s, Van den Corput had a second residence built for himself at Chaussée de Waterloo no. 876, called Villa La Clairière (https://monument.heritage.brussels/fr/Bruxelles_Extension_Sud /Chaussaee_de_Waterloo/876a/16307). Located beside the Bois de La Cambre, about 4.5 kilometres to the southeast of the Avenue de la Toison d'Or, the villa functioned as a weekend and summer residence. In 1903, Anna Boch left the house at Avenue de la Toison d'Or no. 73 to move to her newly built villa at Rue de l'Abbaye no. 30, about 2 kilometres to the southeast of her former residence. She now lived, as it were, between the two houses of Van den Corput. While not direct neighbours, it is remarkable how both collectors explicitly chose to move within the then-flourishing suburban neighbourhood in the southeast of Brussels.

Van den Corput's and Boch's choices as to where to settle and set up their collections mirror some of the general trends indicated earlier about the (sub)urbanization of Brussels's southeastern areas, and Ixelles in particular. At the same time, the location of their residences, as well as the ways in which they presented and used their collections there, reveal how, for these collectors, the suburb functioned as an ideal place to creatively stage their idiosyncratic, at times opposing, ideas of the arts and of cultural pre-eminence, precisely because of the emancipatory potential of these suburban areas.

Édouard Van den Corput was one of the first collectors to settle in Ixelles. In 1865, he had a new mansion built for himself at Avenue de la Toison d'Or no. 21, designed in a historicist style by architect Désiré De Keyser (1823–97). At the edge of the city centre, on the former second surrounding wall of Brussels, the Avenue de la Toison d'Or was then the geographical point of departure for the extension of the Belgian capital towards the southwest. The avenue experienced a construction boom of mainly neoclassical and eclectic bourgeois mansions between 1860 and 1870, endowing it with its typical residential character. Due to the proximity of the old and new palace of justice as well as the newly created Avenue Louise (constructed as of 1860), the Avenue de la Toison d'Or was an attractive place of residence for members of the upper bourgeoisie, especially those active in the liberal professions such as lawyers and doctors – as was Van den Corput (Deschaumes, 2007–9).

At the beginning of the nineteenth century, Ixelles had still been a rural area with a village core and only a few country houses owned by wealthy aristocratic families. From the 1850s to the 1860s onwards, the area attracted an increasing number of inhabitants, since it offered new spatial possibilities for bourgeois residents as well as the vicinity of green areas (Herla, 2016–17). Around 1850, the urbanization of Ixelles started from the southeastern portion of the boulevards replacing the former wall. First stretching along the Chaussée d'Ixelles, the suburban expansion then proceeded notably along the Avenue Louise (Douillet and Schaack, n.d.-a). The area around the new avenue – including the Rue de l'Abbaye where Anna Boch built her new villa in the early 1900s – was designed by architect and urban planner Victor Besme (1834–1904) on the basis of his *Plan général pour l'extension et l'embellissement de l'agglomération bruxelloise* (1862–6). Besme's ambitious and influential plan for the extension and embellishment of Brussels emphasized the monumental aspect of the capital of the then-flourishing industrial state of Belgium (Smets, 1977: 38–40; Zitouni, 2010). At the southeastern tip of the suburban extension, a new park – the Bois de La Cambre – was created between 1861 and 1866 (Douillet and Schaack, n.d.-b). It was next to this park that Van den Corput had his Villa La Clairière built between 1871 and 1873. Designed in an exuberant neo-Flemish Renaissance style, the villa functioned as a stage where Van den Corput could showcase his cultural ideals as well as his vision on the future of art, collecting, and patronage.

Van den Corput's residence and collecting activity clearly mirrored his public aspirations. Not only did he send artworks from his collection to public exhibitions such as the prestigious show *Les Primitifs Flamandes* in Bruges in 1902 (Weale, 1902: 124–5), but he was also interested in the broader public dissemination of images of his collection and its display. In 1900, for example, the Brussels-based editor Philogène Wytsman (1866–1925) published a number of interior views of Van den Corput's residences in the series *Intérieurs et mobiliers de styles anciens*, in which the collector's houses were praised as a model of fine taste, "une merveille de bon goût" (Wytsman, 1900: 27). In the published photographs, Van den Corput's suburban villa appears as the home of the ideal modern connoisseur, with clear hints to the tradition of Southern Netherlandish *Kunstkamer* of the seventeenth century, while at the same time also referring to the nineteenth-century concept of the period room (Figure 4.3).

In his own writings, Van den Corput (1897) gave expression to his ideas about the position and role of the private collector in the public sphere. He believed that private collectors played a crucial role in cultural and artistic life, among others, because they preserved important objects from the past for the future, maintained aesthetic standards, and

Figure 4.3. C. Aubry fils, the *Salle à Manger* in Édouard Van den Corput's Villa La Clairière at Chaussée de Waterloo no. 876, photograph. Source: Photo published in Wytsman, 1900: 54.

passed on cultural values. In so doing, the contemporary amateur aesthete, according to Van den Corput, had a considerable social responsibility: he functioned as a cultural role model and educator, and thereby made a significant contribution to the general public good and the glory of the resident city and nation.[3] Van den Corput's writings about art and collecting mirror the more traditional ideals of the early nineteenth-century connoisseur-collector as an important figure of public significance. In this way of thinking, he clearly stood at odds with many of his contemporary private collectors like Anna Boch, who, as we will see, shifted her attention more to individual preferences and focused more closely on her own (elitist) social networks rather than on a broader educational or societal mission (Müller, 2017, 2020).

Interestingly, Van den Corput's motivation for defending these older cultural values aligns with his ideas about the process of urbanization. As a physician and politician – his areas of professional activity – he was strongly engaged in matters of public health, social and cultural progress, and how they related to the ongoing urban development. His writings on these topics demonstrate his belief in the process of urbanization and its positive impact on society, showing how he reconciled the bourgeoisie's new suburban cultural identity with more traditional, urban values and national ideals (Van den Corput, 1886, 1899). Further building on Besme's urban development plans, Van den Corput defended the benefits of the urban extension and promoted the great economic as well as cultural potential of the suburbs. Instead of dull, homogeneous, and conventional areas, he characterized the newly built urban area as of essential value to the project of societal progress. The suburbs, he argued, played a crucial role in giving shape to an ideal new city that was better organized, healthier, and, with its novel architecture and stunning collections, also more beautiful. Among the measures he proposed was the distribution of "de petits manuels … des nombreuses curiosités qu'offrent aux voyageurs notre ville et ses environs, avec les indications des excursions à faire … ou des monuments, théâtres, musées, bibliothèques ou collections particulières à visiter" (Van den Corput, 1886: 20).[4] Thereby, the suburban extension contributed considerably to the international attractiveness and reputation of the capital and the country as a whole. In Van den Corput's opinion, there was eventually no difference between the "old" city centre and the "new" suburbs – for him, "Bruxelles-capitale" was one whole (Van den Corput, 1886). This vision materialized in the symbolic location of his suburban Villa La Clairière at the Chaussée de Waterloo near the Bois de La Cambre park. Initially, the geographical area around the park belonged to Ixelles. In 1864, however, in the midst of the booming suburban construction works, the

Avenue Louise, Bois de La Cambre, and surrounding area – including the Chaussée de Waterloo – were annexed by the capital and became known as the Southern Extension (Bruxelles-Extension Sud). In 1873, when Van den Corput had his villa built beside the park, he had thus chosen a spot that had the benefits of the capital and the suburb alike.

For Anna Boch too, the suburb was a place of new, emancipatory potential, from where she could closely follow the urban art world while, at the same time, keep a distance from it to ensure her social and creative freedom. From her suburban residences, she had easy access to the main areas of activity of Brussels's lively avant-garde movement. As a painter and prominent member – the only female one – of the artistic circles Les XX and La Libre Esthétique, she regularly exhibited her own works and pieces from her collection at the exhibitions organized by these circles, which took place at the Palais des Beaux-Arts in the heart of the city. Not only the exhibition halls but also other venues frequented by members of these avant-garde circles for socializing purposes – such as galleries, cafés, bookshops, and the like – were located in the city centre (Brogniez and Debroux, 2017).

On the other hand, the homes of several avant-garde artists and their patrons also became important places of sociability of a more personal and semi-public character. Being an important society figure within her network, Anna Boch held regular salons and concerts at her home during the 1880s and 1890s, which were frequented by her artist-friends, musicians, and other amateurs. When moving to the Rue de l'Abbaye, she settled in a neighbourhood where she lived in close proximity to numerous renowned artists such as Théo van Rysselberghe (1862–1929) and Constantin Meunier (1831–1905; Brogniez and Debroux, 2013: 46). Her house was designed by architect Paul Hermanus (1859–1911) in Art Nouveau style, according to the latest fashion. It was conceived as a place entirely dedicated to art and its enjoyment in a highly aestheticizing atmosphere. Boch's large art collection extended to every room, from the entrance hall, to the salons on the ground floor, to her studio on the first floor, as well as to her private rooms on the upper floor. The central hall, with its staircase by Victor Horta (1861–1947), gave direct access to the comfortable *Salon jaune* (Figure 4.4), a room decorated with light, yellow wall hangings that formed a decisive contrast with the dark colours and heavy wooden furniture and wall panelling of Van den Corput's neo-Flemish Renaissance interiors. Boch's hospitable salon, with its elegant mantelpiece likewise designed by Horta, presented some of the masterpieces of her collection of modern paintings, including Paul Gauguin's *Conversation in the Meadows at Pont-Aven* (visible in the photograph hanging on the wall behind Anna Boch), as

Figure 4.4. The *Salon jaune* in Anna Boch's villa at Rue de l'Abbaye no. 30, photograph. Source: Reproduction of an original photograph from ca. 1903, photographer unknown, Villeroy & Boch AG company archive, Mettlach, Germany.

well as Georges Seurat's *The Seine near Grande-Jatte* and Paul Signac's *The Bay* (Thomas et al., 2000: 156), all three of which she would bequeath to the Museum of Fine Arts of Brussels.[5] The *Salon jaune* directly connected to the music room, which also housed many key pieces from her collection. A central place of sociability, the latter was where Boch held regular musical salons on Monday evenings and other artistic gatherings. During these events, visitors had ample opportunity to socialize while enjoying, examining, and discussing Boch's own creations and the works from her collection displayed in the different rooms on the ground floor.

Certainly, these social events at Anna Boch's home were not accessible to a large audience; neither was her collection publicized to a general public. Rather, her villa was a sphere reserved for like-minded artists, musicians, amateurs, and aesthetes, who were – like Boch herself – members of a close-knit and elitist network circling around the artistic avant-garde. As such, Anna Boch's house and collection functioned not only as a central place for the creation, promotion, and appreciation of cultural and artistic advancement. It was also a fertile ground for the development of a new kind of exclusive sociability that flourished especially in the turn-of-the-century suburb of Ixelles (Brogniez and Debroux, 2013, 2017). It was this new sociability that opened up new possibilities for female artists and collectors to become active in the art world. While the sphere of connoisseurship and the collecting of Old Master paintings, as well as the public exhibition and discussion of art in general, was generally considered a task for male amateurs, the outspoken semi-public character of the artistic salons that flourished in the avant-garde circles was particularly beneficial for the greater involvement of female artists and patrons.

The liberating atmosphere of the suburban environment certainly played its part within this context. In fact, throughout her activity as an artist and patron, Anna Boch concentrated less on her potential impact on the general public or on an established base of inner-urban connoisseurs, authorities, institutions, museums, and so on. Neither did she cultivate an involvement with national ideals or goals. Instead, she focused mainly on the very personal and intimate connotations of her own artistic production and collecting. Her main aim – to have a direct contact with contemporary artists and to encourage the avant-garde – is reflected in her decision to settle in Brussels's southeastern suburb of Ixelles, a free-thinking artists' and liberal-elite neighbourhood, where her collection primary functioned as a space of semi-public sociability.

Conclusion: The Suburb as a Creative Battleground

In this chapter we have taken an explicit socio-geographic approach in analysing the history of private collections in late nineteenth-century Belgium. By combining historiographies on expanding European collecting cultures with those focusing on suburban expansion, we have established how both historical processes are more closely tied to each other then has previously been understood in literature. The creative practice of setting up a material collection and using such displays for expressing core cultural values and notions of sociability was inherently enmeshed with specific spatial contexts. Art and antique collections were set up and became part of the interior of private homes, and these, in turn, formed part of broader spatial surroundings invested with changing historical meanings and symbolic connotations. However, due to their inherently fluid temporal and spatial nature (McManus and Ethington, 2007; Van Damme and Oosterlynck, 2021), suburbs have often gone unnoticed and unquestioned in relation to collecting practices. While suburbs were inherently always in transition in time and space, their classic, stereotypical depiction has been one of stasis, "emptiness" – a stereotypical history devoid of cultural and creative practice (Harris, 2018; and chapter 1 of this volume).

Nevertheless, while art and antique collecting was originally tied to the socio-spatial binary of noble rural villas and upper-class inner-city mansions, the newly emerging suburbs in the second half of the nineteenth century merit closer study and attention. Suburbs opened up new possibilities to old and established collecting practices, originally dating back to the Renaissance and Enlightenment period. Suburbs became the *terra nova* or "creative battleground" from which new generations of private collectors – among whom increasingly were women – could start to etch out different societal roles, aesthetic and material interests, and cultural values.

Within the Belgian context, this process became ever more visible in the later decades of the nineteenth century and was best articulated around the capital of Brussels. In comparison to the other artistic centres of the time, Antwerp and Ghent, the city of Brussels experienced the most feverish suburban growth in the period under examination. Moreover, Brussels's suburban growth was to a large extent driven by a more elitist and upper-class demand for new villa-park neighbourhoods, modern sanitized streetscapes, and new spacious and convenient homes with ample freedom to express domesticity and individual privacy. These were precisely the sort of suburbs that best fitted the needs and wishes of the type of art and antique collectors to be found in our empirical database. Given the bias of our research towards the more well-off and renowned art and antique collectors, it would be interesting to see more future studies on the socio-spatial milieu of the newly arising groups of middle-class collectors, focusing on cheaper and ephemeral forms of collecting. Chances are high that middle-class groups increasingly preferred similar suburban contexts as the collectors studied here, albeit in a more scaled-down and affordable fashion. However, studying collectors lower down the social scale would potentially reveal other pathways into the city that would no longer necessarily pass through the centre first or, historically, originate there. Although more research is needed here, it can already be hypothesized how, from at least the 1870s onwards, Belgian suburbs increasingly became a springboard to creatively construct new and specific suburban subjectivities.

What was it after all that made the newly arising Belgian suburbs so attractive to these late nineteenth-century generations of collectors? While these suburban contexts normally lacked the classic urban infrastructures of creative interaction and face-to-face sociability – traditional picture galleries, museums for ancient art, cafés, shopping arcades, and so on – they made up for that deficiency by offering more freedom and privacy of interaction. This liberty in expressing more personal sensibilities and artistic judgements – away from, indifferent to, or in conscious opposition to official and mainstream norms and beliefs – summed up the emancipatory potential of suburban spaces. The Brussels-Ixelles example perfectly illustrates how collectors with sometimes very different temperaments, age, gender, and conflicting ideas like Van den Corput and Boch nevertheless shared the same preference for acting from and within the suburban space.

The new suburban extensions around 1900 sheltered an increasing splintering of identities, sensibilities, and tastes in ways that were less evident in an urban centre dominated by an officially accepted collecting culture and practice. Thus, suburban collections in many ways better represent the broadening range of creative associations, functions, and purposes for which the turn-of-the-century expanding art world and material culture could be mobilized by its owners than the established, path-dependent collecting practices to be found in inner-city cores. While different in content and function, both Van den Corput's and Boch's collections were, through their suburban locus, intimately tied to an ongoing modernization process. Finally, Belgian and broader European suburbs thus emerge at the turn of the century not only as spaces of a new creative engagement with an expanding art world and material culture. Equally, they are important to understand changing positions and debates about the evolving public role and societal potential of this engagement itself.

NOTES

1 Müller, 2019: vol. 2, database.
2 The "city centres" are here defined as the historic centres within the early modern city walls and enclosures that formerly surrounded the cities and were transformed into large avenues in the nineteenth century. For Brussels, this transformation concerns the so-called "pentagon," for Antwerp the "halfmoon" surrounded by the "Leien," and for Ghent the designated area commonly recognized in research as the historic core (see "Histoische stadskern van Gent" [Historic city centre of Ghent], https://inventaris.onroerenderfgoed.be/erfgoedobjecten/140021). As "suburbs," we define here the urban extensions outside of these areas that became urbanized from the nineteenth century onwards. Collections in the rest of the province are those that existed outside and at some distance from the studied

cities and were mainly located at country estates. Collections that were distributed over different locations (for example at a city palace and at a country estate) were recorded once at each location.

3 See also a number of unpublished writings by Édouard Van den Corput preserved in the Archive and Cultural Centre Arenberg (ACA) in Enghien/Edingen, Belgium, and at the Ghent University Library (UG): Van den Corput's "lettre-préface" for Alexandre Tillot, *Les collectionneurs de Belgique* (unpublished manuscript, ca. 1904), in Papiers de Alexandre Tillot, ACA, BA 29/15; and *Les collectionneurs de Belgique, répertoire, prospectus et bulletin de souscription*, Bruxelles 1904, UG, BIB.VLBL. HFI.C.184.04.

4 This quotation translates as "small handbooks ... of the numerous curiosities offered to travellers by our city and its surroundings, with indications of the excursions to be made ... or the monuments, theatres, museums, libraries, or private collections to be visited."

5 Paul Gauguin, *Conversation in the Meadows at Pont-Aven*, 1888, KMSKB-MRBAB, inv. no. 5092; Georges Seurat, *The Seine near Grande-Jatte*, 1888, KMSKB-MRBAB, inv. no. 5091; Paul Signac, *The Bay*, 1906, KMSKB-MRBAB, inv. no. 5090.

REFERENCES

Aron, P., and C. Vanderpelen-Diagre. 2013. *Edmond Picard (1836–1924). Un bourgeois socialiste belge à la fin du dix-neuvième siècle. Essai d'histoire culturelle.* Brussels: Musées royaux des Beaux-Arts de Belgique.

Avery-Quash, S. 2018. "'The Lover of the Fine Arts Is Well Amused with the Choice Pictures That Adorn the House': John Julius Angerstein's 'Other' Art Collection at His Suburban Villa, Woodlands." *Journal of the History of Collections*, 30(3): 433–52. https://doi.org/10.1093/jhc/fhx039

Baetens, R. 2013. *Het 'soete' buitenleven. Hoven van plaisantie in de provincie Antwerpen. 16de-20ste eeuw.* Antwerp: Petraco-Pandora NV.

Bilston, S. 2019. *The Promise of the Suburbs: A Victorian History in Literature and Culture.* New Haven, CT: Yale University Press.

Boone, M., and G. Deneckere, eds. 2010. *Gent: stad van alle tijden.* Brussels: Mercatorfonds.

Brogniez, L., and T. Debroux. 2013. "Les XX in the City: An Artists' Neighborhood in Brussels." *Artl@s Bulletin*, 2(2): 38–51. https://docs.lib.purdue.edu/artlas/vol2/iss2/5/

Brogniez, L., and T. Debroux. 2017. "Une exposition à l'échelle de la ville. Sociabilités des espaces complémentaires aux Salons des XX et de La Libre Esthétique." *COnTEXTES. Revue de sociologie de la littérature*, 19. https://doi.org/10.4000/contextes.6327

Buyck, J.F. 1994. "Het Vingtisme en zijn opponenten te Antwerpen." In M.A. Stevens and R. Hoozee, eds., *Impressionisme en symbolisme. De Belgische avant-garde 1880–1900*. London: Royal Academy of Arts, 59–69.

Cowan, B. 2004. "An Open Elite: The Peculiarities of Connoisseurship in Early Modern England." *Modern Intellectual History*, 1(2): 151–83. https://doi .org/10.1017/S1479244304000113

Crane, S. 2000. *Collecting and Historical Consciousness in Early 19th-Century Germany*. Ithaca, NY: Cornell University Press.

Debroux, T. 2013. "Dans et hors la ville: Esquisse d'une géographie des artistes plasticiens à Bruxelles (19e–21e siècles)." *Brussels Studies*, 69. https:// doi.org/10.4000/brussels.1177

Denis, B. 2016. "In Search of Material Practices: The Nineteenth-Century European Domestic Interior Rehabilitated." *History of Retailing and Consumption*, 2(2): 97–112. https://doi.org/10.1080/2373518X.2016 .1194712

Deschaumes, C. 2007–9. "Ixelles. Avenue de la Toison d'Or." *Inventaire du patrimoine architectural région Bruxelles-Capitale*. https://monument.heritage .brussels/fr/streets/10502830

Douillet, I., and C. Schaack. n.d.-a. "L'Avenue Louise et les rues adjacentes." Service public régional de Bruxelles, Direction des Monuments et des Sites. https://monument.heritage.brussels/files/cities/1001/documents/louise _bruxelles-extensions_sud.pdf

Douillet, I., and C. Schaack. n.d.-b. "Bois de La Cambre." Service public régional de Bruxelles, Direction des Monuments et des Sites. https:// monument.heritage.brussels/files/cities/1001/documents/bois_cambre _bruxelles-extensions_sud.pdf

Duverger, E. 1974. "Het legaat van barones Van den Hecke-Baut de Rasmon aan het Museum van Antwerpen." In A. Monballieu, ed., *Jaarboek Koninklijk Museum voor Schone Kunsten Antwerpen*. Antwerp: KMSK, 211–84.

Emery, E., and L. Morowitz. 2003. *Consuming the Past: The Medieval Revival in Fin-de-Siècle France*. Aldershot, UK: Ashgate.

Evans, R.J.W., and Marr, A. eds. 2006. *Curiosity and Wonder from the Renaissance to the Enlightenment*. Aldershot, UK: Ashgate.

Goddard, S.H. 1998. "Investigating and Celebrating the "Golden Age" in Nineteenth-Century Antwerp, 1854–1894." In L.S. Dixon, ed., *New Studies of Northern Renaissance Art in Honor of Walter S. Gibson*. Turnhout, BE: Brepols, 151–63.

Goldman, N. 2012. "Un monde pour les XX. Octave Maus et le groupe des XX: analyse d'un cercle artistique dans une perspective sociale, économique et politique." Unpublished PhD diss., Université Libre de Bruxelles.

Greefs, H., B. Blondé, and P. Clark. 2005. "The Growth of Urban Industrial Regions: Belgian Developments in Comparative Perspective, 1750–1850."

In J. Stobart and N. Raven, eds., *Towns, Regions and Industries: Urban and Industrial Change in the Midlands, c.1700–1840*. Manchester: Manchester University Press, 210–28.

Guichard, C. 2008. *Les amateurs d'art à Paris au XVIIIe siècle*. Seyssel, FR: Éditions Champ Vallon.

Harris, R. 2018. "Suburban Stereotypes." In B. Hanlon and T.J. Vicino, eds., *The Routledge Companion to the Suburbs*. London: Routledge, 29–38.

Herla, M. 2016–17. "Ixelles – Histoire du développement urbanistique – Partie 1." Service public régional de Bruxelles, Direction des Monuments et des Sites. https://monument.heritage.brussels/files/cities/1050/documents/Ixelles_developpement_urbanistique_1.pdf

Iskin, R.E., and B. Salsbury, eds. 2020. *Collecting Prints, Posters, and Ephemera: Perspectives in a Global World*. New York: Bloomsbury.

Jackson-Stops, G., E. Blair MacDougall, H.A. Millon, L. Cowen Orlin, and G. J. Schochet, eds. 1989. *The Fashioning and Functioning of the British Country House*. Washington, DC: The National Gallery of Art.

Kuhrau, S. 2005. *Der Kunstsammler im Kaiserreich. Kunst und Repräsentation in der Berliner Privatsammlerkultur* Kiel: Ludwig.

MacGregor, A. 2007. *Curiosity and Enlightenment: Collectors and Collections from the Sixteenth to the Nineteenth Century*. New Haven, CT: Yale University Press.

MacGregor, A. 2014. "Aristocrats and Others: Collectors of Influence in Eighteenth-Century England." In I. Reist, ed., *British Models of Art Collecting and the American Response: Reflections across the Pond*. Farnham, UK: Ashgate, 73–85.

Machen, A. 2018 [1895]. *The Great God Pan and Other Horror Stories*. Edited by Aaron Worth. Oxford: Oxford University Press.

May, L. 2020. *Suburban Place-Making: Political Economic Coalitions and "Place Distinctiveness" (Antwerp, c.1860–c.1940)*. Unpublished PhD diss., University of Antwerp.

McManus, R., and P.J. Ethington. 2007. "Suburbs in Transition: New Approaches to Suburban History." *Urban History*, 34(2): 317–37. https://doi.org/10.1017/S096392680700466X

Morowitz, L. 2006. "A Home Is a Woman's Castle: Ladies' Journals and Do-It-Yourself Medievalism in Fin-de-Siècle France." *Nineteenth-Century Art Worldwide*, 5(2). http://www.19thc-artworldwide.org/autumn06/49-autumn06/autumn06article/159-a-home-is-a-womans-castle-ladies-journals-and-do-it-yourself-medievalism-in-fin-de-siecle-france

Müller, U. 2015. "Particuliere kunst- en antiekverzamelaars in negentiende-eeuws Gent, door de lens van internationale reisliteratuur." *Handelingen der Maatschappij voor Geschiedenis en Oudheidkunde te Gent*, 69, 103–27. https://doi.org/10.21825/hmgog.v0i0.5844

Müller, U. 2017. "The Amateur and the Public Sphere: Private Collectors in Brussels, Antwerp and Ghent through the Eyes of European Travellers in the Long Nineteenth Century." *Journal of the History of Collections*, 29(3): 423–38. https://doi.org/10.1093/jhc/fhw032

Müller, U. 2019. "Between Public Relevance and Personal Pleasure: Private Art and Antique Collectors in Brussels, Antwerp and Ghent, ca. 1780–1914." Unpublished PhD diss., Ghent University/Antwerp University, 2 vols.

Müller, U. 2020. "Les collectionneurs privés. Acteurs d'un monde de l'art en mutation." In J. D. Baetens, E. Berger, I. Goddeeris, N. Goldman, D. Laoureux, and U. Müller, eds., *Adjugé! Les artistes et le marché de l'art en Belgique (1850–1900)*. Paris: Mare & Martin, 100–9.

Müller, U., and M. Sterckx. 2022. "Art and Domestic Space: Continuity and Change in Private Collectors Interiors in Belgium, ca. 1830–1930." In C. Moran, ed., *Domestic Space in France and Belgium: Art, Literature and Design (c.1850–1920)*. London: Bloomsbury, Academic, 49–76.

Nys, L. 2012. *De intrede van het publiek. Museumbezoek in België 1830–1914*. Leuven: Leuven University Press.

Sachko Macleod, D. 1996. *Art and the Victorian Middle Class: Money and the Making of Cultural Identity*. New York: Cambridge University Press.

Sachko Macleod, D. 2008. *Enchanted Lives, Enchanted Objects: American Women Collectors and the Making of Culture, 1800–1940*. Berkeley: University of California Press.

Silverman, W.Z. 2008. *The New Bibliopolis: French Book Collectors and the Culture of Print, 1880–1914*. Toronto: University of Toronto Press.

Smets, M. 1977. *L'av ènement de la cité-jardin en Belgique: histoire de l'habitat social en Belgique de 1830 à 1930*. Brussels: Mardaga.

Sweet, R. 2004. *Antiquaries: The Discovery of the Past in Eighteenth-Century Britain*. London: Hambledon and London.

Thomas, T., C. Dulière, and É. de Jacquier de Rosée. 2000. *Anna Boch 1848–1936*. Tournai: Renaissance du livre.

Thomas, T., M. Lenglez, and P. Duroisin. 2005. *Anna Boch. Catalogue raisonné*. Brussels: Éditions Racine.

Tollebeek, J. 1998. "Geschiedenis en oudheidkunde in de negentiende eeuw. De Messager des sciences historiques 1823–1896." *BMGN – Low Countries Historical Review*, 113(1): 23–55. https://doi.org/10.18352/bmgn-lchr.4617

Tollebeek, J. 1999. "Het verleden in de negentiende eeuw. Arthur Merghelynck en het Kasteel van Beauvoorde." *Verslagen en mededelingen van de Koninklijke Academie voor Nederlandse taal- en letterkunde*, 109(1): 107–47.

Van Damme, I. 2015. "Recycling the Wreckage of History: On the Rise of an 'Antiquarian Consumer Culture' in the Southern Netherlands." In A. Fennetaux, A. Junqua, and S. Vasset, eds., *The Afterlife of Used Things: Recycling in the Long 18th Century*. London: Routledge, 37–48.

Van Damme, I., and S. Oosterlynck. 2021. "Seeing through the Darkness of Future Past: 'After-Suburbia' from a Historical Perspective." In R. Keil and F. Wu, eds., *After Suburbia: Urbanization on the Planet's Periphery*. Toronto: University of Toronto Press, 257–76.

Van den Corput, E. 1886. *Bruxelles-Capitale, son rôle scientifique et artistique. Avantages de sa situation. Projets d'installations nouvelles et d'embellissement.* Brussels: Maladry & Soemer.

Van den Corput, E. 1897. *Collectionneurs et collections*. Brussels: Lamertin.

Van den Corput, E. 1899. *Utilité des embellissements de Bruxelles. Nécessité de l'agrandissement territorial de la capitale de la Belgique*. Brussels: Lamertin.

Vanderpelen, C. 2011. "L'émergence des salons littéraires." *Les Cahiers de la fonderie*, 43: 66–70.

Van de Woestyne, G. 2020. *Karel en ik. Memento*. Edited by J. De Smet, L. Jansen, P. Theunynck, and H. Vandevoorde. Antwerp: Davidsfonds.

Verlaine, J. 2014. *Femmes collectionneuses d'art et mécènes de 1880 à nos jours*. Paris: Hazan.

Weale, W.H.J. 1902. *Exposition des Primitifs flamands et d'art ancien. Bruges, première section: tableaux. Catalogue*. Bruges: Desclée, De Brouwer et Cie.

Whelan, L.B. 2010. *Class, Culture, and Suburban Anxieties in the Victorian Era*. New York: Routledge.

Wytsman, P. 1900. *Intérieurs et mobiliers de styles anciens. Collection recueillie en Belgique*. vol. 1. Brussels: Wytsman.

Zitouni, B. 2010. "La croissance d'une agglomération urbaine: Bruxelles vue à travers ses archives (1828–1915)." Unpublished PhD diss., Paris, Institut d'études politiques/Vrije Universitieit Brussel.

5 The Art of Living in the Australian Suburb: Creative and Cultural Production at Home in Suburban Melbourne, 1910s–1960s

SUSAN REIDY

Introduction

In 1911, a Melbourne poet and playwright, Louis Esson (1878–1943), wrote scathingly that the suburban home "stifles ... the creative spirit ... denounces Art, enthusiasm [and] heroic virtue" (Esson, 1980 [1911]). That same year a recently married couple, who had met at art school, moved into their first home, a new house built in a small suburban subdivision known as Fairy Hills on the northeastern fringe of metropolitan Melbourne. The house of Norman Macgeorge (1872–1952) and his wife May (1882–1970) contained a studio that Norman used for his profession as an artist and a writer and lecturer on fine art. As Fairy Hills became solidly suburban, other artists moved there – Napier Waller (1893–1972), Christian Waller (1894–1954), and Lina Bryans (1909–2000) – where they too practised as professional artists in home-based studios. During the twentieth century, these people were able to shape a suburban way of living and working that embraced creative production while also actively participating in the cultural milieu of their locale, the metropolis, and the nation. While studies have been done on the suburb as the meaningful subject of Australian artists (McAuliffe, 1996; Slater, 2004), this chapter is a reflection on the life experiences of artists who not only made their homes in the suburbs but also produced their art at home, as well as constructed a way of living that embraced creative collaborations and friendships.

The suburban experiences of these artists were at odds with Esson's derisive newspaper piece, which casts the suburb as a place that would suppress creativity. This type of hostile generalization about suburban life has remained a persistent commentary on urban living and a preoccupation of scholars in Australia (Gilbert, 1988; Forsyth, 2012; Davison, 2013) and internationally (Burton, 2015). Such attention has been to the detriment of suburbia's reputation. But is it deserved? Or is the Australian suburb a positively useful place for cultural production of the artistic kind? This chapter first examines the negative idea that Australian suburbs are mindless and tedious places, as promulgated by Esson and another critic of Australian suburbia. This position is then set against the suburban experiences of five people who lived in Melbourne. My aim is to delve into the associations between their cultural production and ways of life, how they utilized their homes to sustain professions as artists and their creative interests, and the individuality they brought to these endeavours over a period of six decades. I draw on sources including newspapers and magazines of the times, diaries and personal papers, institutional records, and biographical monographs. This historical case study is deliberately descriptive. My approach as a social historian is to bring into the open aspects of the past that shine a light on ideas about human experience and, in this case, to examine what living a creative life in the suburbs encompassed, to reveal how it could provide capacity for creativity, including the use of home as a workplace, and to show that suburban women could participate in creative work and practices just as well as men. The case study will also reveal that, in this setting, the fruits of cultural production also resonated within friendships, collaborations, and contributions to society beyond suburban boundaries. But first, it is necessary to explain the Australian concept of the city.

By the beginning of the twentieth century, Australia had become and remains a highly urbanized society. The predominant urban form is suburban, and metropolitan suburbs are where most Australians live. The binary idea of city and suburb as separate entities (Harris, 2020: 29) does not hold in Australia, where "city" as both word and concept is used interchangeably to mean both the central core and the whole metropolis.[1] Australians conceive of the city as an agglomeration of its centre and its suburbs, where neither is discrete nor separate from the other. The centre contains many cultural, government, financial, and business institutions but relatively few homes, and it cannot be envisaged as a dominating core. Neither are suburbs lesser places. The centre and the suburbs together constitute the metropolitan whole. The lives of the artists described in this chapter will show the extent to which "city" in Australia means the entire metropolis. What is it, therefore, about the positions taken by critics of suburbia that is at odds with the Australian city as a complete entity? Is criticism of suburbia an illusion?

Creating a Fantasy: Critical Commentary on the Australian Suburb

The modern cultural reputation of the suburb as a "hard place to love" has its roots in nineteenth-century urban expansion and the unease it caused about the developing nature of cities (Davison, 2013). In 1868, the English cultural commentator John Ruskin gave a lecture ("The Mystery of Life and Its Arts"), in which he described his country's industrial slums as "festering and wretched" (Ruskin, 1893 [1868]: 184). Critiques such as this one, about congested, shoddy urban places where people supposedly lived sordid lives, constituted a new type of public commentary about urban lifestyle. In Australia, from the 1870s, the older residential suburbs close to urban centres began to slide into decay, and eventually they also became marked as "slums" and attracted adverse comments (Anderson et al., 2004). Judgmental attitudes towards "slums" perhaps constituted the first version of critical complaining about suburban living.

At the same time as slums were being denigrated, there was an alternative. In the later decades of the nineteenth century and into the twentieth century, Australian cities swelled with an ever-increasing number of new suburbs, which featured many private free-standing houses with gardens, giving rise to urban agglomerations with a distinctly residential way of life. This development was not far removed from an alternative to the slums that John Ruskin had proposed in his 1868 lecture, an ideal place where people would live in "beautiful" houses surrounded by gardens, spaciously arranged in clean streets with fresh air and views of the countryside (Ruskin, 1893 [1868]: 184). Just a few years after Ruskin's lecture was published, the English town planner Ebenezer Howard was so inspired by Ruskin's ideal that he included its description as a quotation in *Garden Cities of To-Morrow*.

In his book, Howard presented an urban typology, the "garden city," that was close to Ruskin's proposal, although Howard added the idea that residential, industrial, and civic areas should be separated (Howard, 1902: 50–1). In the first two decades of the twentieth century, Howard, in turn, was an influence (along with the American "city beautiful" ideology) on the Australian town planning movement's support for residential urban suburbs full of houses with gardens (Garnaut, 2000: 47–8) where people could live a comfortable, healthy life. Ever since, Australian urban expansion has occurred mostly in this manner, the creation of suburbs full of residential individuality.

It was in the period of nascent town planning that Esson took aim at Australia's suburban residential environment, and while it appeared to come out of nowhere, Esson may have followed the lead of a contemporary English journalist Thomas Crosland, who published *The Suburbans*, in which he took a harsh and critical look at new suburbs in London (Crosland, 1905). Crosland's book was immediately available in Australia, and it attracted a few newspaper reviews, all negative and noting the cynicism of his views (see, for example, *Age*, 16 September 1905, 4). Crosland's attitude was tough concerning the lives of suburban women, finding both them and the suburbs they inhabited to be empty of culture. So began the idea of the modern suburb as a dreary wasteland that was "inimical to creativity" (Bilston, 2013: 621, 632).

Crosland's and Esson's type of disapproval persisted throughout much of the twentieth century in Australia as a notable if occasional commentary on urban society (see Rowse, 1992; Kinnane, 1998). Even novelists, including Lennie Lower (1930) and Patrick White (1956), found suburbia wanting. One of the most widely read critics of Australian suburbs was the well-known architect and social commentator Robin Boyd, who wrote extensively on public taste in the 1950s and 1960s. One of Boyd's main targets was suburbia's swathes of detached suburban houses and gardens. In *The Australian Ugliness*, he describes the transformation of a place of natural beauty into a suburb as a process of degradation: "The subdividers arrive, and behind them the main wave of suburbia ... [S]oon rows of cottages and raw paling fences create a new land scape ... [O]nce any man sets his eyes on any pretty place in Australia the inexorable process of uglification begins" (Boyd, 1960: 35). Boyd's type of post-war anti-suburban critique fits into what has been called a "sneer" from urban intellectuals (Davison and Dingle, 1995: 3).

By positioning suburbs as boring or unattractive, complainers had the effect of elevating their own positions at the same time as they narrowed thinking about suburban living, which resulted in the creation of a rhetorical, anti-suburban language (Slater, 2004: 177–8; Peel, 2002) and, in certain cases, the belittling of women's lives (Johnson, 2012: 219). Esson's and Boyd's superiority centred on ideas that the suburb was alienating and unsophisticated. For Esson, the suburban home – including as a woman's realm – was a dull place of trivialities like "afternoon tea." For Boyd, the suburban way of life encouraged bad taste and conformity. Yet these commentaries did not necessarily reflect the realities of urban life – they were opinions, just as Ruskin's was. They were not reportage; instead, they presented a moralizing point of view in which suburbia was presented as a singular and faulty environment.

Yet not all suburbs are blasted plains; they can be "fertile terrain" for the creative mind (Gibson, 2015: 50). The critics ignored the reality of the whole Australian metropolis as a place of dynamic urban relationships and suburbs as places of potential. Hugh Stretton observed: "You don't have to be a mindless conformist to choose suburban life. Most of the best poets and painters and inventors and protesters choose it too. It reconciles access to work and city with private, adaptable, self-expressive living space at home" (Stretton, 1975: 21). In Robin Boyd's case, it was certainly true that creativity could be cultivated in the suburbs: His parents were artists, as were many members of his extended family (Niall, 2002). He grew up and spent his adult life in suburban Melbourne, where he became a notable architect and widely read social critic. Indeed, while Boyd was disdainful about certain types of design, he was also a champion of suburban living through his own modernist designs for family homes and his crucial role as director of the Institute of Architects' Small Homes Service in the 1950s (Serle, 1995; Dingle and O'Hanlon, 1997: 37–8).

The suburbs contain many strong and complex community relationships, sophisticated creative productivity, and cultural connections that extend not only to the local but reach into the life of the city and beyond. These are to be found not in an "undifferentiated rubric of suburbia" (May, 2009: 79), not in the ghastly and fantastical landscape of the critics but in the specificities of place and individual personal experiences.

Creating Art in a Melbourne Suburb: Cultural Production in the Home

During the twentieth century, four artists lived in a small corner of a Melbourne suburb in a place known as Fairy Hills, where they utilized their homes to practise their art and to participate in and contribute to the cultural life of metropolitan Melbourne. Fairy Hills is a small locality forming the southern corner of the Melbourne suburb of Ivanhoe. The most prominent physical feature of Fairy Hills is the riverine bushland of the Yarra River and Darebin Creek, waterways that define, respectively, the locality's southern and western boundaries. Fairy Hills takes its name from a small farm that, in the 1880s, lay within the rural Shire of Heidelberg, close to Alphington (a suburb then on Melbourne's northern fringe) and two nearby rural villages, Ivanhoe (as it was then) and Heidelberg; all were connected to the city centre via a metropolitan train service. As elsewhere in Melbourne, suburban development and subdivision in the district followed the railway: in 1889 the Ivanhoe Advancement Society was formed with the aim of promoting the area to potential suburban homemakers (Garden, 1972: 172).

The Fairy Hills farm was sold in 1896 and subsequently subdivided into allotments suitable for suburban "villas." The subdivision's new streets, which meandered around the slopes above the picturesque riverine landscapes, began to fill with detached houses with their own gardens, a progressive transformation of the land from rural to suburban. During the twentieth century's first decade, Fairy Hills was keenly promoted as a modern and scenic place for suburban living, and the district and metropolitan Melbourne were cast in a bright and dynamic light: "The marvel of modern civilization is the growth of great cities and their environs" (Ivanhoe and Alphington Progress Society, 1910). By 1911, the Fairy Hills subdivision contained 11 streets, 274 suburban allotments, and 53 houses (Public Record Office Victoria, Rate Books for 1911–12). One of the newest houses that year was built for Norman and May Macgeorge on an allotment with a frontage on the Yarra River. The house contained a studio, Norman's workplace, where he would paint and write.

During the twentieth century, there were three clusters of artists who lived in suburban Melbourne and have become well known: "Monsalvat," an artist colony established in the 1930s by artist and mentor Justus Jorgensen in the outer northeastern suburb of Eltham; "Heide," the home from 1934 of art patrons Sunday and John Reed in Bulleen (a few kilometres northeast of Fairy Hills); and "Open Country," the home of a family of artists in the suburb of Murrumbeena (southeast of the city centre). From 1913 to the 1950s, Open Country nurtured the creative Boyd family of parents, children, spouses, and other relatives, Robin Boyd among them (Niall, 2002). By contrast, the situation at Fairy Hills was not a consciously artistic group but rather a place where the resident artists fostered associations into friendships and occasional collaborations. The Macgeorges would make their home a social hub as well as a workplace. The river landscape of Norman Macgeorge's suburb was a recurring subject of his art, and he described it as "a perfectly lovely spot as far as natural beauty goes" (Pigot, 2001: 12). Although Norman's artistic talent remained competent rather than inspired (one review described his work as "sincere"; Bell, 1933), he nevertheless exhibited and sold paintings throughout his career. In his way, Norman was continuing the long-standing association of the district of Heidelberg with the art of landscape painting that began with Louis Buvelot in the 1870s and was consolidated in the early 1890s by the famous Heidelberg School of *plein air* painters, Tom Roberts and Arthur Streeton among them (Clark and Whitelaw, 1985).

Norman Macgeorge was a significant and well-known figure in Melbourne's art scene for thirty years, not only as an exhibiting painter. During the 1920s and 1930s, he presented very well-attended art appreciation lectures at Melbourne's National Gallery and the University of Melbourne, and in rural Victoria. He was an active member of several art organizations, including the Victorian Artists Society (VAS) and the Contemporary Art Society (CAS), and in the 1930s he publicly championed the cause of modern art and reviewed art exhibitions for the *Herald*. May Macgeorge settled into life as a homemaker, hostess, and community volunteer. Although she had abandoned the artistic ambitions that had taken her to art school, she found self-expression in the everyday creativities of cultivating the garden, making her own stylish clothes (as many women did throughout the twentieth century), and decorating the house interior and Norman's exhibitions. Hospitality was central to the lifestyle of the Macgeorges, and they regularly welcomed into their home other residents of Fairy Hills and visitors from Melbourne's art community.

Fairy Hills was busy with local development during the 1920s, and the Macgeorges were involved. Residents formed a social and progress association in 1921, and its first activity was to build a community hall with the resources and labour of the residents. Tennis courts were soon added, and from 1928, the hall accommodated a small government school for the vicinity's ever-increasing number of young children. Norman Macgeorge was one of three trustees who managed the hall for the community (Fairy Hills Social and Progress Association, 1921), a typical example of his unpaid community service. The locality gained its own station in 1922 (named Darebin after the creek) on the existing train line, thus increasing the convenience of metropolitan travel. By 1927, a small commercial precinct had been established along the main road next to Darebin Station, with eight shops and a motor garage (Public Record Office Victoria, Rate Books for 1926–7).[2] Fairy Hills also gained two churches in the 1920s. May Macgeorge was an active member of the convivial and community-minded St Stephen's Church Ladies' Guild from its inception in 1927.

Two more artists, Napier and Christian Waller, moved to Fairy Hills in 1922, just one street away from the Macgeorges. The house was built to the Wallers' specifications, with a large studio as its centrepiece. Napier Waller had met Christian Yandell in 1913 while they were both studying at the National Gallery art school. They married in 1915 before Napier enlisted in the army. During the First World War, Christian worked in Melbourne as a graphic artist and book illustrator (for example, Edwin Brady's [1918] *Australia Unlimited*), and exhibited with the VAS. In 1916, Napier was badly wounded in the Battle of Bullecourt in France, and his right arm was amputated. He was soon able to draw and paint with his left hand, and among his first efforts on returning to Australia was a book and touring exhibition of his war sketches (pre- and post-injury; Waller, 1918; for a digital copy of the whole book, see the artworks list in the appendix of this chapter). After the war, both Napier and Christian exhibited at the VAS, where they would have met Norman Macgeorge (*Argus*, 1 September 1919, 6; *Herald*, 23 May 1921, 6).

Napier and Christian Waller had lifelong successful careers as artists. They were both exceptionally talented and in the 1920s expanded their interests beyond painting and drawing into other mediums, including the new technique of linocut, plus woodblock printing and, unusually for fine artists, large-scale painted and mosaic murals for public and commercial buildings and stained glass, particularly for churches. Napier remained Australia's pre-eminent artist of public murals for thirty years; surviving examples include the Melbourne Town Hall (1926–7) and the State Library of Victoria (1929). The couple found interest in allegory, myth, and symbolism and fashionably embraced the Art Deco style. Napier combined its modern aspects with a neoclassical approach in his depictions of the human body, as encapsulated in a portrait of his wife (1932) and the mosaic mural *I'll Put a Girdle Round About the Earth* (1933) on the façade of Melbourne's Newspaper House on Collins Street (for a digital image, see the artworks list in the appendix of this chapter). Christian used Art Deco's geometric, curvilinear, and naturalistic forms to great effect in her depictions of human figures and the spaces around them (Draffin, 1978; Thomas, 1992).

In 1926, Napier created some designs for William Montgomery's stained glass studio in Melbourne, and after Montgomery died in 1927, Napier and Christian completed Montgomery's full-scale drawings (cartoons) for the enormous Stevens Window in the University of Melbourne's Wilson Hall (Hughes, 2007: 231–2). This project was the beginning of a growing body of stained glass work by Napier and Christian, but in these years, they could only create the designs in their home studio; a glass company painted and cut the glass and assembled the windows. These experiences piqued the Wallers' interest in executing the glass painting, cutting, and window assembly themselves, for in 1929 they left Australia to expand their knowledge of printmaking and stained glass design and techniques of production in Europe. They spent time at a London glass workshop, scrutinized stained glass windows in French and British cathedrals, and in Italy they were inspired by the famous medieval mosaics in Venice and Ravenna.

Christian and Napier returned to Melbourne in 1930, bringing with them an electric furnace for annealing stained glass (*Argus*, 10 March 1930, 6). They immediately embarked upon changes to their Fairy Hills home to create additional workspace. A new studio for glass work was built adjoining the house, while the original studio was converted to a living room but had its ceiling raised to a double height with a gallery, making it possible to hang full-scale cartoons required for making murals and stained glass windows. Another studio was built in the garden to house a hand printing press. These changes made the Waller property into quite an intense place for the making of artworks, where cultural production had a powerful physical presence. Within a month of the couple's return to Melbourne, Napier gained a commission from the University of Western Australia in Perth for a mosaic to decorate its new Winthrop Hall. Napier completed the design work and assembly of *The Five Lamps* mosaic in the Waller house studios. Then it was packed and transported across the continent to Perth, where Napier supervised the installation (*Daily News*, Perth, 3 September 1931, 6). The Wallers also began taking commissions for stained glass windows. They collaborated on one of the first of these, three windows for St John's Church of England in the Melbourne suburb of Camberwell (*Argus*, 7 August 1930, 5).

For Christian, this period was an exceptional time, for in 1932 she produced *The Great Breath*, a special artist's book of linocut prints (Waller, 1932; for a digital copy of the whole book, see the artworks list in the appendix of this chapter). Christian's interests in northern European mythology were graphically incorporated into the illustrations, and she completed every aspect of the book at the Waller house studios: designing and cutting the plates, preparing the papers, printing all the pages on the hand press. She hand-bound each book with silk cords and covers of green cloth embossed with gold ink. The book attracted high praise; Melbourne art critic Blamire Young wrote of Christian's "intensely imaginative genius" and of the "impeccable taste" in the book's production, from the "velvety richness" of the black inking to the "perfection of the hand-printing and the choice of paper" (*Herald*, 9 April 1932, 24). Young was also an artist. It was not unusual in the first half of the twentieth century for Australia's art critics to be artists, who could bring professional rigour to their reviews; other artist/critics included George Bell, the famous Arthur Streeton, and Norman Macgeorge himself.

Napier began teaching visual art at the Melbourne Technical College in 1932, and Christian created a mural (*Hymn of the Robe of Glory*) for a chapel at a new cemetery in the outer Melbourne suburb of Fawkner. Norman Macgeorge obligingly modelled for one of the mural figures (for a digital copy of Christian's portrait sketch, see the artworks list in the appendix of this chapter). But from this period, Christian's main occupation was designing, painting, and making stained glass windows, mostly for churches, an activity that sustained her artistry for the rest of her life. As well, working in that medium perhaps reflected her

Christian beliefs, and it gave her the opportunity to make a living as an artist in a profession dominated by men. Her work certainly justified the incorporation of the glass studio into the Wallers' home. Among Christian's works in the 1930s was the design and assembly of seven stained glass windows for St James Church in nearby Ivanhoe (1938), and she closely supervised their installation. The commission for these windows had come from the architect Louis Williams, who designed numerous buildings for the Anglican Church (including St Stephen's in Fairy Hills), and the Wallers collaborated with him on a number of projects (Correspondence, 1938). Williams admired their stained glass work; he wrote that, "of all the [stained glass] designers in Australia, I consider that there are none comparable with Napier Waller and Mrs. Waller: I would even place Mrs. Waller first" (Thomas, 1992: 47).

Norman Macgeorge was also industrious during the 1930s in metropolitan Melbourne. Apart from occasional one-man exhibitions of his art, he also chaired the organizing committee of an art show held in Ivanhoe in 1934 to celebrate the Shire of Heidelberg's centenary and its rise in municipal status to "City." This event was another example of Norman's voluntary community service. He persuaded artists who had connections to the Heidelberg vicinity to participate in the Ivanhoe exhibition; among those who agreed were William Frater, Ada Plante, William McGuiness, Napier Waller, and Christian Waller. Norman publicized the exhibition and drew attention to Napier and Christian's "generous gift" of an exhibition poster and catalogue cover (*Argus*, 3 November 1934, 20). Christian's design for the catalogue cover was a striking black and white linocut composition of paint emerging from several paint tubes to form the long hair of two standing female figures (*Heidelberg Art Exhibition November 1934*). The following year, Norman joined the newly formed Contemporary Art Group, which promoted modernist artists in an exhibition held at Melbourne's National Gallery. Norman participated in the show, even though his work would have looked old fashioned among the show's modernist works by Arnold Shore, Margaret Preston, and others (*Argus*, 9 July 1935, 5).

Napier Waller gave up teaching in the late 1930s, as he gained a commission in 1938 from the Commonwealth Government of Australia to create a series of stained glass windows and interior mosaics for a Hall of Memory in a national war memorial planned for the nation's capital city, Canberra. Napier was an obvious choice for these artworks, not only because he was considered to be the country's leading creator of large-scale public murals and well versed in stained glass design, but also because he had served in the First World War. That same year Norman Macgeorge, at the age of sixty-five, held his last solo exhibition, and, as she had done previously for his exhibitions, May Macgeorge decorated the gallery. The women's pages of the Melbourne press often reported on social aspects of exhibitions; in this case, May's decorations were described in the pages of the *Argus*: "Peonies, deep red and palest pink … were arranged in quaint low holders and in squat pottery jars … [B]owls of deep wine purple wallflowers, pure white fruit blossom, and bush flowers and tea tree were placed here and there in delightful effect" (*Argus*, 21 September 1937, 6). There were future implications for May's interest in artful arrangements of colours and objects, as these would lead to an abiding friendship with one other artist who was soon to live in Fairy Hills. Fortuitously, also in 1937, Norman Macgeorge joined the organizing committee for another Ivanhoe art show. Among its exhibits were works by two accomplished modernists, William Frater and Ada Plante, and three paintings by a young woman just beginning her career in art, Lina Bryans.

Lina had met William Frater in the mid-1930s, and her artistic development benefitted from their friendship. William lived in Alphington, the suburb next to Fairy Hills, and he knew the Wallers, who were his contemporaries. William may have introduced Lina to Ada Plante, who at this time was living and painting in Fairy Hills in what had been a colonial inn, known as Darebin Bridge House. In 1940, when Lina inherited money, she bought the old house and moved in with her young son. The house had plentiful space for living accommodations and studios for Lina, Ada, and occasionally other artists, among them William Frater and Ian Fairweather (Forwood, 2003). Part of Lina's enduring fame is due to the friendship she extended to Fairweather. She provided a home and a studio for the brilliant but reclusive artist for two years, during which he produced a large body of work, and Lina bought several of his paintings. Today Fairweather remains one of Australia's most admired modernist artists.

In the years Lina lived in Fairy Hills, she became a skilful painter and portraitist, and was an active member of the Melbourne Society of Women Painters. She also made Darebin Bridge House into a salon for Melbourne artists and the *Meanjin* magazine literary set and thus became known for her stylish way of living and as a gregarious host (Bail, 1981; Forwood, 2003). Lina herself described the house as "a centre of activity and art" (Albiston, 1986: 36). The manner in which Lina lived in Fairy Hills is a reminder of the Australian characterization of "city" as a whole metropolis; there is no sense in which such a salon could only flourish in the "centre" – an Australian suburb was an eminently suitable location. In addition, another aspect of that lifestyle would endure for Lina because, during her time in Fairy Hills, she established a lifelong friendship with the Macgeorges and particularly with May Macgeorge.

Creating Lifestyle in a Melbourne Suburb: Friendships and Residential Creativity

When Louis Esson criticized the suburbs, he took aim at the lifestyle of the suburban woman, writing that "the suburb exists for the villa, and the villa for the drawing-room. The drawing-room exists for the lady of the house ... It is here that she expresses herself, holds her court, wears curious costumes, dispenses gossip and afternoon tea" (Esson, 1980 [1911]). The friendship between Lina Bryans and May Macgeorge grew because of geography, but its qualities expose the reductionism at the core of Esson's idea of suburban conviviality. His conception does not recognize that rich meaning can be found in civility and the bonds forged between people. The friendship between the two women was nurtured by hospitality, and it flourished because of a shared artistic sensibility and a suburban lifestyle in which connections reached beyond the personal satisfactions of self-expression and personal creativity. When Lina Bryans moved to Fairy Hills, she was a young artist at the beginning of her career, but although she had financial means, she had also left her marriage and was raising her son alone. As such, she was a person slightly outside society's cultural norms. For her, friendship could provide community and social stability. For the rest of her life, Lina kept letters she had received from May Macgeorge. She also kept a photograph in which May is seated in a studio in front of a large portrait of her husband Norman, with a few of his small paintings at her feet (Bryans, Lina, Papers). The setting of this photographic tableau is the glass studio in the Waller house, and the portrait (which was unfinished) was the work of Napier Waller. The letters and photograph that Lina kept were mementos of May Macgeorge, and they give us an understanding of the relationship between the women.

Among the interests that May and Lina shared were the hospitable home, a love of beautiful objects, and a fascination with the use of colours, textures, and forms. Beautiful objects could be as large as a home or as small as a vase of flowers, and they were valued not only for their aesthetic qualities but because they could be shared. A clue to this notion can be found in a gift of flowers from May, and Lina's response: "You are always giving me such lovely surprises – today the parcel with the most exquisite garden flowers. I have them in a deep greenish-blue pottery vase. I think you would approve" (Macgeorge Collection, letter from Lina Bryans to May Macgeorge, 29 October 1963). May and Lina's interests of this kind were centred on the appreciation of interior spaces and artful combinations of materials to create beauty. For twenty years, the two women exchanged correspondence and gifts in the knowledge that the recipient would respond as a connoisseur. May's decoration of the Macgeorge's home attracted journalistic attention in the 1940s, which revealed the richness of the house interior. Throughout the rooms, there were book-lined alcoves, hangings of printed linen, old blue and white china, many pictures on the walls, Persian rugs on the "softly glowing" waxed floors, and "antique brassbound cupboards" (Johnston, 1948). Lina was also attentive to décor, for she created new interiors for Darebin Bridge House. Lina decorated the rooms and painted the timber floors in colours of "dull green, rust, or deep red" and then waxed them to produce shine ("Artist's Home Was Once a Hotel," *Herald*, 29 May 1944, 7). A sitting room had its woodwork and floors toned to match the "moss green" colour of the walls, and the room's décor was completed with a striped Dhurri rug, pewter vessels, an old French armoire, and one of Ian Fairweather's paintings (Wilson, 1947). May described the result as "quite perfect" (Macgeorge, May, Diaries, 5 September 1947).

Ballet was another other interest that Lina shared with May (and Norman), and together with the Macgeorges she attended the Princess Theatre in the city in October 1947 for the opening night of the Ballet Rambert's first Australian season (Macgeorge, May, Diaries, 17 October 1947; *Herald*, 18 October 1947, 7). This occasion took place almost a year after Norman Macgeorge's first book, *Borovansky Ballet in Australia and New Zealand*, was published (Macgeorge, 1946). The Macgeorges were enthusiastic patrons of ballet, having attended many performances during the 1930s when a number of small European dance companies came to Australia and performed to public acclaim. Edouard and Xenia Borovansky toured Australia with the Royal Covent Garden Russian Ballet in 1938, then settled in Melbourne and established a dancing school and a ballet company (Salter, 1980: 86). The European dance companies also brought to Australia their practice of associating with artists. In France after the First World War, contemporary ballet companies such as Sergei Diaghilev's Ballet Russes commissioned painters, including Leon Bakst and Pablo Picasso, to design sets and costumes. In the 1940s, Melbourne artists also made associations with ballet companies.

As he had done with all his projects, Norman used his home as workplace. He called on his artworld friends and associates to contribute essays about the Borovansky Ballet Company along with photographs, paintings, and sketches of the company's dancers and ballets. The Macgeorges paid for the book's publication. Norman persuaded his neighbour, Napier Waller, to design the dust jacket, which they discussed at the Waller house (Macgeorge, May, Diaries, 21 May and 8 June 1946). This book seems an unusual project for Norman, who was in his seventies, but the trajectory of his life clearly indicates that he was energetic and ever the collaborator, keen to involve his friends within Melbourne's art community. The book sold very well and gained good reviews and congratulations: "Mr Macgeorge is well known in his own field of art, and his tribute to the young Australian ballet is typical of the influence this lively fusion of dance, drama, music, and design can exercise beyond the four walls of the theatre" (*Argus*, 9 November 1946, 12).

Norman followed this success with *The Arts in Australia* (Macgeorge, 1948), a book that contained twenty-three essays and illustrations by contributors on painting, sculpture, music, dance, and theatre in Sydney and Melbourne. As was the case with the ballet book, it featured work by several of the Macgeorges' friends, including two paintings by Lina Bryans and numerous works by Napier Waller. This book was the last of Norman's public efforts. He and May continued to live quietly in their home in Fairy Hills, the house where they had made the best of the suburban way of life, with each in their own way contributing to the art of the city.

Conclusion: Creating Culture from the Suburban Home to the Nation

As Norman's life as a public figure drew to a close, the Macgeorges made plans to ensure their commitment to the arts would remain. In the mid-1940s, they discussed the idea that their house could be "left for a sort of hostel for artists and musicians" (Macgeorge, May, Diaries, 22 September 1946). They chose to bequeath the property to the University of Melbourne for use as a centre for artistic production. This gift would have long-term cultural resonances. Following May's death in 1970, the bequest to the university has continued to fund residential fellowships in the Macgeorge House and other scholarships in the creative and performing arts. This period was also a busy time in the Waller household, where work continued on the many stained glass and mosaic designs for the National War Memorial's Hall of Memory. Although Napier is the named creator of the complete Hall of Memory artworks, it is likely that Christian contributed to the window designs; while unacknowledged officially, her style is evident in the finished works, in particular the patterning and use of colours. Otherwise, throughout the 1940s, the bulk of Christian's work was making stained glass windows for Protestant churches in suburbs and country towns in three states (see, for example, *Herald*, 12 February 1944, 3). She estimated her body of work in this medium as totalling over fifty windows (Westhoven, 1948).

Napier went to Canberra in 1950 to oversee the installation of the Hall of Memory's stained glass windows at the Australian War Memorial (*Canberra Times*, 27 July 1950, 4). The long time frame for this project was partly due to war delaying progress but also because of the work's great scale. There are fifteen large stained glass windows, and the hall's tall interior walls and dome are completely lined with a mosaic of heroic figures and patterning. Every aspect is redolent with symbolism, representing quintessential human attributes related to military service and sacrifice. The Wallers were no strangers to the use of symbolism in art and incorporated into the window and wall designs an abundance of motifs from crowns to flowers and animals (for digital depictions of the Hall of Memory interior and explanations of the symbolism, see the artworks list in the appendix of this chapter). Christian died in 1954, so she never saw the Hall of Memory completed, the centrepiece of one of the most important institutions associated with Australia's fascination with the memorialization of war. It is a powerful example of how suburban creativity can have reverberations within national identity.

While Napier's attention was on the Hall of Memory in the 1950s, May Macgeorge continued her friendships with him and Lina Bryans. There was to be one final collaboration between the Macgeorge and Waller households. In 1962, Napier designed several stained glass windows for St Stephen's Church in Fairy Hills, a commission he received from May Macgeorge. It would also be her final donation to the church where she had worshipped since 1927. One window was dedicated to the memory of her husband Norman (*Parish Notes*, 1962; Harridge, 2000).[3] The commission for these windows brought to a close the era of Fairy Hills as a mid-century suburban place of art. But they also represent the sensibilities of that group of people, their embrace of practical collaborative associations between neighbours and participation in the life of local and metropolitan communities.

For all the adverse writing about suburbia, one might conclude that suburbs are elusive places without the kind of vigour that conceptualizes city or country. Yet, the fact that most Australian urban dwellers live in the suburbs makes suburban criticism into a kind of absurdity. The complaining critics of the suburbs were expressing opinions that created an *idea* of the suburb in which everything was the same (and wanting), but they mistook physical appearances for conformity and disregarded the experiences of those who lived there. Norman Macgeorge, May Macgeorge, Christian Waller, Napier Waller, and Lina Bryans chose to live in Fairy Hills, where they were able to experience lives of great substance, mostly of their own making. They utilized their homes to devote themselves to cultural production, live creatively, make connections with each other and the local community, and participate meaningfully in the cultural life of the city and the nation. In opposition to the complainers' ideas, their homes were not sterile. Suburban location and home ownership gave the Macgeorges, the Wallers, and Lina Bryans freedom and space to produce their art according to their own tastes and preferences of practice, whether it was Norman Macgeorge painting nature on the river, Napier Waller and Christian Waller making artworks for locations throughout the country, or Lina Bryans providing studios and amenity in her home for herself and other artists.

The complainers also ignored the capacity of the suburb to support human interactions of cooperation and friendships. Living near other artists encouraged community and the embrace of creative household practices. By collaborating with others, each of these people honoured the processes of work and dealt with the complexities that came with them. Lina Bryans not only found her artistic feet in Fairy Hills but also her taste for the atelier and salon, and the opportunity to surround herself with those who also chose modernism. May Macgeorge was in many ways the ultimate collaborator, supporting the practices of art and embracing community. Her life demonstrates that conceptions of cultural production need not be limited to the exercising of professional creativity and that women could find opportunities for producing an aesthetic landscape within their suburban homes. For these people, cultural production was not only enabled by the suburban way of life; it also became an integral part of their suburban experiences. As a result, suburbia and cultural production were not mutually exclusive; rather, together they made a way of life that was both creative and suburban. For the Macgeorges, the Wallers, and Lina Bryans, their home-based neighbourhood of artists allowed them to experience the suburb as a place of creative possibilities that extended beyond local boundaries into the cultural life of the metropolis and the nation.

Appendix

Artworks of Discussed Artists Available Online

Bryans, Lina. The National Gallery of Victoria holds several works; see the website, https://www.ngv.vic.gov.au/explore/collection/artist/705/

Waller, Christian. 1932. *The Great Breath: A Book of Seven Designs*. Digitized images of the complete original edition available on the Art Gallery of NSW website, https://www.artgallery.nsw.gov.au/collection/works/250.1975/

Waller, Christian. 1937. Portrait drawing of Norman Macgeorge for Fawkner Cemetery. State Library of Victoria website. Digital identifier H2000.63/4. https://www.slv.vic.gov.au

Waller, Napier. 1918. *War Sketches on the Somme Front by M. Napier Waller*. Digitized copy available on the State Library of Victoria website, https://viewer.slv.vic.gov.au/?entity=IE6075848

Waller, Napier. 1932. Painting, *Christian Waller with Baldur, Undine and Siren at Fairy Hills*. National Gallery of Australia. Accession no. NGA 84.845. Available on the website, https://searchthecollection.nga.gov.au/object/49895

Waller, Napier. 1933. *I'll Put a Girdle Round About the Earth*. Oil painting prepared for the Newspaper House mosaic. Image available on National Gallery of Australia website, https://nga.gov.au/exhibitions/art-deco/ (Images of the mosaic on Newspaper House, Collins Street, Melbourne, can also be seen online.)

Waller, Napier. 1950s. Hall of Memory, Australian War Memorial. Images of the stained glass windows and wall mosaics, and an explanation of their symbolism available on the National War Memorial website, https://www.awm.gov.au/visit/visitor-information/features/hall-of-memory

NOTES

1 Municipal governments are also known as cities (as in the City of Banyule, which is the municipality of Melbourne where Fairy Hills is located).
2 One unforeseen effect of Darebin Station was that its name would eventually overtake Fairy Hills as the locality name.
3 When St Stephen's was closed in the 1980s, the windows were removed and installed at St James, Ivanhoe.

REFERENCES

Archival Sources

Bryans, Lina. Papers. State Library of Victoria. MS9420.
Correspondence with Christian Waller, 1938. State Library of Victoria. Louis Williams Papers. MS10990.
Fairy Hills Social and Progress Association. 1921. *Grand Fete Souvenir Programme*. Pamphlet. Heidelberg Historical Society Collection.
Heidelberg Art Exhibition November 1934. Exhibition Catalogue. State Library of Victoria, Christian Waller Art and Artists File.
Ivanhoe and Alphington Progress Society. 1910. *Beautiful Ivanhoe: The Suburb of Model Homes and Scenic Charm*. Booklet. Heidelberg Historical Society Collection.
Macgeorge, May. Diaries. Macgeorge Collection. University of Melbourne Archives.
Macgeorge Collection. University of Melbourne Archives.
Parish Notes. 1962, August. Newsletter of St James' Ivanhoe and St Stephen's Darebin. Heidelberg Historical Society Collection.
Public Record Office Victoria. Ivanhoe Riding Rates Books for years 1911–12 and 1926–7. VPRS/2870.

Published Sources

Albiston, V. 1986. "The Pink Hotel." *This Australia*, 5(4): 32–6.
Anderson, F., C. Coney, and E. Nelson. 2004. "Haunt of Convicted Thieves and Prostitutes." In P. Yule, ed., *Carlton: A History*. Carlton Vic: Melbourne University Press, 430–44.
Bail, M. 1981. *Ian Fairweather*. Sydney: Bay Books.
Bell, G. 1933. "Sincere Work by Artist." *Sun*, 25 July 1933.
Bilston, S. 2013. "'Your Vile Suburbs Can Offer Nothing but the Deadness of the Grave': The Stereotyping of Early Victorian Suburbia." *Victorian Literature and Culture*, 41(4): 621–42. https://doi.org/10.1017/S1060150313000144

Boyd, R. 1960. *The Australian Ugliness*. Melbourne: Cheshire.

Brady, E.J. 1918. *Australia Unlimited*. Melbourne: George Robinson.

Burton, P. 2015. "The Australian Good Life: The Fraying of a Suburban Template." *Built Environment*, 41(4): 504–18. https://doi.org/10.2148/benv.41.4.504

Clark, J., and B. Whitelaw. 1985. *Golden Summers, Heidelberg and Beyond*. Sydney: International Cultural Corporation of Australia.

Crosland, T.H.W. 1905. *The Suburbans*. London: John Long.

Davison, G. 2013. "The Suburban Idea and Its Enemies." *Journal of Urban History*, 39(5): 829–47. https://doi.org/10.1177/0096144213479307

Davison, G., and T. Dingle. 1995. "Introduction: The View from the Ming Wing." In G. Davison, T. Dingle, and S. O'Hanlon, eds., *The Cream Brick Frontier: Histories in Australian Suburbia*. Clayton, Vic: Monash Publications in History, 2–17.

Dingle, T., and S. O'Hanlon. 1997. "Modernism versus Domesticity: The Contest to Shape Melbourne's Homes, 1945–1960." *Australian Historical Studies*, 27(109): 33–48. https://doi.org/10.1080/10314619708596041

Draffin, N. 1978. *The Art of M. Napier Waller*. South Melbourne: Sun Books.

Esson, L. 1980 [1911]. "Our Institutions IV: The Suburban Home." In *Ballades of Old Bohemia: An Anthology of Louis Esson*. Ascot Vale, Vic: Red Rooster Press, 90–1

Forsyth, A. 2012. "Defining Suburbs." *Journal of Planning Literature*, 27(3): 270–81. https://doi.org/10.1177/0885412212448101

Forwood, G. 2003. *Lina Bryans: Rare Modern 1909–2000*. Carlton Vic: Miegunyah Press.

Garden, D. 1972. *Heidelberg: The Land and Its People, 1838–1900*. Melbourne: Melbourne University Press.

Garnaut, C. 2000. "Towards Metropolitan Organisation: Town Planning and the Garden City Idea." In S. Hamnett and R. Freestone, eds., *The Australian Metropolis: A Planning History*. Sydney: Allen & Unwin, 46–64.

Gibson, M. 2015. "Creativity and Attenuated Sociality: Creative Communities in Suburban and Peri-Suburban Australia." In J. McDonald and R. Mason, eds., *Creative Communities: Regional Inclusion and the Arts*. Bristol: Intellect, 49–61.

Gilbert, A. 1988. "The Roots of Anti-Suburbanism in Australia." In S.L. Goldberg and F.B. Smith, eds., *Australian Cultural History*. Cambridge: Cambridge University Press, 33–9.

Harridge, W. 2000. *The Stained Glass Windows of St. James' Anglican Church, Ivanhoe, Victoria*. http://stjameschurch.org.au/wp-content/uploads/2020/01/Stained-glass-windows-booklet-Feb-2019.pdf

Harris, R. 2020. "Using Toronto to Explore Three Suburban Stereotypes, and Vice Versa." In J. Nijman, ed., *The Life of the North American Suburbs: Imagined Utopias and Transitional Spaces*. Toronto: University of Toronto Press, 23–44

Howard, E. 1902. *Garden Cities of To-Morrow*. London: Swan Sonnenschein.

Hughes, B. 2007. *Designing Stained Glass for Australia: 1888–1927: The Art and Professional Life of William Montgomery*. PhD diss., University of Melbourne.

Johnson, L.C. 2012. "Creative Suburbs? How Women, Design and Technology Renew Australian Suburbs." *International Journal of Cultural Studies*, 15(3): 217–29. https://doi.org/10.1177/1367877911433744

Johnston, E. 1948. "The River Yarra Is Their Friend." *Australian Home Beautiful*, 27(2): 16–17, 44.

Kinnane, G. 1998. "'Shopping at Last!' History, Fiction and the Anti-Suburban Tradition." *Australian Literary Studies*, 18(4): 41–55. https://doi.org/10.20314/als.36e5699bc7

Lower, L.W. 1930. *Here's Luck*. Sydney: Angus & Robertson.

Macgeorge, N. 1946. *Borovansky Ballet in Australia and New Zealand*. Melbourne: F.W. Cheshire.

Macgeorge, N., ed. 1948. *The Arts in Australia*. Melbourne: F.W. Cheshire.

May, A. 2009. "Ideas from Australian Cities: Relocating Urban and Suburban History." *Australian Economic History Review*, 49(1): 70–86. https://doi.org/10.1111/j.1467-8446.2009.00250.x

McAuliffe, C. 1996. *Art and Suburbia*. Sydney: Craftsman House.

Niall, B. 2002. *The Boyds: A Family Biography*. Carlton: Miegunyah Press.

Peel, M. 2002. "The Ends of the Earth." In T. Bonyhady and T. Griffiths, eds., *Words for Country: Landscape and the Language in Australia*. Sydney: University of New South Wales Press, 177–89

Pigot, J. 2001. *Norman Macgeorge: Man of Art*. Melbourne: Ian Potter Museum of Art, University of Melbourne.

Rowse, T. 1992. "Heaven and a Hills Hoist: Australian Critics on Suburbia." In G. Whitlock and D. Carter, eds., *Images of Australia*. St Lucia Qld: University of Queensland Press, 240–50.

Ruskin, J. 1893 [1868]. "The Mystery of Life and Its Arts." In J. Ruskin, *Sesame and Lilies: Three Lectures*. London. George Allen. https://www.gutenberg.org/files/1293/1293-h/1293-h.htm

Salter, F. 1980. *Borovansky: The Man Who Made Australian Ballet*. Sydney: Wildcat Press.

Serle, G. 1995. *Robin Boyd: A Life*. Carlton Vic: Miegunyah Press.

Slater, J. 2004. *Through Artists' Eyes: Australian Suburbs and Their Cities 1919–1945*. Melbourne: Miegunyah Press.

Stretton, H. 1975. *Ideas for Australian Cities*. Melbourne: Georgian House.

Thomas, D. 1992. *The Art of Christian Waller*. Bendigo, Vic: Bendigo Art Gallery.

Waller, C. 1932. *The Great Breath: A Book of Seven Designs*. Melbourne: Golden Arrow Press.

Waller, N. 1918. *War Sketches on the Somme Front by M. Napier Waller*. Melbourne: A. Vidler.

Westhoven, M. 1948. "She Copes with a Five Year Queue." *Argus Women's Magazine*, 27 July 1948, 1–2.

White, P. 1956. *The Tree of Man*. London: Eyre & Spottiswoode.

Wilson, G. 1947. "The Painters' Pub." *Australian Home Beautiful*, 26(13): 26–9.

6 Ideal Homes and Haunted Houses: Twenty-First-Century Irish Suburban Art and Writing

Introduction

In many ways, aspects of contemporary Irish suburbia, particularly the spaces of Ireland's post-crash ghost estates, would seem to confirm many of the worst anxieties about suburbanization. Contemporary Irish suburban development has been viewed as synonymous with dysfunctional financial structures, infrastructural mismanagement, ecological calamity, and socio-political marginalization. However, what is remarkable about the territory of contemporary Irish suburbia, particularly in its more marginal aspect, is that it has inspired and engendered a rich oeuvre of art and writing characterized by radical and dynamic forms of creativity. Whatever negative associations suburban artists and writers have reflected about their experience within or regarding Irish suburbia, the innovative and subversive nature of their work contests and gainsays more conventional, normative notions of suburbia as a creatively stagnant, artistically inert space. In their work, suburbia often becomes a site in which themes and cultural anxieties, both historical and contemporary, that have been partially disavowed in the cultural mainstream can be disclosed and interrogated.

This chapter, then, is deeply engaged by the socio-economic and geo-historical contexts relevant to contemporary Irish suburbia; yet its primary focus rests on suburban creativity as it manifests within the literary and visual art fields. The chapter seeks to understand how contemporary Irish literary and artistic cultures have challenged and adapted the dominant tropes evinced within what has been usefully termed the "suburban imaginary" – "the discursive domain in which images of and stories about the suburbs circulate" (Dines, 2020: 6). This focus on specifically Irish suburban artistic culture will also contribute towards the increasing diversification of suburban cultural studies as it pivots away from its traditional focus on North American or British suburbia. The widening of the geographical field away from national cultures in which conventional discourses of suburbia have been engendered has produced a more complex understanding of how suburban art relates to its subject matter. Recent studies of suburban artistic creativity (in a range of media) in countries such as Germany, France, Australia, and Croatia have generally contested rather than confirmed conventionally pejorative conceptions of suburban culture; this chapter accords with and contributes to this problematization of long-established negative conceptions of suburbia as a site for cultural and artistic production (Merkel, 2013; Prančević, 2013; Tarr, 2013; Rooney, 2018). In line with more recent critics of artistic suburban culture, such as Jo Gill (2013), Timotheus Vermeulen (2014), and Martin Dines (2020), this chapter will also closely attend to the stylistic aspects of the artworks under discussion, moving beyond critiques that tend to see the art and literature of suburbia as simply a vehicle for social commentary. Through greater scrutiny of the formal properties of suburban artistic representation, suburbia can be recognized as an aesthetic principle within the artwork itself. As Dines argues, an attentiveness to the formal and stylistic textures of suburban art helps to reveal "how suburbs not only feature in but also structure the stories in which they feature ... [and] the ways that suburban representation and suburban spatiality are mutually constitutive" (Dines, 2020: 23).

In order to explore these issues, this chapter will examine three artworks focused on Irish suburbia that are taken as paradigmatic of their respective media: Aideen Barry's multimedia visual art installations *Subversion and the Domestic* (2007) and *Possession* (2011), Valerie Anex's photographic essay *Ghost Estates* (2013), and Conor O'Callaghan's novel *Nothing on Earth* (2016). These texts have been directly and fundamentally inspired by Irish suburbia and are profoundly concerned with the forms, effects, and ontologies of the suburban. They also display a particular preoccupation with the contemporary, which is related to their interest in the urban periphery. As Keil has observed, it is at the "margins of the metropolis, more than in the increasingly uniform and normed inner cities, [that] new urban forms and ways of life emerge. Those tend to be indicative and directive for our existence in the twenty-first century overall" (Keil, 2017: 25).

The three works under consideration here exemplify how the Irish suburb has offered a singular lens through which to comprehend the transforming economic dispensation in twenty-first century Ireland; they are indicative of the new forms of social existence and experience that are emerging in the new century. Through these works, the ostensibly moribund, even ruined, spaces of Irish suburbia will be recognized as nutritive of some of the most searching and pertinent artistic responses to the Celtic Tiger and its aftermath. They also reveal how the spaces of contemporary Irish suburbia have functioned to engender new, more experimental modes of artistic expression as writers and artists have developed aesthetic strategies commensurate with the often strange, uncanny territories of Ireland's suburban hinterland. As such, this chapter seeks to partake of the increasingly dominant approach within suburban cultural studies to "challenge an ordering of urban space that produces the suburban as 'sub-creative' and a denigrated spatial 'Other'" (Bain, 2013: 4).

The Irish Case: An International Perspective

From the early nineties until the global financial crisis of 2008, the Republic of Ireland experienced a period of unprecedented economic growth, which has been defined by the epithet the "Celtic Tiger," a term intimating a parallel between Ireland's economic success and Asian "tiger" countries whose economies grew rapidly in the late eighties and early nineties. During this period, Ireland transformed from a relatively impoverished society at the periphery of Europe to a highly globalized, liberalizing, and more prosperous nation. Demographically, Ireland's population increased by a fifth between 1991 and 2006, and in 2004 its population breached four million – the highest number of people living in the country since 1871.

It is important to note, however, that over the total period of the Celtic Tiger, Ireland's economic growth model did not remain uniform or consistent. Rather, as Ó Riain (2014), Donovan and Murphy (2013), and others have argued, the fundamental mechanisms of economic development shifted significantly at the turn of the millennium: as growth based on increased productivity and rising exports began to falter, capital was gradually channelled away from export-oriented industry and redirected into property, building construction, and domestic consumerism. The result was a residential and commercial property market that grew precipitously and perilously, facilitated by the increased lending capacity of poorly regulated Irish banks, whose ability to supply finance to the construction industry was enabled by greater access to a larger European banking market after Ireland's adoption of the euro in 2002. Owing to this influx of capital, competition grew within the banking sector to provide credit to prospective homebuyers, who were happy, even anxious, to avail of such easily available loans in order to secure a rung on the property ladder. Aside from increased demand, construction was also stimulated by relaxed planning rules, limited building controls, and a variety of tax interventions and incentives such as the 1986 Urban Renewal Act, the cutting of the capital gains tax in 1998, Section 23 Relief, and the Upper Shannon Rural Renewal Scheme (1998–2006), which promoted property development in the economically disadvantaged rural Irish counties of Leitrim, Longford, Roscommon, Sligo, and Cavan (O'Callaghan, 2013). Aside from these policy drivers, building was also catalyzed by a growing property fetishism spurred by cultural discourses of speculation, accumulation, and consumption that became pervasive in Ireland at this juncture. Ultimately, as Ó Riain has cogently summarized, during this period an economic approach that had been rooted in "public and private productive investments" was "sidelined by an asset bubble rooted in financial and property speculation ... 'Development' and 'financialization' competed as projects within the political economy, resulting in financialization winning out with disastrous results" (Ó Riain, 2014: 5). As O'Toole has argued, these economic changes engendered a shift in the national culture: "Consumption would replace production ... The nation was to think of itself as a lottery winner, the blessed recipient of a staggering windfall. It was spend, spend, spend" (O'Toole, 2009: 22).

The remarkable effects on the property sector of Irish economic policy, combined with the emerging neoliberal socio-cultural habitus, are starkly revealed in the statistical data. Between 1996 and 2005, over half a million houses were built in Ireland (for a population of just over four million). During the same period, house prices saw precipitous increases as did land prices, while mortgage debt ballooned: between 2002 and 2007 alone, it trebled in size from 47.2 billion to 139.8 billion (O'Callaghan, 2013: 19). By the late noughties, Ireland's demand for new homes had reached a fever pitch; between 2006 and 2009, 244,000 homes were built, even though the 2006 census revealed that 266,000 homes stood vacant (Marcinkoski, 2016: 30). The absurd oversupply of housing, combined with the huge accumulation of mortgage debt, meant that the catastrophic immobilization of the international monetary market in 2008 (after the bankruptcy of Lehmann Brothers) dealt a fatal blow to the property industry in Ireland. With Irish banks no longer capable of providing finance to service the purchase of new property, property prices collapsed, and the banking and property sectors entered into a death spiral, resulting in thousands of vacant homes in "ghost estates" throughout Ireland as well as an unemployment rate that peaked at 15 per cent. To ensure the survival of its financial system, the Irish state bailed out the nation's financial institutions through the Bank Guarantee Scheme in 2008. In 2009, the government established the National Asset Management Agency (NAMA), a "bad bank" created to take on property development loans from Ireland's banks. Ultimately, however, the state had to enter a larger bailout program between November 2010 and December 2013 that was negotiated with the Troika (European Commission / International Monetary Fund / European Central Bank), which required a commitment to enact a series of severe and unpopular austerity measures.

One of the most striking and iconic manifestations of Ireland's failed property sector during this period was the prevalence of so-called "ghost estates" throughout the country. The term "ghost estate" was coined in 2006 by the economist David McWilliams and has become a mainstay of the post–Celtic Tiger cultural lexicon (McWilliams, 2006). It denotes housing developments left abandoned or unfinished in the wake of the crash. More precisely, a "ghost estate" can be defined as a development of ten or more houses, built post-2005, where more than 50 per cent of units are either vacant or under construction. These failed residences were a particularly potent manifestation of the "return of the real" within Ireland's national consciousness, evidence of the traumatic rupturing of the chimeric, pre-crash vision of Ireland. As Boyle and colleagues have observed, these residences constitute "the material and symbolic apotheosis of Ireland's economic crisis" (Boyle et al., 2014: 122). In the post-crisis period, there were ghost estates in every county in Ireland, with many situated on the outward periphery of large urban centres such as Dublin and Cork. Perhaps understandably, those ghost estates situated in more rural locales have been the focus of artistic and media representation. These starkly dilapidated sites seem to powerfully solicit allegorization and aesthetization; yet it is crucial to remember that these estates were at the extreme edge of a more general phenomenon and that "housing development and housing vacancy permeates both urban and rural contexts" in Ireland and "should be viewed as relational components of the same system" (O'Callaghan, 2013: 23). However, it is worth noting that those Irish housing estates constructed in particularly isolated rural areas stretch the definition of suburbanity to its limit, or perhaps beyond it. As McManus has argued, these estates might be understood as indicative of a process of "'rurbanisation,' with the spread of suburban-style housing estates to rural locations" (McManus, 2018: 27). The notion of "rural suburbs" concisely intimates the warped, paradoxical logic of building suburban-style residences in distinctly non-suburban locales without sufficient infrastructure or economic justification. Indeed, it could be argued that these housing estates represent an extreme manifestation of an ideal suburbia, a suburbia of the mind, which largely exists in an ontological vacuum within its surrounding environment (27).

The modes of recent suburbanization in Ireland are paradigmatic of transnational trends in modern (sub)urbanization, echoing, in particular, forms of development in other Western nations, such as Britain, the United States, Canada, and Australia, in which "the construction of the suburbs was instrumental in the 'suburban solution' of capitalist overproduction crises: the massive erection of single-family homes with their associated shopping centres was an ideal platform for the shift of capital from the (glutted) production sector into societal consumption" (Keil, 2017: 28). This process, as Marcinkoski notes, might be understood as a variant of what Harvey (1985) understood as capital switching and represents a gradual trend "from urbanization as a *response* to economic growth to urbanization deployed as a *driver* of economic growth" (Marcinkoski, 2016: 18).

In broad terms, then, Ireland's increasing reliance upon (sub)urbanization and housing finance systems as primary facets of its economic expansion is a manifestation of a larger trend in urban and economic development in the West, but the particular nature of Irish urbanization – in its extremity and fragility – can be most productively and pertinently compared to Spain, where similar political and socio-economic factors converged. There were, of course, dissimilarities in the rate, scale, and type of urban development; however, Spain and Ireland display striking points of correspondence in terms of their respective growth journeys. Spain, like Ireland, went through a remarkable voyage of transformation, evolving from one of the most impoverished nations in Europe in the 1980s to an increasingly prosperous nation at the turn of the millennium; it was a period in which the unemployment rate dropped considerably, foreign investment flowed in, and private wealth soared. This growth was characterized by unprecedented levels of urbanization activity, which was facilitated and lubricated by European structural and cohesion funding, government tax incentives, the adoption of the euro, and more easily available and cheaper credit in the European banking system. Domestic and foreign demand combined to produce a housing bubble of absurd proportions in Spain. During the country's speculative urbanization cycle, more houses were being built per annum than in Germany, France, and the United Kingdom combined, despite Spain only having a fifth of the combined population of those countries (Barbero-Sierra et al., 2013: 100). When the global financial crisis struck, the underlying weaknesses in Spain's growth model were catastrophically exposed. The collapse of the construction sector contributed to Spain's unemployment rate rising to 26 per cent in 2012, with 55 per cent youth unemployment, while in the same year almost 2 million unsold homes remained on the market ("Spain Unemployment," 2013). As in

Ireland, half-completed housing estates haunted the nation as banks and property developers collapsed or filed for bankruptcy. These nearly deserted wastelands were called *ciudades fantasma*, or "ghost towns," and became a "kind of tourist anti-attraction" in the post-crash landscape (Chakrabortty, 2011).

In registering such parallels and contexts to Ireland's suburban journey, a more complex picture of the property bubble and bust can be established in which Irish suburbanization is understood as profoundly influenced by (and expressive of) deleterious exogenous economic dynamics, as well as more well-documented endogenous governmental, business, and regulatory processes. This analytic approach falls in line with more "systemic" critiques of suburbia, as defined by Keil, which see "suburbanization as an outgrowth of contradictions caused by capitalist urbanization processes and intrinsically entwined with them" (Keil, 2017: 80). In responding to the complex recent transformations in Ireland's suburban milieu, visual artists and writers have not only produced works of remarkable innovation and originality, but many have also used their particular experience of suburbia to contribute to a "systemic" cultural analysis of Irish suburbanization. In doing so, their work has helped to clarify the complex matrix of individual, national, and transnational forms of behaviour that engendered the particular forms of contemporary Irish suburbia.

Outside In / Inside Out: Visual Art in the Suburbs

Contemporary visual artists focusing on the suburban have tended to exhibit and explore a fundamental contradiction, even paradox, regarding Irish suburbia and creativity. On the one hand, their work appears to valorize more conventionally negative images of suburbia as a site of social conformity, psychological repression, and cultural isolation. However, their artwork – inspired by and often constituted within Irish suburban locales – also reveals Irish suburbia as a site conducive to creativity, especially in relation to modes of artistic expression that are socially engaged and politically salient. The spaces of contemporary suburbia, particularly the ghost estate, have also permitted and engendered the artistic exploration and acknowledgement of cultural anxieties and subject matter – mental illness, hysteria, contemporary female domestic experience, new modes of familial organization – often occluded from more mainstream cultural production. The work of artist Aideen Barry exemplifies these contradictions and is paradigmatic of the forms of response suburbia has engendered within the contemporary Irish art scene. Barry is a visual artist whose work integrates and incorporates a remarkable variety of genre and media, including performance, sculpture, film, animation, musical composition, and drawing. She grew up in Cork, but after studying in the Galway-Mayo Institute of Technology, she chose to stay in Galway to remain close to her artistic peers, becoming involved in several artistic collectives designed to contribute to the cultural and economic life of the city. She has been integral to the emergence and consolidation of the visual art community in Galway and has also established herself as a national figure within contemporary visual art. She is an elected member of Aosdána (an affiliation of creative artists who honour artists whose work has made an outstanding contribution to creative arts in Ireland) and an associate member of the Royal Hibernian Academy.

Much of Barry's work emerges directly from her lived experience of suburbia, particularly her seven years spent in a house in an unfinished estate in Claregalway, outside Galway. In 2006, Barry was diagnosed with obsessive-compulsive disorder (OCD), and in an interview in 2009, she frames her condition as both a symptom and a reflection of modern Irish suburban experience:

> Sometimes I look out the window on a Sunday and I see the neighbor from number 43 cleaning the inside of the hubcaps of her 08 Lexus hybrid 4×4 monster car and think "Bloody Hell, she is definitely more OCD than me!" ... I mean you only have to look around at the new Ireland with the thousands of cloned housing estates, and the "stepfordzombiness" that has settled in to what were Irish Villages and market towns up and down the country and wonder, what kind of nightmare have we woken up to in our country? (Barry, 2009)

These attitudes patently echo or draw out more orthodox conceptions of suburbia as a culturally conformist, architecturally banal space characterized by social atomization, psychological anomie, and creative vacuity. Yet when Barry is asked in the same interview what is the most "inspiring place or space" she has been in, she replies: "My Housing Estate." This oppositional duality – suburbia as simultaneously vital artistic muse and source of oppressive anxiety – is explored in Barry's collaborative project entitled *Subversion and the Domestic* (2007) in which she used her own home, which was built during the Celtic Tiger property boom, as a site to creatively reflect upon contemporary Irish suburbia. As the title implies, the work was designed to transform and reorganize Barry's residence in a manner that problematized and resisted notions of domesticity, especially as manifest in contemporary Irish culture. For the project, Barry drew together a range of Irish and international artists including Ben Roosevelt, Niall Moore, Dominic Thorpe, Jackie Sumell, and Nuisance Bears. One of the most striking contributions was Dominic Thorpe's *It's Not My Place*, which placed the phrase "It's Not My Place" in oversized black plastic lettering underneath the upper-story window of Barry's semi-detached home. The deeply ambiguous and cryptically contextualized phrase is redolent with a range of meanings connected to concepts of home, belonging, and possession (whether legal, metaphysical, or ontological). Like other works produced for the project, Thorpe reckons with the seemingly disjointed nature of Irish suburban existence concomitant with rapid suburban expansion.

Barry herself echoes these concerns in her 2007 stop-motion film *Levitating*, in which she becomes the protagonist in a Gothic domestic drama that involves the completion of a range of quotidian chores – ironing, hoovering, putting out the bins, shopping – while levitating six inches above the ground. Manically propulsive, increasingly arrhythmic xylophone music provides the soundtrack to the video and amplifies the feeling of existential unease and circumambient pressure that the woman appears to feel. As Barry makes clear, *Levitating* is a personal study of the hysteria resulting from a perceived pressure to conform to behaviours and standards that pertain to the cultures within new suburban communities. In 2006, a year before the piece was produced, she recalled feeling "under" enormous emotional distress trying to fit in with the new "modern living" of Celtic Tiger suburbia: "I wanted to fit in with everyone in my housing estate in Claregalway and would be up all night cleaning and manicuring my house" (Andrews, 2011). In *Levitating*, this incessant need to publicly enact and demonstrably attain a certain preordained ideal suburban lifestyle is instantiated by the woman's state of constant levitation: it appears not as a reflection of some form of preternatural transcendence but as a breathless, belaboured performance, as though achieved under extreme stress having been compelled by an unseen force. Through such effects, Barry's work reveals itself as patently attuned to the corrosive illusions and aspirations attendant upon contemporary domestic life in certain forms of modern Irish suburbia; yet it also seeks to reflect how older Irish cultural and constitutional structures of containment, which have historically circumscribed female expression and agency, still resonate into the present. In this manner, the suburban periphery becomes a space in which profoundly pertinent issues regarding contemporary Irish female experience and the domestic are productively explored, disclosed, and illuminated

While Barry's *Levitating* and the other artworks in *Subversion and the Domestic* exude and delineate anxieties related to suburbia as a space of psychological dislocation, social hegemony, and fractured community, the project itself is a testament to the emergence of suburbia as a site of collective cultural enterprise in a space not conventionally understood as productive of subversive artistic expression. What is particularly remarkable about Barry's work for *Subversion* is that virtually every aspect and stage of the artistic process – from inspiration, to creation, through to exhibition and public engagement – occurred within the bounds of her suburban house on the outskirts of Galway. While there are obvious limitations to making and mediating art in this way, there are also undoubtedly artistic freedoms and creative frequencies of expression that are perhaps not available within more traditional art venues. Through the art's location and presentation in an outer-urban residential zone, as opposed to a more conventional inner-urban gallery space, it challenges and inverts the more orthodox models of visual art exhibitionism and mediation in Ireland. It draws cultural attention (and a potential audience) to a relatively marginalized urban locale as well as challenges the spatial aesthetics of the crisp white modernist cuboid exhibition space that has come to dominate the display of contemporary art in Ireland. As Jane Humphries argues, installations such as these play "on spatiality to contest and subvert traditional gendered spatial binaries of male (exterior, white cube space) and female (interior, domestic space) and display interchanges or inter-spatiality" (Humphries, 2010: 4). The spatially disruptive strategy of Barry's artistic project resonates with other European suburban artists such as KVART, a collective of Croatian artists that was founded in 2006 in Trstenik, a suburban region (*kvart*) of Split. KVART aimed to expose the exploitation of Split's suburbs by "profiteering developers keen to capitalize the privatization of public space" after the liberalization of the Croatian economy in recent years. Like Barry's work, KVART's activism also works to dislocate conventional notions of where art and culture happen within the city and "seeks to challenge the presumption of dependence of the periphery on the centre" (Dines and Vermeulen, 2013: 12).

Subversion and the Domestic is part of what Barry has referred to as her "dystopic domestic investigations," and a second stop-motion film, *Possession*, made in 2011, revisits the themes explored in the former project, though from the other side of the global financial crisis. The video is set in a ghost estate in Galway, and Barry again assumes the persona of the manic, harassed, lonely suburban woman, who appears afflicted with a compulsion to complete a series of domestic tasks, but in a bizarre – often hysterically bizarre – manner: she cuts toast with a garage door, becomes a hoover attempting to suck up dust, consumes vast amounts of cake at her kitchen table, and turns into a lawnmower with scissors embedded in her hair. Both inside and outside environments are rebarbative and estranging, and the woman's antic behaviours increasingly appear to be a function of her possession or instrumentalization by the house itself. Despite her strenuous efforts, at the denouement of the film the protagonist is eventually consumed and devoured by the apparently demonic house. As the title makes clear, the work is a mediation on the variegated forms and meanings of possession, and the narrative amplifies the slippage – experienced by many Irish suburbanites in this period – between home as a possession and house as a possessor. As Fahey observes, houses in Barry's work are "treacherous possessions" in the post-crisis dispensation, with the home becoming "a millstone," an equity disaster, an expensive and precarious possession whose actual ownership hovers uncertainly between that of buyer, owner, bank, and state (Fahey, 2018: 222).

Barry's art self-consciously and patently recycles and reconstitutes a range of tropes from the Gothic – the victimized or imprisoned woman, the apparent animism of inanimate objects, the *unheimlich* home – and draws particularly on Irish literary progenitors such as Sheridan Le Fanu, Bram Stoker, and Charles Maturin. Yet her aesthetic also echoes and reproduces surrealist filmic techniques adopted by mid-century Czech and Polish artists such as Jan Švankmajer, Walerian Borowczyk, and Jan Lenica. Barry's use of this surrealist technique is partly designed to illuminate and spatially instantiate subterranean anxieties regarding the domestic within Irish society and to translate "the unspeakable fears and frustrations of contemporary Irish women into a rich lexicon of distorted images" (Fahey, 2018: 196). This hybridization of postmodern Gothic modes and neomodernist surrealist tones is also part of what distinguishes Barry's work as especially innovative in its procedures. As Fintan O'Toole has remarked, it registers as a "new note" in the contemporary art scene, particularly in the way that its black humour renders and revitalizes the Gothic to produce art that is "surreal, absurd, technologically inventive and wildly energetic" (O'Toole, 2012). While the originality of Barry's work must be understood as a product of her technical brilliance and radical insight, it might also be seen as a function of the particular forms of suburban environment she chooses to focus on and in which she constitutes her artistic vision. What was it, one might ask, about these brand-new unfinished (subsequently ghost) estates in Galway that provoked her particular mode of aesthetic response? It might be argued that Barry's recourse to the strange collocations and non-sequiturs characteristic of surrealist art – the juxtaposition of the real and the simulacrum, the rational with the irrational, the quotidian with the oneiric – potently evoke the complex and singular affect engendered by these locales: the sense that these estates have been jarringly interpolated into a vernacular landscape and social milieu without any authentic attempt at assimilation and then abandoned as absurd vestiges of an unfathomably dysfunctional socio-economic dispensation rendered increasingly incompatible with the real.

Ghost Estates in Fine Art Photography

In addition to engendering site-specific artwork such as Aideen Barry's *Possession*, ghost estates have also inspired a plethora of fine art photography work by artists living in Ireland and abroad. The potency of the Irish ghost estate as a site for photographic examination of the contemporary was underlined in March 2011 when Kim Haughton's picture of horses in a ghost estate in Leitrim was selected as one of eleven images to feature in the *Guardian*'s "A History of Europe in Pictures 1945–2011." The picture was part of Haughton's *Shadowlands* (2011) series, one among many photographic explorations of the ghost-estate phenomenon, which include David Farrell's *An Archaeology of the Present* (2010–13), Anthony Haughey's *Settlement* (2011), Eoin O'Conaill's *Reprieve* (2012 – present), and Ruth Connolly's *If You Lived Here, You'd Be Home by Now* (2017). These works are representative of what Justin Carville cogently defines as the "topographical turn" in Celtic Tiger and post–Celtic Tiger photography, an approach that does not simply register the aesthetic and formal changes to Ireland's suburban landscapes but also understands "topographical change as the physical manifestation of much deeper cultural and political processes and crises to which society has become blinded" (Carville, 2018: 251). Photography is a particularly salient medium in terms of the consideration of these processes and crises, since it was often the "'post-photographic' digitally generated" idealized images of suburbia created for property advertisement that helped precipitate Ireland's rapid and ill-fated suburban expansion in the first instance. Yet if digital photographic imaging became an instrument creatively deployed to intensify and facilitate speculative development, fine art photography offered a remarkably fertile medium to creatively analyse the logics underpinning that development and its attendant socio-cultural and economic effects.

A particularly achieved example within the genre of ghost-estate photography is Valerie Anex's (2013) photo essay *Ghost Estates*.[1] Anex is a critically acclaimed photographer from Lausanne, Switzerland. Her mother is from Leitrim in Ireland, where Anex spent most of her summer holidays as a youth. Anex's interest in Ireland's changing urban form was engendered by her shock at the dramatic changes that occurred in the surrounding landscape every summer she returned to Leitrim from Lausanne: "It started with new roads, then came new supermarkets, and then very big ones, suddenly people lived in new houses, and new cars were everywhere." Ireland's suburban expansion was particularly striking for Anex because of how it contrasted to her own native Switzerland, where "there is a scarcity of building land, landscapes are protected, and it is very difficult to get new building permits" (Anex, 2014). For Anex, Switzerland appears to represent a more ingrained, settled, and therefore oblique form of capitalist modernity as reflected and constituted in its relatively static urban infrastructure and development. Whereas in Ireland, the contradictory, and often incoherent, processes of late capitalist modernization are eminently more explicit, given how tangibly they are expressed in the transformed suburban topography and concomitant infrastructure. *Ghost Estates* is composed of a series of photographic journeyings through ghost estates in counties Cavan, Roscommon, Longford, and Leitrim in 2011; these failed housing developments are representative of some of the most egregious instances of property development in Ireland at this time. The Irish counties in which they were built had the highest rate of ghost estates in the country when standardized by population (O'Callaghan, 2013: 23), and part of the reason there was so much oversupply in this region was that it was covered by the Upper Shannon Rural Renewal Scheme, which provided tax incentives to developers to build. As noted earlier, such schemes are underpinned by an approach that uses urbanization as a driver of economic growth rather than a function of it. As a consequence, no deep-rooted economic benefit was achievable from the residential building in this region as "most of the employment generated involved short-term, construction-related jobs, rather than more sustainable employment in businesses" (McManus, 2018: 21).

In terms of Anex's approach in *Ghost Estates*, the photographs generally present houses front on and at a distance. Often the book will present a diptych of images in which two almost identical houses are presented side by side: the method simultaneously draws attention to the prototypical architecture of the estates and the subtle, barely perceptible differences between houses. Though Anex adopts a seemingly objective, even documentary style perspective to her subject matter, the work generates a cool, sombre, uncanny tonal effect. The photographs echo the aesthetic of Dusseldorf School artists and photographers such as Thomas Ruff, Andreas Gursky, and Candida Höfer. Höfer in particular seems a pertinent influence; like Anex, she presents scrupulously observed public spaces devoid of human activity but not necessarily of any form of presence. For Anex, the absence of any photographs of people in her series was also designed to intimate that "human concern was away from the preoccupations of the companies who built the houses" (Teicher, 2014). In her attentiveness to the enigmatic but palpable absence of the human, Anex's work is paradigmatic of much ghost-estate photography in that it presents ghost estates as dysfunctional landscapes haunted by a spectral humanity. This stance might be understood as a defining feature of this medium's approach in the sense that other cultural forms depicting ghost estates tend to present dysfunctional human subjects haunted by a spectral landscape.

The ghostly suburban community that seems to linger within the spaces of Anex's photographs is made most manifest in the final photograph in the book, which captures an advertised digital image of an idyllic suburban scene printed on a poster draped over the bare wall of an unfinished house in a ghost estate (Battery Court) in Longford. Anex's central framing of the simulated image of a future suburban dream within the abandoned ruins of a dystopic suburban ruin succinctly points up the forms of cognitive dissonance at play in the planning of the estate and intimates the irrationality of Ireland's larger suburban development in the later stages of the Celtic Tiger. Yet despite Anex construing these digitalized simulacra of suburbia as symptomatic of a vacuous, hyper-consumerist culture, perhaps counterintuitively, they also act as a form of inspiration in terms of developing her artistic perspective and approach. She relates how the 3D software modelling used by the architects for the estates "inspired" her and that she found these "3D imagery models interesting; there is something morbid in the way they aim to reach perfection. I tried to imitate or refer to these models in certain of my photographs" (Anex, 2014). Anex also intended the book of photographs itself to resemble a property catalogue or brochure that might have been used to market the estates in the first instance. In this regard, her work might be said to hybridize two kinds of apparently contradistinctive forms of creative enterprise related to contemporary Irish suburbia. Even as her work adduces and ironizes more utilitarian modes of commercial creativity through its visual allusions to hyperreal marketing imagery, slick advertising slogans, and property brochure formats, these forms and formats also underpin Anex's own aesthetic and thus become creatively transmuted into fine art. Part of the effect of this blending of forms is to confuse the housing estate as an ideal form, in a metaphysical sense, with its uncanny physical manifestation, provoking the viewer to reflect on the manner in which, for a time, neo-liberal and commercialist discourse completely overwhelmed the real, becoming a kind of Baudrillardian "map that precede[d] the territory" (Baudrillard, 1994: 1).

For Anex, despite her compelling aesthetization of the houses she photographed, *Ghost Estates* is primarily intended to expose how Ireland's economic and suburban calamity "is not a transitional crisis but a structural one"(Anex, 2013: 74). Her project, which is underpinned by a Luttwakian critique of "turbo-capitalism,"[2] is designed to engender new modes of thinking regarding the processes of urbanization in Ireland and the "limits of a system based on continual progress and growth" (74). Her hopes for such suburban reimagining were partially fulfilled in 2015 when a ghost estate called "Waterways" in Leitrim, a locale that features in *Ghost Estates*, became a site of well-publicized protest by a "guerrilla garden" group called Nama to Nature, who planted nearly a thousand native tree varieties on the abandoned site of the ghost estate. Frank Armstrong, a spokesperson for the group, explained that the action was designed to highlight the consistent failure of property developers to protect biodiversity on their land and to show that "the tragedy of the Irish commons where tax breaks and lax planning allowed developments with no regard to infrastructure could be challenged and that ownership only stretched so far. Without proposing anarchy, we were asserting that communities hold rights over all property in their vicinity" (Armstrong, 2015). In this small, but significant example, it is possible to view Anex's work as part of a larger politico-cultural enterprise to use the ruins of ghost estates as sites of radical social and political potential. Anex pithily states: "If these houses presage the imminent end of a system, they also constitute the opportunity to envisage living beyond it" (Anex, 2013: 74). It is salient to note that this form of socially engaged photography has been deployed elsewhere in Europe in response to the ruins born of speculative urbanization. An exhibition entitled *Unfinished*, which was displayed in the Spanish Pavilion at the Venice Biennale 2016, also focused on half-built buildings left in the wake of the financial crisis and included seven photographic series, multiple design graphics and plans, and lectures by relevant practitioners and academics. This varied content not only drew attention to the unfinished structures, many of which were suburban, resulting from the crash in Spain but also showcased architectural solutions that creatively respond to the degraded built environment. According to co-curator and architect Iñaqui Carnicero, the overall aim of the work was to "reflect on Spain's unfinished architectures to stimulate a discussion around how to move from a discourse of past ruins to one of future opportunities" (Amaya, 2016).

Suburbia in Contemporary Irish Fiction

While photography and visual art performance have been quick to analyse and document the evolving territories of Ireland's suburban geography, Irish literary fiction, particularly the novel (perhaps owing to the aesthetic demands of the form), has been slightly less expeditious in its response to contemporary Irish suburbia. Part of the reason is perhaps the relative lack, or perceived lack, of an established tradition of suburban Irish writing. In 1992, Fintan O'Toole observed: "One finds that the great tradition of Irish writing is silent on the subject of the suburbs, so you can slip out from under its shadow. No one has ever mythologized this housing estate, this footbridge over the motorway, that video rental shop" (O'Toole, 1992: 1–2). Just over a quarter of a century later, Joe Cleary observed of the Irish suburbs that they "may have had their minor singers, but, mostly, they have seemed too featureless to offer much to 'literature,' too stripped of a distinct identity certainly to qualify as subject matter for 'Irish Literature'" (Cleary, 2018: v).

Cleary is patently aware of the problematic nature of rigid demarcations of Irish literary canonicity. He does add the caveat that the situation is beginning to change regarding the recognition of Irish suburban writing, but what he and O'Toole are emphasizing is a critical reflex within Irish cultural criticism that tends to understand canonical or "great" Irish writing as largely (and legitimately) indifferent to the Irish suburbs. Their commentary, then, is as much to do with the historic mapping of the dominant sites of Irish writing as it is to do with a dearth of writing about suburbia. As Cleary notes, Irish literature has been conventionally understood as emerging out of non-suburban topographies and locations – the western Atlantic seaboard, the tenements and streets of Ireland's inner cities, Irish Big Houses and their surrounding parklands, the milieu of Irish midland towns – which are linked to the dominant aesthetic modes in Irish writing, such as romanticism, realism, pastoralism, and the Gothic. Suburbia is construed to exist somewhere between these realms, a liminal space, lacking the cultural excitement or inspiring natural environment of more orthodox sites of Irish art (Cleary, 2018).

However, it is possible now to point to a growing corpus of Irish novels that reveals suburbia as the context for some of Ireland's most politically salutary and socially incisive contemporary fiction, including Anne Enright's *The Gathering* (2007), Donal Ryan's *The Spinning Heart* (2012), Claire Kilroy's *The Devil I Know* (2012), Tana French's *Broken Harbour* (2013) and *The Secret Place* (2014), and Paul Murray's *The Mark and the Void* (2015). This recent proliferation of contemporary fiction focused on the Irish suburbs, particularly the ghost estate, represents a distinct challenge to conventional formulations of the established territories of Irish literary culture and provokes a recalibration of dominant conceptions of where creative literary practice is engendered.

Conor O'Callaghan's novel *Nothing on Earth* is exemplary of the growing genre of Irish suburban fiction, especially the subgenre of ghost-estate Gothic (O'Callaghan, 2016). This short novel weaves a strange, compelling tale about the unexplained disappearance of a young family of four who live in a show house on the grounds of a largely abandoned ghost estate outside a nameless small town in Ireland. The family rent the house from a rural property developer who, rather ominously, goes by the name of Flood. From early on in the novel, there is a sense of a spectral futurity pervading the residence: "The things of the show house belonged to the lives that should have happened but never did. They gave off no noise at all, and that was more deafening than anything" (34). The depiction of the estate in the novel was inspired by O'Callaghan's own journeying through ghost estates in Ireland after the break-up of his marriage in 2008. During this traumatic period, the author, who was living in Manchester at the time, would often return home to County Louth in Ireland and drive the "'hinterland' of Dundalk" in the evening, where he would "happen upon umpteen half-finished housing developments: middle of nowhere, uninhabited, overgrown" (O'Callaghan, 2017). O'Callaghan became a kind of flâneur of the extreme suburban periphery, wandering arbitrarily through ghost estates and acutely observing their remarkable features, forms, and affects: "I remember bare breeze-blocks, graffiti, ragwort and poppies in among the rubble, scraps of glass and scaffolding. I remember occasionally calling 'Hello,' for company's sake, into hollow space. Nothing. Not even an echo. There were moments ... when I felt fear like never before or since, and legged it to the car, and sped back to a world marginally more populated" (O'Callaghan, 2017). While this eerie flânerie of Ireland's forsaken suburban frontier is the immediate inspiration for the novel's world, the text is also haunted by the memory of "the family of four" O'Callaghan "broke up." The house that the family move into in the novel is based on the first suburban house O'Callaghan lived in with his own family in 1995, which was situated on a new development "that remained a building site for a whole summer after we moved in" (O'Callaghan, 2017). In creating the novel, therefore, O'Callaghan amalgamates various suburban locales and experience as part of his process of engaging (sometimes unconsciously) with different and competing strata of memory linked to his family and its dispersion.

The ghost estate, it might be argued, is especially conducive to the exploration of past suffering and present upheaval, since it is a site of lingering trauma in which time and memory coalesce disturbingly. These uncanny sites powerfully manifest a sense of disjunctive temporality, a dislocated present that is haunted both by memories of past historical catastrophes linked to land mismanagement and future visions of a new quasi-utopian social dispensation that never came to be. In the novel, O'Callaghan's private familial tragedy finds a correlate with a national socio-urban tragedy through the symbolic significance of the ghost estate: the disappearance of the family and the collapse of the ghost estate in the text are not only a redolent figuration and concretization of O'Callaghan's previous psychic dislocation but might be read as a fragmentary parable of contemporary national discord and politico-economic collapse.

The novel is narrated by a local priest who may or may not be implicated in the disappearance of the youngest member of the family living on the estate: a twelve-year-old girl who comes to his door seeking help when her father disappears. She is dishevelled, grimy, and visibly malnourished when she first enters the priest's home, and her appearance clearly manifests her recent trauma and destitution:

> It was also marked in places, her skin was: scratches, creases, streaks of dirt, and words. There were actual words scrawled around her skin, dozens in blue, frayed at the edges, blurred by sweat and largely illegible. The more blurred ones resembled bruises. The more intact were like little darns meant to mend those points where the fabric of her flesh had worn threadbare. (O'Callaghan, 2016: 9)

The emaciated and disturbingly bizarre appearance of the girl bespeaks the degraded circumstances in which the family have been living, while her corporeal attenuation and insubstantiality reflect the entropic world of the increasingly derelict ghost estate. The girl's "threadbare" flesh augurs the fate of the estate and reveals the text's central fascination with the process and consequence of disappearance.

Aside from the central plot device – the mystery of the disappearing family members – the novel also hints at older more cataclysmic forms of disappearance. The ghost estate is built on land that used to belong to an Anglo-Irish landed estate, and we might feel that it is haunted by the ghosts of Irish victims of English colonial rule and, most patently, the era-defining loss of rural population during the Great Irish Famine, during which whole villages, hamlets, and communities disappeared through death or migration. The novel's explicit references to Stanley

Kubrick's *The Shining*, a film depicting a haunted hotel built on the site of old Native American burial grounds, would seem to amplify this notion of the spectral return of an imperially repressed subaltern populace. The novel also invokes other forms of historical trauma linked to population loss and the devastation of rural communities. Slattery, the large landowner who sold the land to Flood, refers to the estate as "Flanders Fields," and it transpires later that he has conducted a survey of all the men of the parish who fought and died in the First World War. All of these histories appear to accrete on the site of the estate in the novel, or are uncovered by its effect, and it might be argued that this inference reflects the manner in which ghost estates appear to give expression to historical pathologies and traumas that continue to haunt contemporary Irish society. In this sense, O'Callaghan's writing could be said to explore how Ireland's failed suburban estates are powerfully overlaid with histories and trauma that are unique to the Irish experience and the changing forms of land use in Ireland. In drawing out these submerged pasts, however, O'Callaghan deploys a similar strategy to other contemporary suburban writers who also contest the notion of suburbia as a space characterized by the absence of history. For example, Georg Klein's German novel *Roman unserer Kindheit* (2010) follows a group of children living in a post-war housing estate in Germany who discover an underground medieval system of tunnels. The children's adventures in the underground labyrinth serve to "evoke certain metaphors relating to memory and oblivion … [and] can be read as acts of recovering repressed parts of a collective identity and history" (Merkel, 2013: 45).

O'Callaghan's novel, of course, is also very much invested in the contemporary and in analysing the larger connections between economic development and failed suburbanization that are pertinent beyond Ireland's shores. If we read O'Callaghan's text as an allegory of early twenty-first-century Ireland, it could be argued that the novel invokes questions about who, ultimately, is responsible for the failure of the property market and the Celtic Tiger crash. Certainly, it references some of the more obvious factors that left the Irish economy vulnerable to collapse: the property developer Flood is convicted of fraud and can readily be understood as the personification of the corruption, greed, and incompetence that characterized much failed housing development in Ireland during this period. The novel also offers a larger critique of the underlying dynamics at play in Irish society and its economy at this time. Two family members are employed by a large multinational software company on the outskirts of town. This software plant, which eventually moves its operation to the "East," underlines Ireland's imbrication within transnational economic networks that leave the country and its property sector perilously exposed to fluctuations in the international market place. This larger context is also underlined early on in the novel during a conversation between the narrator and the girl. When the narrator asks her why she doesn't simply phone her missing father, she replies:

> "We have no more credit."
> It struck me, at the time, as one of several peculiar phrases she used. She said it as if credit were something held and frittered collectively, all at once. I couldn't altogether get away from the phrase's implication of belief or lack of it. *We have no more credit.* (O'Callaghan, 2016: 12)

O'Callaghan concisely leverages the multiple implications of the word "credit" here in order to underline the connection between the collapse of the estate in the novel and the larger international economic catastrophe of which it is a symptom. The girl and her family are victims of the "credit" crunch, and the narrator's fixation with the peculiarity of the girl's phrase intimates the strangeness of their experience within the ghost estate: the reality that obscure and deeply mysterious movements of capital in distant European and American cities and countries can profoundly determine the fate of individuals living in estates in the rural provinces of Ireland. In this sense, the novel appears to reflect what Mark Fisher understands as the eeriness of capital, a force that "does not exist in any substantial sense, yet is capable of producing practically any kind of effect" (Fisher, 2016: 13). Capital might be thought of as a kind of ghostly influence, therefore, and the family's apparent possession by the spectral forces at work in the estate understood as a Gothicized expression of the ways in which many Irish families were overwhelmed (or possessed) by large, unfathomable, transpersonal (almost supernatural) economic forces that engendered behaviours that may only retrospectively appear irrational or absurd. In this way, the novel intimates that many individuals in Ireland at this time were not the self-aggrandizing so-called "Tiger cubs" whose greed and excess debilitated a functioning economy. Rather, as with many homebuyers within Spain, Greece, the United States, and other highly financialized economies, they were unwitting victims within a structurally flawed speculative growth model over which they had little agency and whose underlying logics were profoundly obscured.

The patent Gothicism of the novel's world underlines how O'Callaghan reaches into the past in order to formulate a new mode for the contemporary. As Jarlath Killeen has pointed out, the author "takes the supposedly outmoded and creaky ingredients of the Gothic ... and brings them into direct contact with contemporary Ireland, which turns out to be as peculiar and unsettling as Count Dracula's Transylvania" (Killeen, 2017). Part of O'Callaghan's updating of the Gothic is not simply his change of mise-en-scène, however. The novel also draws on a postmodern aesthetics, particularly in its metafictionality and ontological instability, both of which serve as a formal reflection of the world of the ghost estate, whose status *qua* residential housing estate becomes increasingly tenuous, even absurd, as the narrative progresses. By the end of the novel, the estate (like numerous estates throughout Ireland at the time) is more ideal construct – a figment of advertised aspiration – than a fully realized suburban community. This particular quality of the ghost estate, then, has engendered its own effects in the text. In order to fully replicate the specific atmospheric textures and uncanny realities of the ghost estate, O'Callaghan develops a style that hybridizes older Gothic tropes with more recent postmodern textual strategies, yet without slavishly imitating either. In the process, the author achieves a literary style that is experimental and artistically invigorating, which moves beyond more dominant naturalist modes that have been so characteristic of modern Irish fiction.

Conclusion

What is consistent across all three artists discussed in this chapter is that their creative responses have been intricately and fundamentally linked to their experience of living in or attentively travelling through Irish suburban locales. The result of these creative explorations has been art that is culturally radical and aesthetically varied: O'Callaghan's *Nothing on Earth* precisely discloses the discomfiting experience of those living in peripheral, failed suburban spaces and the highly complex network of causes that resulted in their socio-economic marginalization; Barry's work exposes problematic aspects of contemporary Irish domesticity while simultaneously destabilizing and disrupting the way in which visual art is exhibited and mediated in Ireland; Anex's *Ghost Estates* reveals ghost estates as immanent with emancipatory potential and especially provocative of the impulse towards economic and (sub) urban reimagining. These artworks also resonate with suburban artistic cultures emerging elsewhere in Europe. O'Callaghan's writing, while exploring the unique significance of the ghost estate within Ireland, is also attendant to the temporal dimensions of suburbia in a manner similar to contemporary German literary fiction focusing on suburbia. Barry's and Anex's work indicate how Irish suburban visual art shares aesthetic and thematic similarities with contemporary suburban art-works produced in other countries such as Spain and Croatia. These affinities intimate that Irish suburban artists are part of an emerging cultural response in Europe that reveals suburbia as a site of radical artistic and cultural creativity.

In toto, the work of these artists, and the larger artistic output that they represent, produce and demand a more granular compre-hension of contemporary suburban Irish experience and suburban realities more generally: their work discloses suburban culture as profoundly more variegated, metamorphic, and dynamically lay-ered than conventionally understood. It is not that contemporary Irish suburban art and writing completely reject and dispel the qual-ities and states emphasized in more reflex notions of suburbia – as spaces of atomization, conformity, ennui. It is that they reveal another dimension to the suburban, which complicates and contests these tropes. At their most productive, these artworks explode reductive binaries regarding suburbia and the city, and bring their opposi-tional terms into creative synthesis: suburbia can be *both* repressive *and* emancipatory, *both* socially isolating *and* generative of commu-nity, *both* politically disenfranchised *and* a site of resistance and pro-test, *both* stifling of individual expression *and* provocative of radical, avant-garde art.

NOTES

1   Anex has exhibited in a number of galleries throughout Europe, and her work has appeared in, among other publications, *Lens* (*New York Times* photo blog), *Slate*, as well as in the *Huffington Post*.
2   This analogy is a reference to the work of Edward N. Luttwak, *Turbo-Capitalism: Winners and Losers in the Global Economy* (New York: Harper Collins, 1999).

REFERENCES

Amaya, L. 2016. "Previewing the 2016 Venice Biennale: Spain's 'Unfinished.'" *Archinect*, 12 May 2016. https://archinect.com/features/article/149944447/previewing-the-2016-venice-biennale-spain-s-unfinished
Andrews, K. 2011. "Aideen Barry – Exploring Gothic Terror in Suburbia." *Galway Advertiser*, 14 July 2011. https://www.advertiser.ie/Galway/article/41704/aideen-barry-exploring-gothic-terror-in-suburbia
Anex, V. 2013. *Ghost Estates*. Geneva: Uqbar.

Anex, V. 2014. "Questions: Ghost Estates." Email correspondence with the author, 4 August 2014.

Armstrong, F. 2015. "Blight of Ghost Estates Remains, Three Years After Nama to Nature Tree Planting." *Irish Times*, 9 April 2015. https://www.irishtimes .com/opinion/blight-of-ghost-estates-remains-three-years-after-nama-to -nature-tree-planting-1.2168955

Bain, A.L. 2013. *Creative Margins: Cultural Production in Canadian Suburbs*. Toronto: University of Toronto Press.

Barbero-Sierra, C., M.J. Marques, and M. Ruíz-Pérez. 2013. "The Case of Urban Sprawl in Spain as an Active and Irreversible Driving Force in Desertification." *Journal of Arid Environments*, 90: 95–102. https://doi .org/10.1016/j.jaridenv.2012.10.014

Barry, A. 2009. "Ex Tenebris Lux: An Interview with Aideen Barry." *An Cathach*, 26 May 2009. http://www.ancathach.com/2009/05/ex-tenebris -lux.html

Baudrillard, J. 1994. *Simulacra and Simulation*. Translated by S.F. Glaser. Ann Arbor: University of Michigan Press.

Boyle, M., R. Kitchin, and C. O'Callaghan. 2014. "Post-Politics, Crisis, and Ireland's Ghost Estates." *Political Geography*, 42: 121–33. https://doi.org /10.1016/j.polgeo.2014.07.006

Carville, J. 2018. "The Narrow Margins: Photography and the Terrain Vague." In E. Smith and S. Workman, eds., *Imagining Irish Suburbia in Literature and Culture*. Basingstoke, UK: Palgrave Macmillan, 249–74.

Chakrabortty, A. 2011. "Nightmare for Residents Trapped in Spanish Ghost Towns." *The Guardian*, 28 March 2011. https://www.theguardian.com /world/2011/mar/28/residents-trapped-spanish-ghost-towns

Cleary, J. 2018. "Sing, Muse, of Irish Suburbia." In E. Smith and S. Workman, eds., *Imagining Irish Suburbia in Literature and Culture*. Basingstoke, UK: Palgrave Macmillan, v–ix.

Dines, M. 2020. *The Literature of Suburban Change: Narrating Spatial Complexity in Metropolitan America*. Edinburgh: Edinburgh University Press.

Dines, M., and T. Vermeulen. 2013. "Introduction: New Suburban Stories." In M. Dines and T. Vermeulen, eds., *New Suburban Stories*. London: Bloomsbury, 1–14.

Donovan, D., and A.E. Murphy. 2013. *The Fall of the Celtic Tiger: Ireland and the Euro Debt Crisis*. Oxford: Oxford University Press.

Fahey, T. 2018. "'And This Is Where My Anxiety Manifested Itself …': Gothic Suburbia in Contemporary Irish Art." In E. Smith and S. Workman, eds., *Imagining Irish Suburbia in Literature and Culture*. Basingstoke, UK: Palgrave Macmillan, 209–26.

Fisher, M. 2016. *The Weird and the Eerie*. London: Repeater Books.

Gill, J. 2013. *The Poetics of the American Suburbs*. Basingstoke, UK: Palgrave Macmillan.

Harvey, D. 1985. *The Urbanization of Capital: Studies in the History and Theory of Capitalist Urbanization*. Oxford: Blackwell.

Humphries, J. 2010. "Crossing the Threshold: The Domestic House/ Home in Irish Art Installations." Paper presented at "Writing Irish Art Conference," Trinity College, Dublin, 20 November 2010.

Keil, R. 2017. *Suburban Planet: Making the World Urban from the Outside In*. Oxford: Polity.

Killeen, J. 2017. "How Celtic Tiger's Death Led to a Gothic Revival." *The Irish Times*, 28 April 2017. https://www.irishtimes.com/culture/books/how-celtic-tiger-s-death-led-to-a-gothicrevival-1.3065069

Marcinkoski, C. 2016. *The City That Never Was: Reconsidering the Speculative Nature of Contemporary Urbanization*. New York: Princeton Architectural Press.

McManus, R. 2018. "Brave New Worlds? 150 Years of Irish Suburban Evolution." In E. Smith and S. Workman, eds., *Imagining Irish Suburbia in Literature and Culture*. Basingstoke, UK: Palgrave Macmillan, 9–38.

McWilliams, D. 2006 "A Warning from Deserted Ghost Estates." *Sunday Business Post*, 1 October 2006. http://www.davidmcwilliams.ie/a-warning-from-deserted-ghost-estates

Merkel, C. 2013. "Entering No-Go Areas: Suburbs in Contemporary German Literature." In M. Dines and T. Vermeulen, eds., *New Suburban Stories*. London: Bloomsbury, 41–50.

O'Callaghan, C. 2013. "Ghost Estates: Spaces and Spectres of Ireland after Nama." In C. Crowley and D. Linehan, eds., *Spacing Ireland: Place, Society and Culture in a Post-Boom Era*. Manchester: Manchester University Press, 17–31.

O'Callaghan, C. 2016. *Nothing on Earth*. Dublin: Doubleday Ireland.

O'Callaghan, C. 2017. "My Novel's Roots? A Rootless Half-Life Roaming Ghost Estates." *Irish Times*, 7 April 2017. https://www.irishtimes.com/culture/books/my-novel-s-roots-a-rootless-half-life-roaming-ghost-estates-1.3039357

Ó Riain, S. 2014. *The Rise and Fall of Ireland's Celtic Tiger: Liberalism, Boom and Bust*. Cambridge: Cambridge University Press.

O'Toole, F. 1992. "Introduction: On the Frontier." In D. Bolger, *A Dublin Quartet*. London: Penguin, 1–6.

O'Toole, F. 2009. *Ship of Fools: How Stupidity and Corruption Sank the Celtic Tiger*. London: Faber & Faber.

O'Toole, F. 2012. "Gothic Realism in the Here and Now: Haunted Houses of a Dead Boom." *Irish Times*, 1 September 2012. https://www.irishtimes.com

/culture/tv-radio-web/gothic-realism-in-the-here-and-now-haunted-houses
-of-a-dead-boom-1.525515

Prančević, D. 2013 "Kvart KVART: Contemporary Art Activism in Suburban
Split, Croatia." In M. Dines and T. Vermeulen, eds., *New Suburban Stories*.
London: Bloomsbury, 175–86.

Rooney, B. 2018. *Suburban Space, the Novel and Australian Modernity*. London:
Anthem Press.

"Spain Unemployment Rate Hit a Record: Youth Rate at 55%." 2013. *BBC
News*, 24 January 2013. https://www.bbc.com/news/business-21180371

Tarr, C. 2013. "From Riots to Designer Shoes: *Tout ce qui brille/All that Glitters*
(2010) and Changing Representations of the Banlieue in French Cinema."
In M. Dines and T. Vermeulen, eds., *New Suburban Stories*. London:
Bloomsbury, 31–40.

Teicher, J.G. 2014. "A Stroll through Ireland's Eerie Ghost Estates." *Slate*,
10 August 2014. https://slate.com/culture/2014/08/valerie-anex
-photographs-ghost-estates-in-ireland-in-her-series-ghost-estates.html

Vermeulen, T. 2014. *Scenes from the Suburbs: The Suburb in Contemporary US
Film and Television*. Edinburgh: Edinburgh University Press.

PART 3

The Suburban Creative Milieu

7 Halfway between Nature and Culture: Uccle Centre d'Art, a Colony of Artists in Brussels's Suburbs in the Interwar Period

TATIANA DEBROUX

Which retreat is more comforting, peaceful and closer to the big city?
– Alfred Bastien, "Les peintres de la forêt de Soignes" (1920: 278)[1]

Introduction

In 1933, a local newspaper from Brussels celebrated the twenty-fifth anniversary of "pastoral exhibitions" that were organized annually in Uccle, a southern suburban municipality of the city.[2] In 1908, art exhibitions started to take place in an inn whose owner appreciated artworks and the company of artists visiting the vicinity to paint outdoors. For decades, artists had been coming to the area for walks outside the city. They also came to find inspiration in the local, still rural landscape, partly covered with woods at the beginning of the twentieth century and shaped by several small rivers and valleys. An article on the twenty-fifth anniversary of the exhibitions noted:

> Following the example of Paris, which has Montparnasse, Brussels has its art centre. It is Uccle. Indeed, no municipality around the capital has so many painters, sculptors, writers, and musicians; neither Auderghem, Boitsfort, nor Schaerbeek can even claim to have as many artists of notorious talent as Uccle-la-Jolie [Pretty Uccle], where all painters, since Breughel, have so often planted their easel. Uccle is truly the art centre of Greater Brussels, not only because of the large number of artists who live there but above all because of the art movement that has flourished there for twenty-five years now.[3]

The enthusiastic columnist just quoted presented Uccle as a cluster of artists, a place where artists supposedly gathered in even greater numbers than in Schaerbeek. The latter was another former suburban municipality in the northern part of the city, which remained for decades the most important artists' cluster.

At the turn of the twentieth century, Brussels hosted hundreds of visual artists, who were not only attracted by its status as the capital of an early industrialized country but also by its vivid creative scene (Brogniez et al., 2017). Two creative areas attracted artists and art amateurs in the new suburbs of Brussels: from the 1860s, Schaerbeek and Saint-Josse, north of the historical city core; and from the 1880s, Ixelles and Saint-Gilles, south of the city centre. New creative clusters emerged further from the centre during the first decades of the twentieth century in a second wave of suburban creativity. Even if Uccle never developed as intensely as the older clusters, concentration of artistic activity in Uccle was part of this dynamic and was the most important of the interwar artists' settlements around Brussels. Proud of this artistic presence, the mayor of Uccle in 1921 invited artists to present works from the area using the new exhibition place that the municipality had bought (a small castle called Wolvendael in close vicinity to the inn where artists gathered). The narrative associated with the creation of the art circle Uccle Centre d'Art (Uccle Art Centre) indicates that the success of this municipal exhibition encouraged the group of artists who participated in the former pastoral exhibitions to formalize their organization one year later.

Uccle Centre d'Art is an example of an artistic group created around a specific space and local identity. This case study illustrates how creativity developed in the suburbs of Brussels by looking at its geography and its social organization. The story also illustrates the ambiguous feelings of these artists regarding the inner city and the threats caused by urban sprawl into the suburbs. By analysing the settlement of artists in Uccle and the material conditions that made this suburban location possible, I seek to understand to what extent and in what way this marginal position in space also corresponded to a distancing from the Brussels cultural scene. Is it a question of real empowerment of an artistic milieu due to its peripheral position that made the emergence of a singular form of creativity possible? Or, on the contrary, does this new geography correspond to a spatial and artistic retreat associated with the upheavals of the urban society at that time?

The material presented originates from a broader empirical research project on visual artists in Brussels from 1830 to present days. The project adopted the perspective of a geographer and investigated the socio-spatial patterns of a professional group that experienced tremendous change in its status during the nineteenth century (Debroux, 2012). Maps were central to the analysis of the evolving geography of artists, as was research on their social relations and the development of urbanization in Belgium. Based on original archival material documenting the creation and daily life of the circle Uccle Centre d'Art, this chapter sheds new light on the nature of interwar suburban Brussels and its creativity.[4]

The chapter considers three dimensions to build a framework. First, the geographical dimension of this suburban creativity example is placed in the broader context of the development of artistic life in Brussels and in the specific spatial context of Uccle. Second, the chapter focuses on the forms of sociability associated with the art circle and the coexistence of artists and supporters, which rendered possible a marginal position in the art world and market. Finally, it is through the scope of the activities organized by the members of Uccle Centre d'Art and their collective mobilization that their values can be investigated. This framework allows questioning of the specific features associated with suburban creativity around Brussels before the Second World War and of the development of a new cultural geography and sociability. The case study of Uccle Centre d'Art illustrates in a concrete way why groups of artists left the city centre, how they were able to practise art outside of the main art market, and how they defended unique values in this remote urban place.

Geography: Why Artists Settled Outside of the City and What They Found There

Artists in Brussels at the Turn of the Twentieth Century and during the Interwar Period

The increase in numbers of visual artists in Brussels during the second half of the nineteenth century was closely linked to the growing population of the Belgian capital and the elevation of the social and professional status of painters and sculptors (Becq, 1982; Hoogenboom, 1993; La Gorce et al., 1997; Barker et al., 1999; Heinich, 1996, 2005). The geography of visual artists within the expanding city reflected this twofold reality. It was in the developing suburban margins of Brussels where many artists settled and founded their studios, but with a clear tropism towards the eastern suburbs (Debroux, 2013). These new territorial developments were dominantly bourgeois, whereas the western ones, such as Molenbeek and Anderlecht, were dominantly working class.

Following the overall pace of urbanization of the city, artists settled first in the northeastern suburbs of Schaerbeek and Saint-Josse, then twenty years later, in Ixelles and Saint-Gilles (see Figure 7.1). These

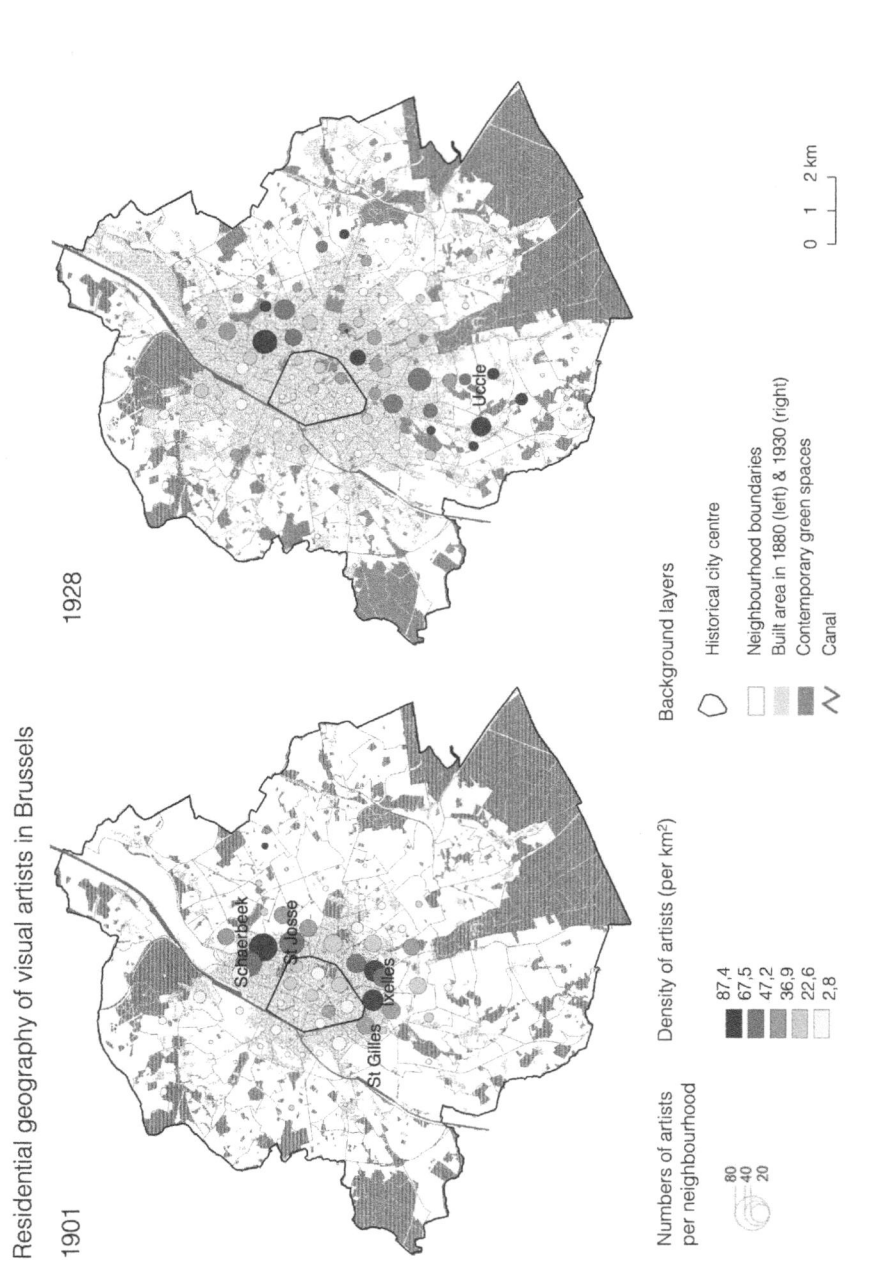

Figure 7.1. Maps of visual artists in Brussels, 1901 and 1928. Data source: Trade directories Mertens & Rosez, 1901 and 1928. Copyright: Author, 2013.

suburbs, which were soon to become neighbourhoods of Brussels, offered to artists essential conditions such as relatively affordable plots or houses at walkable distance from the city centre and its cultural life, as well as daily encounters with affluent households that constituted the artists' clientele. Residential proximity between artists and their clients was long a major explanatory factor of Brussels's artistic geography. The first real art gallery only opened its doors in 1912 (Lewijse, 2005), while it was only after the Second World War that the art market as we know it today developed in Brussels. Therefore, for decades, the role of social encounters between artists, art merchants, and clientele remained an essential element of the local art economy, which was facilitated by its spatial proximity.

These key explanatory factors – affordability of land to build one's own studio or artist's house and spatial proximity with wealthy neighbours – were still valid when Uccle experienced an increase in artist numbers during the first quarter of the twentieth century. Although these numbers always remained modest compared to the two oldest clusters, they were significant during the 1920s and 1930s. The share of artists in Uccle's population increased rapidly at the beginning of the century, whereas the general growth of the population took place at a slower pace (Debroux, 2013). Consequently, from the 1920s, small clusters of artists in Uccle can be noticed in data and maps. Although artists were not completely absent before this period, Uccle only developed a sizable and permanent presence of artists from this point forward.

As mentioned, explanatory factors for the settlement of visual artists in this southern suburb of Brussels were comparable to earlier creative suburbs like Schaerbeek and Saint-Gilles in an earlier urbanization phase. The push factor of higher real-estate prices and competition for studios in the centre explained why vacant land in close proximity to the city centre was perceived as a reasonable economic choice, since it enabled the artists to design their own houses and studios.[5] The fact that wealthy families transformed their secondary residence in Uccle into their main house in order to escape the inconvenience of living in the populated city centre also offered an incentive for artists to follow them. Nevertheless, two additional dimensions need to be taken into account: the remote geographical location of these lands and the chronological context of Uccle's urbanization. Even though pastoral exhibitions in Uccle were illustrative of cultural life outside of the centre, the main artistic activity remained in the central municipalities of the city. It was only when means of transportation facilitated the connection between these suburbs and Brussels (by tramway first, then cars for those who could afford one) that they truly became an option as a place to live for artists. Finally, industrialization had brought massive changes

Figure 7.2. Juliette Wytsman-Trullemans, *Les environs de St-Job (Uccle)*, ca. 1900, oil on canvas, 79 x 112 cm, Musée Charlier. Copyright of the photograph: Olivier De Pauw, 1997, KIK-IRPA, Brussels, Belgium, picture KN007436. Reproduced with permission.

to urban conditions, causing pollution and an extreme densifying process in working-class neighbourhoods as well as in more affluent parts of the city. It generated important health and social problems that led sections of the elite to experience mixed feelings about the city – including the artist population. Everywhere in Europe, since the Industrial Revolution, nature had been perceived by artists as an authentic place and a refuge against the violence of modernity, especially after the First World War (Jacobs, 1985). The surroundings of Uccle still offered natural and agricultural sites that were highly valued by the artists who settled there. They would make use of these landscapes, often in an idealized or anachronistic way.

Suburban Landscapes

The catalogues of the Uccle Centre d'Art annual exhibitions include the titles of all the artworks presented to the audience. These titles are revealing for the kinds of topics favoured by the artists as well as by their clients (Figure 7.2). During the first twenty years of the art circle's existence, representation of landscapes was still common and genre painting clearly dominant. As was the case with representations of nature in artists' colonies (Jacobs, 1985; Lübbren, 2001; Pinson, 2006), paintings and drawings often represented the most picturesque sites of Uccle, believed to be "untouched" by mankind, or an idealized vision of agricultural activities. If such views were still representative of a certain reality, they were specifically selected by artists, while suburban land-scapes in Uccle, in fact, offered much more variety.

Artists who settled in Uccle at the beginning of the twentieth century were looking for a remedy to urban life and its lack of nature. Uccle still provided many natural sites, farms, and suburban production activities such as old sand quarries. Nevertheless, the municipality also consisted of dwellings ranging from modest terraced houses to big mansions along prestigious avenues, villas and castles of the wealthier families of Brussels. In a slightly different geographical context, Uccle corresponded to what Thoreau and American transcendentalists named "borderlands" – places "not too rural or isolated from the germs of urban creativity" (Ghorra-Gobin, 2006: 156n2). This qualifying term "borderlands" indicates the nature of these places: located at the margins of the city, they were therefore fated to evolve and host activities connected to urban life, not only of an economic nature but also related to the cultural life brought in by the new residents. In a short story that included the character of Styne van de Woestyne, daughter of a famous painter member of Uccle Centre d'Art, the writer illustrates the dramatic and rapid changes occurring in Uccle in these times: "At dawn, a storm struck the Van Buuren residence ... [Styne van de Woestyne] would never see ... the little English girl who lived across the street again. The mansion where her friend lived had disappeared during the storm. On its site, construction of two new villas had already begun" (Lopes-Sabino, 1994: 83, translated). In addition to roads that had historically connected the village cores to Brussels (chaussée d'Alsemberg, chaussée de Waterloo), new avenues were constructed that accelerated the urbanization process. During the interwar period, real-estate developments under the form of villas and town mansions increased, which completed the social spectrum of a suburban landscape previously made up of humble ter-raced houses and (very) wealthy secondary residences (Figure 7.3).

Marginal and Elite: Artists in Uccle

The variety of architectural landscapes in Uccle corresponded to a social diversity and specific geographical patterns. Different spaces and hous-ing types sheltered a mix of rural and industrial workers (from the sand

Old village core around church Saint-Pierre

Farm of Crabbegat (farmers)

Chaussée d'Alsemberg
(working classes & petite bourgeoisie)

Villas (bourgeoisie)

Terraced houses (bourgeoisie)

Castle Allard (high society)

Figure 7.3. Postcards illustrating the social gradient of housing types in Uccle (postcards printed during the first quarter of the twentieth century). Source: Author's collection.

industry or those expelled from the city centre), civil servants, petite bourgeoisie, and wealthy families. As mentioned earlier, an important explanatory factor for the geography of artists in Brussels at the turn of the twentieth century was the elevation of their social status, which allowed them to relate more closely to traditional elites – in society but also in urban space. According to French sociologist Nathalie Heinich (2005), visual artists became "elite and marginal," an expression that referred to both their social position and their geographical location (Debroux, 2012, 2013). On one hand, artists needed to distinguish themselves from the working classes, from craftsmen and the petite bourgeoisie; on the other hand, they could not access the highest social spheres (although they were dealing with them regularly in the pursuit of their art). Even if they were living in the same parts of the city, they did not share exactly the same addresses or the same standards of living: artists lived and worked in intermediary spaces at the margins of elite locations. The map showing the residential addresses given for artists in the exhibition catalogue of 1922 and the list of the circle's patrons in 1925 illustrates this reality. Exhibiting artists at Uccle Centre d'Art were required to live in Uccle or in its close vicinity,[6] which was not the case for the wealthy patrons of the art centre. However, within the boundaries of Uccle, the different geographies of the two social types, artists and patrons, can be observed (Figure 7.4).

Patrons of the art circle came from the wealthiest neighbourhoods of Brussels or lived in the higher and eastern parts of Uccle (whether close to the built urban area or in remote properties such as the castle of Vivier d'Oie). By contrast, artists, for their part, preferred areas close to the small valleys that they liked to represent in their works. These spaces also corresponded to more affordable locations that were also adjacent to the historical villages. In the latter, there were premises that had been important for social gatherings of artists, such as inns. What the map (see Figure 7.4) also highlights is the close proximity in which artists were living. Artists concentrated in specific streets (rue Edith Cavell, for instance, or rue Van Zuylen), sometimes near the exhibition spaces of the art circle (the castle of Wolvendael) and other meeting places like the Café des Artistes and Café du Centre. In her recollection of the 1920s, the granddaughter of the painter Henri Quittelier wrote that, where her grandfather lived (close to the sunken road Crabbegat, which he drew so many times), the neighbours included at least eight other painters and sculptors, in addition to writers and musicians who were also involved in the activities organized by the circle (Hammes-Quittelier, 2009). Sociability played an essential role between artists and their patrons, whether to allow artists to sell art or to gravitate around common interests related to the preservation of the suburban landscapes they valued.

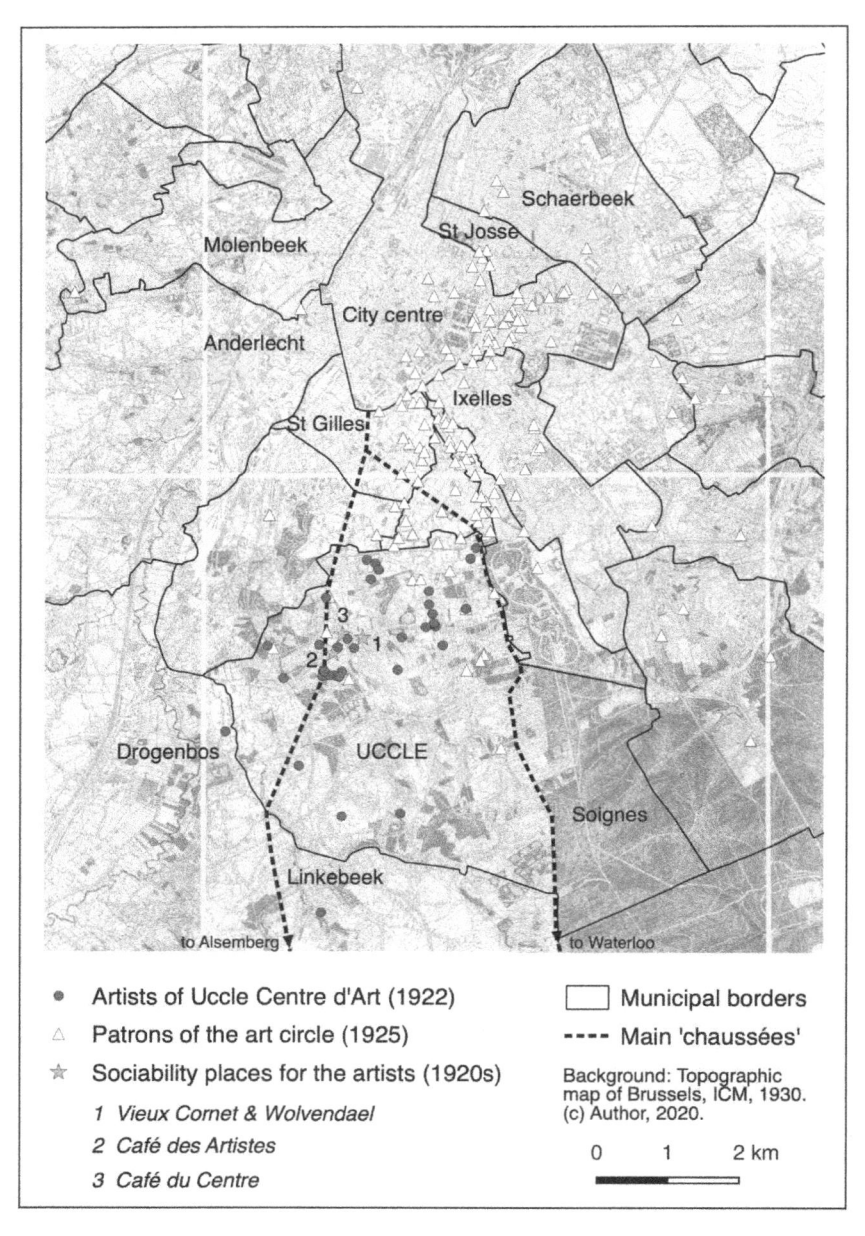

Figure 7.4. Map showing the residential geography of the artists of Uccle Centre d'Art (1922) and their patrons (non-artist members, 1925). Copyright: Author, 2020.

Sociability: Artistic Life Framed by the Suburban Social Environment

Becoming Urban: Emerging Cultural Activities in Uccle

One of the purposes of Uccle Centre d'Art was to collectively improve the visibility of artists' work through the annual exhibition organized in Wolvendael as well as through other kinds of events: readings, concerts, studio visits, dinners, and walks in which artists and non-artist members participated.[7] Following the densification of the suburb due to its growing population, former urbanites started activities mimicking what was done in the city, but in smaller social circles. In a sense, Uccle Centre d'Art offered an extension of a cultural life existing in the city centre and in the older creative clusters. In these smaller art societies, opportunities were probably less diverse but might have been more intense. The archives of the art circle contain pictures of exhibitions, excursions, and banquets that were often echoed in the press. In order to reach a larger audience, and in an attempt to raise the interest of the cultured bourgeoisie of Ixelles and Saint-Gilles, letters of invitation were often sent, and posters and even banners were hung in the streets of these municipalities. The organizers had great ambitions: for the first edition of the annual exhibition, in addition to local and regional elites, they invited the Queen ... who did not show up (Figure 7.5).

Intermediaries Fostering Suburban Artistic Life

A few figures were central to organizing the social activities surrounding the artistic exhibitions. As sociologists Howard S. Becker would theorize a few decades later, "intermediaries" were crucial nodes in the art world (Becker, 1982). Their importance was perhaps even more central in smaller societies where things and projects could develop faster but with limited resources.

The first non-artists to play a central role at the beginning of the pastoral exhibitions were innkeepers, who opened their inns to the painters and sculptors. These places were central for sociability outside of the city centre, especially in the early days when suburbs were mainly rural and could only be reached on foot. The inns and restaurants offered beverages and simple food as well as an occasion to rest (Maziers, 1999). Once these premises had become places to stay for artists living nearby because they offered cheap food and large spaces for groups to meet, inns and their owners were often prone to encourage the presence of artists to increase consumption and reputation but also for the sake of art. Le Vieux Cornet (where the exhibitions took place initially), Le Café des Artistes, and Le Café du Centre (where the circle's committee

Figure 7.5. Example of a social gathering of Uccle Centre d'Art members, a dinner in Beersel, 1937. Source: Administration communale d'Uccle, Archives Service Culture.

meetings were held), or the inn In 't Hoef were part of these anchoring places for Uccle Centre d'Art (Figure 7.6). A journalist recalled:

> A few years ago, the cabaret "In 't Hoef" was run by a sculptor who had married a young lady named Van Cutsem. This was the time of its greatest splendour. Many artists were then meeting there around the heavy tables where one could joyfully taste a Lambic and a Faro [beers], famous for a mile around. Each of these artists contributed to enriching the small art museum, brought together by the "baes" [boss] in the Flemish rooms where modern paintings were displayed alongside engravings, ancient prints, and authentic antiquities from the old Uccle. The small museum is now empty. The last Van Cutsem to run the cabaret moved to Knocke-sur-Mer [sic], where he took all the memorabilia.[8]

Other important protagonists for the settlement of artists in the suburbs were local authorities. The history of the famous artists' colony of Sint-Martens-Latem is linked with the initiative of its mayor: at the very beginning of the twentieth century, he invited artists and offered them decent living and working conditions (Haesaerts, 1968: 40). In Uccle too, Mayor Xavier

Figure 7.6. Postcard of the interior of the inn In 't Hoef, printed in the first quarter of the twentieth century. Source: Author's collection.

De Bue played a decisive role in 1921 by commissioning an exhibition and inviting artists from the municipality – leading to the creation of Uccle Centre d'Art. Even so, the relationship between members of the circle and municipal authorities were complicated: they were allies (in organizing the annual exhibition, buying artworks, and several aldermen were protectors of the circle), but they often argued regarding the future of the municipality. Archives are full of letters between the circle's secretary and the authorities, not only concerning the practical organization of the exhibitions but also criticizing planning decisions, as illustrated in the next sections.

As mentioned, Uccle Centre d'Art not only counted artists among its members (mostly visual artists, as well as a few musicians and writers) but also relied on non-artist members. "Protectors" and "honorary members" were the only ones paying to be part of the circle; hence, their role was crucial to support all the activities organized by the circle. Therefore, their numbers were a constant worry: throughout the circle's history, attempts were regularly made to raise the number of non-artist members by organizing concerts, talks, excursions (for its twenty-fifth anniversary, "*un goûter breughélien*"), visits to the studios, and other similar events. Unfortunately, little is known about the identity of these important sponsors. Some clues can be found in the exhibition catalogues, where the publication of the names of "protector" members appeared as an acknowledgement. In 1939, for instance, in addition to the 57 exhibiting artists, the catalogue mentioned the names of 125 supporting members and 65 honorary members, among whom were aldermen, municipal counsellors, doctors, engineers, and industrialists.

Finally, invisible intermediaries in the archives are women, who had always played a central role in cultural life, whether as artists' wives or as patronesses (since women artists were rare; Creusen, 2007). At a time when direct relationships between artists and their clients were essential, two examples related to the cultural life in Uccle stress the role of women. The first is linked to the closure of the castle of Wolvendael in 1934 due to renovations and the subsequent proposal made to the municipality by Mademoiselle Pipyn, a wealthy single woman, to install an art gallery on her property. This initiative can be associated with patronage since no fees were requested,[9] and the only purpose was to help artists to exhibit and sell their works. By comparison, in the same year, the Gallery Isy Brachot offered to host the annual exhibition but asked for 20 per cent commission on sales. The second example is not directly related to the art circle but benefitted some of its members, for instance, Gustave Van de Woestyne. David Van Buuren, a wealthy banker, and his wife Alice erected their Art Deco house in Uccle in 1924, which soon became a haven for artists and a gallery space for the couple's extensive art collection and intense social life orchestrated by the banker's spouse (Lechien-Durant 2000; Anspach 2007).

The Suburbs as Social "Battlefields"

The analysis of the residential geography of artists illustrated how this specific professional group positioned itself socially through and in space. Professionally, artists were dealing with both sides of the social ladder: the craftsmen providing their canvases and material, and the higher classes ensuring their incomes. This in-between position was not exactly reflected in location, since artists favoured proximity to the elites. Yet artists' relations with higher society were complex and not free from class conflicts. Many disputes centred on governance of the suburbs and management of the natural environment in these urbanizing areas.

Leisure activities of the higher social classes were met with resistance by artists on several occasions: they fought, for instance, against projects aimed at securing access to the forest with safer paths clearly delineated and lit with gas (in a sense, making the wildness more urban). They also criticized and opposed the building of leisure infrastructure within the Forêt de Soignes, such as a hippodrome. Another issue was related to the transformation of the natural landscape by economic activities. To oppose cutting down trees, artists pleaded for the preservation of a landscape that had acquired an aesthetic value for the bourgeoisie (who were acquiring many paintings that pictured nature and rural landscapes). A reader of the

Journal des Beaux-Arts complained: "You have to make money, sure. But isn't this necessity unfortunate at a time when the taste for the arts is developing to such an extent that we are trying to hinder this development by all means, for fear of a real overflow? This taste has invaded the crowd; there are no more grocers like in Courbet's time. The grocers themselves know something about fine arts and often have a quite extensive collection."[10]

If artists in the previous examples fought the bourgeois way of life, it must be noted that artists and the bourgeoisie were also allies and shared local interests to preserve the places where they lived, worked (artists), and spent their leisure time. Alliances existed therefore between artists and sections of the wealthy bourgeoisie to preserve their environment – whether urban or suburban.

Commitments: Making Uccle Stay as Seen in Landscape Paintings

"Blessed Are the Cities Guarded by Trees"

An example of a coalition of interests between artists, intellectuals, and the cultured bourgeoisie at the turn of the twentieth century was the preservation of nature[11] – including in the industrialized cities – and natural sites, which still existed in the urbanizing suburbs (Billen, 2015). In the debates from more than a century ago, signs of an emerging environmental discourse can be found (not in the sense of today's ecological meaning, though, but in the sense of preserving landscapes).

In Brussels, the protests focused especially on the preservation of trees in the city centre and the need to incorporate them into the developing suburbs. If the role of King Leopold II and Mayor Charles Buls in this matter was praised in the weekly progressive arts journal *L'Art moderne*, its columns also contained obvious mistrust against local authorities regarding their lack of proper consideration for the greenness of the urban landscape. "Oh! If only Molenbeek, Schaerbeek, and Saint-Josse had been urbanized forty years later! What a dream of greenery and colours would replace the dismal procession of gloomy and sepulchral streets!" lamented a contributor to the arts journal.[12] Another battlefield was centred on the forest located south of the city (Forêt de Soignes, connected to Uccle), notably regarding its preservation from economic exploitation. For a long time, the forest had attracted people going for walks, but from the second half of the nineteenth century, the urbanization process started threatening its margins. In 1880, the artist Camille Van Camp addressed his concerns in a plea to the Belgian authorities, which raised general awareness among artists and amateurs of landscape painting. In less than twenty years, four associations were created that all included Soignes at

Figure 7.7. A prototypical representation of Uccle's old landscapes favoured by the artists of Uccle Centre d'Art: Henri Quittelier, *Le Kamerdelle – Uccle 1925*, engraving, 15 cm x 21 cm. Source: Private collection of Laure Quittelier. Reprinted with permission.

the centre of their activities: the Société Nationale pour la Protection des Sites (1891; co-chaired by woman artist Euphrosine Beernaert), the Touring Club (1895), the Ligue des Amis des Arbres (1905), and the Ligue des Amis de la Forêt de Soignes (1909; founded by painter René Stevens who also had influence on the Royal Commission for Monuments and Sites). These new lobby groups were mainly composed of business elites and engineers (Bertho-Lavenir, 1997). They shared many common traits and concerns with the protectors of Uccle Centre d'Art and were sometimes directly part of the same social circles (Figure 7.7).

In Uccle, similar battles occurred, to a great extent concerning the Crabbegat, a sunken path passing between the inn Vieux Cornet and the castle of Wolvendael. Rumours about selling plots and potential destruction of trees created strong reaction among artists of Uccle Centre d'Art. They launched an unprecedented mobilization, receiving support from the Touring Club and the powerful Ligue des Amis de la Forêt de Soignes. Their influence persuaded the municipality of Uccle to abandon its plan and even to protect the old trees. In the struggle to save the sunken path, artists intervened in the name of their aesthetic expertise. For them, preserving the path and its centenarian trees was a necessity in order to protect the picturesque side of Uccle (*le pittoresque ucclois*). This phrase was mentioned repeatedly in documents of the municipal archives. Another protest movement concerned a project to demolish the old church of Saint-Pierre. The church was acknowledged for its historical and architectural value, and situated in the main village core of what would become the Uccle municipality. Conservative feelings, or at least strong nostalgia against change, can be read in this draft of a letter dedicated to the mayor and local counsellors (the parts censored are underlined):

> Artists, and all people of taste, are indignant at the all too frequent attacks on the beloved face of their beautiful municipality by particular speculators; it would be very unfortunate if the municipal administration followed such deplorable examples … It is regrettable to note that a municipality known as the "Pearl of Brussels" to which new walkers and residents flocked only because of its rural and picturesque aspect is on the verge of becoming, under the pretext of urbanization (!), a banal and colourless suburb.[13]

In Uccle, as well as in Brussels in general, champions of natural sites realized that one of the major arguments they could use to persuade authorities lay in attractiveness for visitors, now that a growing number of inhabitants and visitors frequented these sites for the sake of health, scientific education, and leisure. Among these defenders of the woods, artists were joined by industrialists who associated the image of the city with the image (and indirect success) of their industry (Billen, 1997). They had different goals, but a common cause!

"Artialisation" of Suburban Landscapes

The success of a small book edited by Uccle Centre d'Art (1925), *Uccle au temps jadis* (Uccle in the old days), illustrates another influence that artists had on the population of their time. The book consisted of an illustrated collection of historical and folkloric notes on history, specific

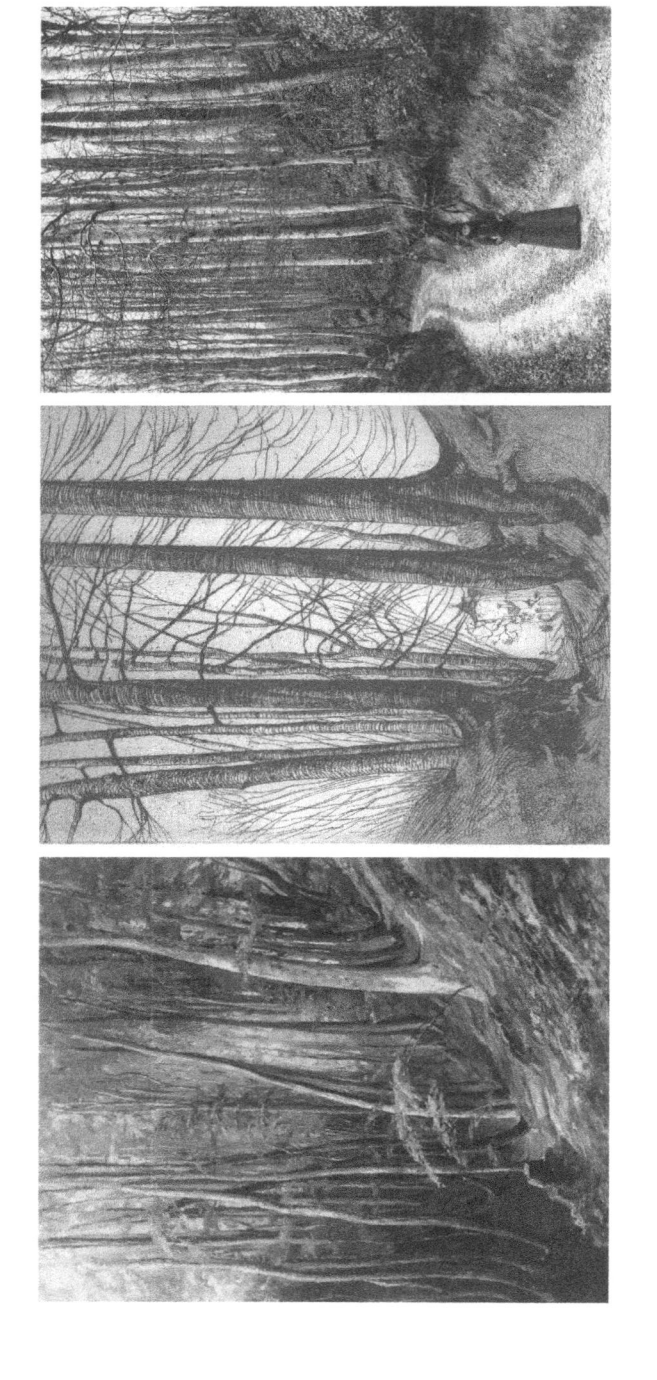

Figure 7.8. The woods, and in particular the sunken path from Crabbegat, were a recurring topic for the artists of Uccle Centre d'Art, who popularized walks in the woods for local strollers. Left: Jean-Louis Minne, *Le Wolvendael (début d'automne)*, photograph of the painting sent by the artist in 1937 to the municipal secretary. Source: Administration communale d'Uccle, Archives Service Culture. Middle: Henri Quittelier, *Uccle, le Crabbegat et le Vieux Cornet*, 1969, engraving, 22.5 x 27.5 cm. Source: Private collection of Laure Hammes-Quittelier. Right: Auguste Van Gele, *Le Crabbegat en direction du Vieux Cornet*, black and white photograph, 1892. Source: KIK-IRPA, Brussels, Belgium, picture A004714. All reproduced with permission.

sites and their legends, and artists working in the municipality. It had to be reprinted, since the audience proved to be highly receptive to such nostalgic recollections of the places where they lived.

The role of artists must be stressed here, and more precisely the role of their works. Once a landscape or natural site had been captured and repeated on canvas, the perception of it changed (Pinson, 2006). Collective imaginaries formed, and the (often outdated) views produced by painters raised new awareness of the changes taking place. This mediating role of artworks in popularizing landscapes and giving them value through their "artialisation" (Roger, 1978) occurred in and around Brussels for sites such as Tervueren, Josaphat, Soignes (and Rouge-Cloître), Uccle, and Linkebeek (and its "artists' valley"). Claire Billen notes: "The forest landscape became particularly interesting because a pictorial movement had made the forest one of its favourite subjects. So much so that walking in the forest and enjoying the views became an artistic activity, and therefore a distinguished one" (Billen, 1999: 62; translated).

The mediating role of the artists themselves is another important dimension in defending their causes. Because of their professional activity and their social status, artists were introduced to a very large social spectrum – from journalists to wealthy art patrons. Therefore, they occupied a privileged position within which to repeat their claims. These "humble Sunday walkers," as members of the group defending Soignes portrayed themselves (Billen, 1999: 65), were in reality more than equipped and skilled to resist and make their claims heard. In addition, once artists took a stand to defend the sites they painted, they could often count on the help of the press to influence views and gain growing support among the general population.

Conclusion – Art or Nature: Situating Suburban Artists' Creativity

Nostalgia cannot only be located in time (modernity looking back) but also in space.

    – Nina Lübbren, *Rural Artists' Colonies in Europe* (2001: 14)

The advent of artists' colonies throughout Europe from the mid-nineteenth century also occurred in Belgium. The country had always been connected to larger cultural centres and their influences during its lively history. Additionally, the rapid development of the Belgian economy allowed the construction of the first railway on the continent, an element that fostered the migration of artists and their settlements outside cities (Jacobs, 1985; Lübbren, 2001). A few localities, for instance, Anseremme or Sint-Martens-Latem,[14] remain famous in Belgian art history for their artists' colonies. However, probably because of the small size of the country and its great variety of landscapes reachable in relatively short distances, artists seemed to favour excursions over long stays in the countryside. The majority of them kept their residence in the city, with Brussels acting as a magnet for the nation's cultural life. It was only later, when modes of transportation diversified within the city, that long-term settlements of artists developed outside the urban boundaries. Improved accessibility between the centre and suburban spaces was an important factor for this changing geography of artists, but their ambiguous relationship with the rapidly industrializing capital was also a factor. Notions of pastoralism and nostalgia are crucial to understanding the migration of artists outside the cities at that time (Cafritz et al., 1988; Alpers, 1996; Salé, 2000; Lübbren, 2001), determining whether they headed to the countryside or selected suburban rural locations. This tropism for suburban environment frequently translated into idealized representations of nature in their artworks but could also express a precocious environmental awareness, as illustrated by the story of Uccle Centre d'Art during its first years.

In *Les peintres de la Forêt de Soignes*, Emmanuel Van de Putte comments on the artists living in and around Brussels:

> If these artists are neither real city dwellers nor real country dwellers, they are, on the other hand, the passageways between the city that is becoming industrialized and a countryside that is becoming urbanized, in short between two worlds in full mutation, being themselves in search of an immutable nature that only seems to fluctuate according to the rhythm of the seasons ... Neither as demanding as the German expressionists nor as rustic as the expressionists of Latem, they have a rather "suburban" soul, that is, they seek both the conveniences of the city and the calm of the countryside. (Van de Putte, 2009: 120, translated)

This specific spatial fabric could only be temporary. These suburbs were likely to be urbanized and their nature domesticated as time passed. These processes accelerated greatly after the Second World War. In addition, by popularizing the suburban landscapes, artists also contributed to the acceleration of these changes. Painters were placemakers of suburbia, thereby transforming the views and landscapes they cherished. Finally, the second half of the twentieth century saw the emergence of a true modern art market in Brussels, and the multiplication of art galleries turned the former relations between artists, merchants, and clients upside down, as well as altered their previous interest in socio-spatial proximity.

New debates in research on creativity question the innate potential of suburbs for creativity, acting as testbeds for experimentation and an aesthetic vanguard away from the urban mainstream (Jones et al., 2016; Grabher, 2018). In this context, the question could be asked whether the suburban location of Uccle has offered emancipatory potential for the artists of Uccle Centre d'Art. Three arguments counter this assumption.

First, in addition to the geographical proximity of Uccle to the city centre (one-hour on foot, faster by tramway), the archives of the art circle contain proof of commissions for several artists that necessitated trips to the city centre. Henri Quittelier, for instance, one of the first and most prominent members of Uccle Centre d'Art, regularly accepted orders for advertising work and developed a series of engravings dedicated to the clientele of the Hotel Albert Ier, located near place Rogier. The fact that most non-artist members lived outside Uccle, and in the city (see Figure 7.4), also illustrates the limited peripherality in which artist members actually evolved and shows their constant interaction with the urban life.[15] In this aspect, those artists were not "authentic outsiders," since their professional activities were integrated in the Brussels art world and upper classes (Fine, 2003).

Second, even if small, Belgium's cultural scene offered room for a diversity of artistic practices: the first two decades of the twentieth century saw the blossoming of (often ephemeral) circles, art journals, and groups, for instance, cubists and constructivists in Antwerp, Flemish expressionists, surrealists, and Brussels animists (Pa and Stappaert, 2000; Laoureux, 2009). Movements or schools never materialized in Brussels with a clear program. Therefore, the "schools of the forest," such as the one described in Uccle, hosted artists from various artistic genres gravitating around a common interest in natural landscapes. They were not unified by a distinctive aesthetic agenda defined by a peripheral identity.

Third, by looking at the artworks exhibited at the Uccle Centre d'Art exhibitions and through the lens of its collective archives,[16] it appears that rather conservative forms of art were produced – not forms that would necessitate developing outside of the mainstream urban art market. However, if they were not acting as a testbed for emerging artistic trends, the suburbs might have been a shelter for these rather outdated genres, which could only persist outside of the city centre, whereas the most up-to-date art production was within the city centre (nostalgia being also located in space, as expressed by Lübbren, 2001). This idea is supported by the violent and bitter responses that artists gave to the invitation of painters and sculptors living outside of Uccle on several occasions. Hanging on the walls of the castle of Wolvendael, artworks pictured anachronistic representations of space, where nostalgia for a disappearing landscape translated into outdated views of the municipality.[17]

The collective history of Uccle Centre d'Art is an interesting example of an art circle with a strong local identity and a territory with a strong visual identity that artists represented abundantly, which was perceived as a sanctuary against urban modernity. If not expressed in their art, it is more in their condemnation of the changes occurring in suburban landscapes that these artists stood among an avant-garde, which shared their concerns about urban conditions and nature preservation, and produced the first environmental discourses. More than in pictorial modernity, these suburban artists used their creativity to promote the beauties of a disappearing landscape, acting in this way as early whistle-blowers for environmental destruction.

NOTES

1   The quote refers to the Forêt de Soignes (Sonian Forest), located south of Brussels and partially included in Uccle. Original French: "Quelle retraite est plus réconfortante, plus paisible et plus proche de la grande ville ?" (Bastien, 1920: 278). All translations from French quotes are by the author, who sincerely thanks Christian Morgner for his careful help with the translation and proofreading.

2   Uccle is the French name of one of the nineteen municipalities that form the Brussels Capital Region nowadays. The Flemish name of this municipality is "Ukkel." However, since the story this chapter narrates is that of a French-speaking group of artists and art amateurs, it is the term "Uccle" that will be used throughout the text. At the beginning of the twentieth century, it was still an independent locality whose mayor was taking part in a "conference" with peers from other localities surrounding Brussels.

3   "25 years of summer exhibitions at Uccle Centre d'Art," an unidentified press article dated 2 September 1933, stored in the archives of the Municipal Department of Culture, files "Uccle Centre d'Art" (translated).

4   Stored at the central administration of Uccle, the archives of the art circle contain letters, documents, images, press clippings, and exhibition catalogues that allow us to access its history and functioning. Lists of the archives' content exist, but no proper inventory of the documents has ever been carried out. For this reason, I refer to these archives simply as "Archives Service Culture Uccle, UCA."

5 Despite the reputation of Brussels as a "paradise for artists" due to more affordable land values (Gower, 1875: 240–1), it must be noted that relatives often helped financially young artists who were following the dream of building their own artist's house. An early example was the Bloemenwerf, an artist's house erected on avenue Vanderaey before Uccle was seen as a new haven for artists. This villa was built in 1895 by young artist and architect Henry Van de Velde with the financial support of his affluent mother-in-law, Maria Sèthe (Thiébaut, 2000: 62). An aesthetic statement for Van de Velde, the house attracted many artists and amateurs outside of the city centre and helped increase the recognition among artists of the advantages offered by the suburbs. Similar examples were the congregation of cultural life around the house of Auguste Oleffe in Auderghem (Schots, 1978; Van de Putte, 2009: 53) and around Constant Montald in Woluwe-Saint-Lambert (Levie and Thiel-Hennaux, 1982).

6 It is interesting to note that, if artists from neighbouring urban locations such as Saint-Gilles or Ixelles were violently excluded from exhibitions and perceived as competitors in the circle's archives, that was not the case for painters or sculptors living in southern locations such as Linkebeek or Drogenbos, which at this time looked very much like the most rural parts of Uccle due to their very low urbanization.

7 The purchase of artworks seldom happened during the exhibitions unless at charity auctions. One hypothesis might be that sales took place behind closed doors or in the artists' studios, which the collective archives of the circle did not record. However, it did not lower the importance of exhibitions and social events with artists, because the exhibitions offered repeated opportunities for encounters that favoured the acquaintance between artists and their patrons.

8 Excerpt from an article published in Le Peuple, titled "In 't Hoef," 1930, Archives Service Culture Uccle, UCA (translated). The wealthy Van Cutsem family was famous for its patronship of artists in Brussels; Henri Van Cutsem, patron and a painter himself, notably included painter Auguste Oleffe in his will (Schots, 1978: 57) and bequeathed his art collection and mansion to sculptor Guillaume Charlier. The building was turned into a museum in 1928, known today as Musée Charlier. As mentioned in the quote, marriage between daughters of cultured bourgeoisie and artists were frequent.

9 The only costs that Uccle Centre d'Art had to cover according to its archives were the expenses to turn the rooms into a gallery. Other art circles followed in the same exhibition space that year – not only originating from Uccle, which caused Uccle Centre d'Art to complain adamantly to the local authorities.

10 Letter from the Journal des Beaux-Arts reproduced in L'Art moderne, 45 (1886): 356 (translated).

11 Quotation in the subtitle is from Verhaeren, 1920: 295.
12 Unsigned opinion piece published in *L'Art moderne*, 35 (1897): 278 (translated).
13 Draft of the letter addressed to the mayor, aldermen, and municipal councillors, dated September 1935, Archives Service Culture Uccle, UCA (translated).
14 Flemish name for Laethem-Saint-Martin, probably Belgium's most famous artists' colony, located in Flanders, 10 kilometres from Ghent (De Ridder, 1945; Haesaerts, 1968; Boyens, 2001).
15 A similar remark applied forty years earlier regarding the geography of the avant-garde artists of the circle Les XX. The need for frequent encounters with the elite commissioning these artists acted as the prime explanatory factor for their geography (Brogniez and Debroux, 2013, 2017). However, financial argument was not the only element: it is obvious that the artists of Les XX clearly favoured the wealthy southeastern suburbs of Ixelles to the ones located East, where a more conservative elite lived who did not support their aesthetic proposals.
16 Individual archives could tell a different story for some of the artists of Uccle Centre d'Art, but they could not be identified during this research.
17 In the same way, the texts and photographs published by the Touring Club deliberately omitted large parts of the landscapes they described (Bertho-Lavenir, 1997; Notteboom, 2007).

REFERENCES

Alpers, P. 1996. *What Is Pastoral?* Chicago: University of Chicago Press.
Anspach, I. 2007. *Musée et jardins Van Buuren*. Brussels: Fonds Mercator.
Barker, E., N. Webb, and K.W. Woods, eds. 1999. *The Changing Status of the Artist*. New Haven, CT: Yale University Press.
Bastien, A. 1920. "Les peintres de la forêt de Soignes." In R. Stevens and L. van der Swaelmen, eds., *La Forêt de Soignes. Monographies historiques, scientifiques et d'esthétique*. Brussels: G. Van Oest et Coe, 273–94.
Becker, H.S. 1982. *Art Worlds*. Berkeley: University of California Press.
Becq, A. 1982. "Expositions, peintres et critiques: vers l'image moderne de l'artiste." *Dix-huitième siècle*, 14: 131–49. https://doi.org/10.3406/DHS.1982.1384
Bertho-Lavenir, C. 1997. "Normes de comportement et contrôle de l'espace: le Touring Club de Belgique avant 1914." *Le Mouvement social*, 178(1): 69–87. https://doi.org/10.2307/3779563
Billen, C. 1997. "Les métamorphoses d'un usage de la nature. Paysages et sites à l'époque de Solvay (1870–1914)." In A. Despy-Meyer and D. Devriese,

eds., *Ernest Solvay et son temps*. Brussels: Archives de l'Université Libre de Bruxelles, 249–70.

Billen, C. 1999. "Conclusion. Le tourisme et la promenade, phénomènes essentiels de l'histoire de la forêt de Soignes." In C. Billen and Collective, *Tourisme en Forêt de Soignes, hier et aujourd'hui*. Catalogue from the exhibition at the Castle Trois-Fontaines, 25 September–28 November 1999. Auderghem: Conseil de Trois-Fontaines, 61–8.

Billen, C. 2015. "De *L'Avocat Richard* aux *Scrupules de Bernus*, le désenchantement bruxellois d'Émile Leclercq (1817–1907)." *Textyles*, 47: 69–81. https://doi.org/10.4000/textyles.2630

Boyens, P. 2001. *Une rare plénitude: les artistes de Laethem-Saint-Martin 1900–1930*. Exhibition catalogue. Ghent: Ludion.

Brogniez, L., and T. Debroux. 2013. "Les XX in the City: An Artists' Neighborhood in Brussels." *Artl@s Bulletin*, 2, Article 5. https://docs.lib .purdue.edu/artlas/vol2/iss2/5/

Brogniez, L., and T. Debroux. 2017. "Une exposition à l'échelle de la ville. Sociabilités des espaces complémentaires aux Salons des XX et de La Libre Esthétique." *COnTEXTES*, 19. https://doi.org/10.4000/contextes.6327

Brogniez, L., T. Debroux, and le Maire, J. 2017. "The Rise of a Small Cultural Capital: Brussels at the End of the 19th Century." In R. Hibbitt, ed., *Other Capitals of the Nineteenth Century*. London: Palgrave Macmillan, 129–57.

Cafritz, R.C., L. Gowing, and D. Rosand. 1988. *Places of Delight: The Pastoral Landscape*. Washington, DC: Phillips Collection.

Creusen, A. 2007. *Femmes artistes en Belgique*. Paris: L'Harmattan.

Debroux, T. 2012. "Des artistes en ville. Géographie rétrospective des plasticiens à Bruxelles (1833–2008)." PhD diss., Université Libre de Bruxelles.

Debroux, T. 2013. "Inside and Outside the City: An Outline of the Geography of Visual Artists in Brussels (19th–21st Centuries)." *Brussels Studies*, 69. https://doi.org/10.4000/brussels.1180

De Ridder, A. 1945. *Laethem Saint-Martin, colonie d'artistes*. Brussels: Éditions Lumière.

Fine, G.A. 2003. "Crafting Authenticity: The Validation of Identity in Self-Taught Art." *Theory and Society*, 32(2), 153–80. https://doi.org/10.1023/A:102394 3503531

Ghorra-Gobin, C. 2006. "La maison individuelle: figure centrale de l'urban sprawl." In A. Berque, P. Bonnin, and C. Ghorra-Gobin, eds., *La ville insoutenable*. Paris: Éditions Belin, 147–58.

Gower, R. 1875. *Handbook to the Art Galleries Public and Private of Belgium & Holland*. London: Sampson Low, Marston, Low, and Searle.

Grabher, G. 2018. "Marginality as Strategy: Leveraging Peripherality for Creativity." *Environment and Planning A: Economy and Space*, 50(8): 1785–94. https://doi.org/10.1177/0308518X18784021

Haesaerts, P. 1968. *Laethem-Saint-Martin: le village élu de l'art flamand*. Brussels: Éditions Arcade.

Hammes-Quittelier, L. 2009. *Henri Quittelier 1884–1980. Peintre, graveur, décorateur et apiculteur. Ucclois*. Brussels: Éditions Echancrure.

Heinich, N. 1996. *Être artiste. Les transformations du statut des peintres et des sculpteurs*. Paris: Klincksieck.

Heinich, N. 2005. *L'élite artiste. Excellence et singularité en régime démocratique*. Paris: Gallimard.

Hoogenboom, A. 1993. *De stand des kunstenaars. De positie van kunstschilders in Nederland in de eerste helft van de negentiende eeuw*. Leiden: Primavera Pers.

Jacobs, M. 1985. *The Good and Simple Life: Artist Colonies in Europe and America*. Oxford: Phaidon.

Jones, C., S. Svejenova, J. Strandgaard Pedersen, and B. Townley. 2016. "Misfits, Mavericks and Mainstreams: Drivers of Innovation in the Creative Industries." *Organization Studies*, 37(6): 751–68. https://doi.org/10.1177/0170840616647671

La Gorce, J. de, F. Levaillant, and A. Mérot, eds. 1997. *La condition sociale de l'artiste XVIe–XXe siècles*. Saint-Etienne: C.I.E.R. Expression Contemporaine.

Laoureux, D. 2009. *Histoire de l'art 20e siècle. Clés pour comprendre*. Brussels: De Boeck.

Lechien-Durant, F. 2000. *Musée David & Alice Van Buuren: Maison de mémoire*. Brussels: Éditions Racine

Levie, F., and D. Thiel-Hennaux. 1982. *Constant Montald, 1862–1944: une vie, une œuvre, une amitié, Emile Verhaeren*. Exhibition catalogue. Brussels: Thiel-Hennaux.

Lewijse, V. 2005. "G.G.G. La Galerie Georges Giroux, monument de l'histoire de l'art belge. Une aventure courageuse, une épopée faite de ténacité et de conviction." *Mémoires – La lettre mensuelle*, juin 2005.

Lopes-Sabino, A. 1994. "Dans le jardin des Van Buuren." In *L'étoile du Nord: Présences portugaises en Belgique et en Hollande*. Brussels: Portugal Embassy, 67–83.

Lübbren, N. 2001. *Rural Artists' Colonies in Europe, 1870–1910*. Manchester: Manchester University Press.

Maziers, M. 1999. "Auberges, laiteries, guinguettes, estaminets …" In C. Billen and Collective, *Tourisme en Forêt de Soignes, hier et aujourd'hui*. Catalogue from the exhibition at the Castle Trois-Fontaines, 25 September–28 November 1999. Auderghem: Conseil de Trois-Fontaines, 41–3.

Notteboom, B. 2007. From Monument to Landscape and Back Again. *Strates*, 13. https://doi.org/10.4000/strates.5673

Pa, J., and B. Stappaerts. 2000. "Le réseau." In New International Cultural Center (NICC), *Mo(u)vements – Mouvements d'artistes en Belgique de 1880 à 2000*. Brussels: NICC, 51–6.

Pinson, D. 2006. "La faute à Cézanne ? À propos de la perception du pays d'Aix par ses nouveaux habitants de villas." In A. Berque, P. Bonnin, and C. Ghorra-Gobin, eds., *La ville insoutenable*. Paris: Éditions Belin, 56–66.

Roger, A. 1978. *Nus et paysages: essai sur la fonction de l'art*. Paris: Aubier.

Salé, M.-P. 2000. "Le mythe de l'âge d'or et le retour à la nature." In P. Thiébaut, ed., *1900*. Paris: Réunion des Musées Nationaux, 341–3.

Schots, H. 1978. *Auderghem et ses peintres*. Brussels: Imprimerie Beirnardt.

Thiébaut, P. 2000. "À la recherche d'un art total. Une figure emblématique: Henry van de Velde." In P. Thiébaut, ed., *1900*. Paris: Réunion des Musées Nationaux, 60–3

Uccle Centre d'Art. 1925. *Uccle au temps jadis*. Uccle: Uccle Centre d'Art.

Van de Putte, E. 2009. *Les peintres de la Forêt de Soignes*. Brussels: Racine.

Verhaeren, E. 1920. *Les villes tentaculaires*. Paris: Mercure de France.

8 Exploring Creativity in Dublin's Suburbs, 1900–2000: Insider, Outsider, Bourgeois, or Bohemian?

RUTH McMANUS

Introduction

This chapter explores the diverse and varied nature of creativity in suburban locations by focusing on several case studies drawn from one metropolitan area, Dublin, across the twentieth century. My approach is rooted in historical geography, which involves investigating the ways in which places change through time, as well as looking at the interaction between people and the landscape or streetscape. In examining a range of different suburbs across a lengthy time frame, I aim to establish some of the different forms that creativity has taken in Dublin over the course of the twentieth century and question whether the socio-economic and physical make-up of suburbs affect the nature of creative activity and the degree to which it is accepted by the mainstream arbiters of taste. In so doing, the chapter will demonstrate a richer and more nuanced suburban creative experience than has been heretofore acknowledged. I also build on earlier arguments I developed with Phil Ethington about the importance of the suburban life-cycle as a research approach (McManus and Ethington, 2007).[1]

The long-standing tendency to portray suburbs in a negative light, as places of monotony, uniformity, and conformity, is persistent in spite of attempts to challenge this portrayal (Harris and Larkham, 1999; McManus and Ethington, 2007). Indeed, research on the "creative class" has tended to reinforce this bias by focusing on inner-city areas. It is only relatively recently that (mostly Australian) academics have begun to counter this perception by highlighting the role that suburbs and fringe areas play as locations for the creative industries (see, for example, Gibson and Brennan-Horley, 2006; Comunian et al., 2010; Felton and Collis, 2012; Collis et al., 2013; Gregory and Rogerson, 2018). Phelps (2012) has highlighted evidence for the creative character of suburban economies in the present and in the past. Increasingly, researchers are recognizing contemporary suburban cultural complexity (see Bain 2016), while Drake (2003) has demonstrated that links between place and individual creativity can be influential in the creative process. Using a broader conception of creativity than the largely economic focus of Phelps and others, this chapter draws on four case studies to uncover the varied nature of creative practices in Dublin's suburbs over the past century or so.

Before introducing the case studies, a brief note about the nature of creativity is necessary in order to contextualize the discussion and start to tease out where the role of the suburb in the creative process might begin. In this chapter, I envisage creativity as having different components. At the inherently individual level, the innate creative impulse exists irrespective of location – in other words, an individual creator (writer/poet/musician/artist/inventor) will be creative irrespective of the location in which they find themselves. This innate impulse could be termed the latent element in creativity. However, it can be argued that external factors may have a bearing on how that creativity finds its expression – whether it is ignited and encouraged or stifled, and whether the tangible or intangible products of that creative process receive validation and recognition or remain ignored or hidden. The creative spark will be nurtured in certain environments more than in others, and it will be directed in particular ways depending on that environment. While a suburban setting may not necessarily have given rise to the creative impulse, it will certainly have affected how, where, and in what form that impulse is expressed. Opportunities afforded by increased levels of education and access to leisure time are other factors affecting creativity. It is also quite likely that the nature and status of the place of origin of the creator will have an effect on how the form of creativity is received by the arbiters of taste.

Creativity can be broadly considered in four ways: (1) artistic expression, such as literature, visual arts, and performing arts; (2) craftsmanship, such as applied arts, pottery, woodwork, tapestry weaving, and fashion makers; (3) industrial inventiveness, such as industrial design, machinery, and computers; and (4) everyday and vernacular creativity, such as clubs, societies, brass bands, and gardening. Three of these different expressions of creativity are identified and explored in the following short case studies for Dublin's twentieth-century suburbs, although there is some overlap between the different elements. The case studies were deliberately selected to cover a broad range of creativity and a diversity of suburban settings in terms of geographical location, time period, socio-economic status, and relationship to the core. The aim is both to establish the diverse range of creative possibilities within suburbs and to tease out the relationship between these activities, their geographical context, and the power structures that may affect their acceptance by metropolitan elites.

The chapter opens with a brief explanation of the historical-geographical context, after which two early twentieth-century case studies are presented. The first of these explores the foundation of a "safe" middle-class musical society, while the second turns to the more avant-garde artistic expression of female artist-craftworkers. The discussion then turns to the changing nature and experience of Dublin's suburbs as a result of twentieth-century policies of mass suburbanization, leading to an examination of two further case studies. These latter case studies look at literary and cinematic creativity emerging from working-class suburbs from the 1970s onwards, which have challenged the dominant representation of the suburb in an Irish context. The concluding section draws upon the range of suburban creative experiences outlined throughout the chapter in order to establish some key features of suburban creativity, as well as posing a number of further questions as to the broader comparative applicability of the discussion.

Dublin's Suburban Geography

To begin, let us look at Dublin city in the early twentieth century. By 1912, Dublin city and county had a population of 477,196, which included the central area renowned for its slum conditions and overcrowding, and prosperous suburban areas outside the city boundaries such as Rathmines, Rathgar, and Pembroke, which were physically close to the centre but socially and politically at a great remove (see Figure 8.1). These southern suburbs remained independently governed until 1930 (McManus, 2019). I argue that the city's evolution during the course of the twentieth century, and the nature of those changes, is reflected in the changing experience of suburban creativity. From the 1920s onwards, the development of low density suburbs on greenfield sites was generally adopted as the solution to the severe problems that beset the central area with its crumbling fabric and overcrowded tenements (McManus, 2002). New single-class estates at the edge of the city emerged, often with limited services initially due to the city authority's "housing first" approach (McManus, 2018). Because of the previous physical development of the city that had seen greater building of middle-class suburbs in the southern area, many of these new working-class estates were developed on the north side of the city and were interspersed with private middle-class suburbs to create a fragmented socio-economic pattern (Brady, 2014, 2016). As we shall see, the nature of suburban creativity changed over time as the physical nature of the suburbs and their social make-up also changed.

In Ireland, as elsewhere, the term "suburb" has been seen as a byword for lack of creativity. In July 2018, a columnist in a leading national

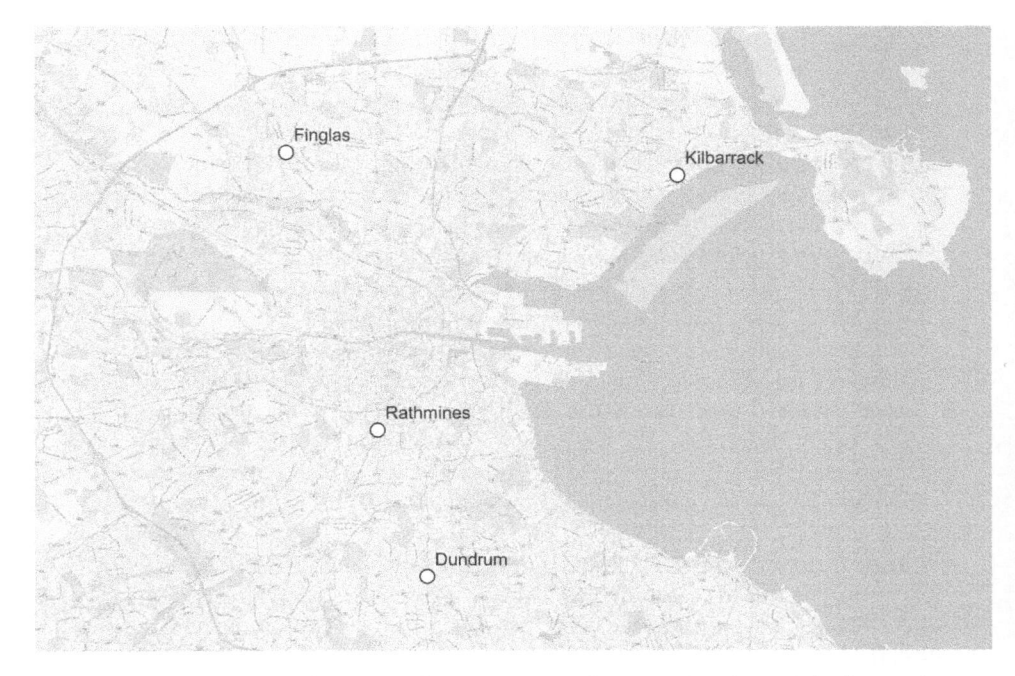

Figure 8.1. Map with case study locations: Rathmines, Dundrum, Finglas, and Kilbarrack/Barrytown. Source: Dr. Finnian Ó Cionnaith / OpenStreetMap.

newspaper, the *Irish Times*, presented a negative account of the west of Ireland city of Galway as it prepared for its stint as European City of Culture in 2020, using the terms "suburban mindset" and "suburban mentality" (McDonald, 2018). This term reflects a mode of thought that sees those living in suburbs as being small-minded, inward-looking, middle-brow passive consumers. More than a century previously, Irish commentator Charles Dawson similarly used the word "suburban" in a derogatory sense to indicate lack of cultural appreciation and, by extension, an anti-creative bias. However, the interesting – perhaps even surprising – observation of Dawson writing about Dublin in 1912 is the way in which he makes a distinction between the term used for its "literary" connotations and its geographical implications. The quote is rather lengthy, but it is worth reproducing in full:

> Is there a cultured literary mind in Dublin, or is your mental state altogether "suburban"? I use the word "suburban" in its literary, not in its literal sense. I use it as denoting "Philistinism." It is one of the plentiful paradoxes Dublin

> affords that the "suburban" temperament is more manifest in the city than in the suburbs. I do not hesitate to say that the "intellectuals" of Dublin live, for the most part, beyond the canal boundaries. Nothing is really more "suburban" than Fitzwilliam Square. Unquestionably, nothing is more provincial. If you desire to hear talk of pictures, books or plays; if you would rub shoulders with socialists, poets, feminists, vegetarians, or dramatists; if, in a word, you would mingle with the thinking few, you must penetrate into Phibsborough or Rathmines. Mind you, I do not guarantee that you will hear a debate on Art or Philosophy in every house in Rathmines. Heaven Forbid. I do not say that Bridge is not played beyond the waters of the Canals, or that the after-dinner snore may not be heard in those regions. I do say, however, that you run more risk of meeting "culture," in the intellectual sense, in the suburban dwelling than in the town house. The National Library, the National Literary Society, the centres of thought connected with UCD and TCD,[2] perhaps the United Arts Club, and one or two other places may be taken as the city rendezvous of these out-dwellers. (Dawson, 1971 [1912]: 255–6)

Rather than indicating that the suburbs were a cultural wasteland, Dawson turned the idea on its head by suggesting it was the central area (Fitzwilliam Square was renowned for its doctors and surgeons at this time) that was parochial or provincial. Indeed, to enjoy intellectual or artistic stimulation, one needed to venture beyond the central area and into the encircling suburbs. These remarks suggest that, even at the start of the twentieth century, the cultural and creative life of the suburbs was being recognized as having greater complexity than the clichéd stereotypes suggested. This is an intriguing point at which to introduce two contrasting experiences of cultural creativity in Dublin's suburbs in the first decades of the twentieth century, the time when Dawson was writing: the Rathmines & Rathgar Musical Society and the Dún Emer Guild / Cuala Press.

A "Safe" Form of Creative Expression: The Rathmines & Rathgar Musical Society

The Rathmines & Rathgar Musical Society, or "R&R" – which still exists today – was established in 1913 in the staunchly Unionist, upper middle-class suburb of the same name. Rathmines immediately adjoined the city to the south, the administrative boundary being marked by the line of the Grand Canal, but was considerably more affluent than the city and was independently governed by its own Urban District Council (Figure 8.2). Just over 38,000 people lived there in 1911, including both wealthy Protestants and middle-class Catholics, most of whom crossed

Figure 8.2. Rathmines in the early twentieth century. Source: Lawrence Collection, image courtesy of the National Library of Ireland.

the canal bridges on a daily basis to work or to avail themselves of the city's retail, leisure, and health services.

The Rathmines & Rathgar Musical Society was intended as an amateur society that aimed to study and perform "operatic, choral and other high-class musical works." The city already boasted a range of professional and semi-professional music, drama, and variety companies, but this new society was to be located in an affluent suburb, where it would offer opportunities for artistic expression and everyday creativity to local residents. The impetus for the formation of the musical society came with the appointment in February 1913 of a new organist and choirmaster, Christopher P. FitzGerald, to the local Roman Catholic Church. The initial meeting to consider the feasibility of forming a musical society probably came about at his suggestion (Dungan, 2013). While the arrival of FitzGerald was significant as a motive musical force, a further stimulus may have come from Englishwoman Violet Hoyle following her attendance with her husband at a Rathmines soirée one evening in 1913. A group of guests had gathered around a piano for a sing-song, and Hoyle remarked on "how beautifully their voices blended," before enquiring if they had ever considered staging light opera (7). Clearly, there existed an environment in which such a musical society could potentially flourish, as the founders of the society "came from a social class where the piano was often a focal point within the house and where musical talent was something to be nurtured" (6). The foundation of such a society, then, was dependent on the socio-economic nature of this suburb, which provided a fertile ground for musical endeavours. The inaugural meeting was held on 20 April 1913 at 48 Summerville Park, off Upper Rathmines Road, "to consider the feasibility of forming a musical society in the township of R&R" (4). According to the minutes of the first annual general meeting (AGM) on 24 September 1914, the meeting was attended largely by residents of the suburb.

A musical society could only come into existence where the necessary musical knowledge and interest existed. However, another factor in the success of the new organization lay in the socio-economic backgrounds of the founders and their successors. The first chairman, Edwin Lloyd, was a local solicitor from a musical family, while Charles Jackson and civil servant W. Gerald Mulvin became joint honorary secretaries. Coming from managerial backgrounds, they brought useful business and administrative skills that would prove vital to the survival and success of the society they inaugurated. These individuals were not just inspired by the possibilities of artistic expression but gave careful thought as to the practicalities of this new endeavour. It was only after "the details as to membership, management, subscription, etc. having been arranged, and having secured the premises [81 Rathmines Road]" (Dungan, 2013: 6) that arrangements began for the first production – Gilbert and Sullivan's *The Mikado*.

The R&R began, and remains, amateur in its ethos, offering a leisure-time musical pursuit for middle-class suburbanites. There has been an enduring association with the Gilbert and Sullivan comic opera repertoire, which is often perceived as being "safe" and "middle-class." Indeed, despite frequently satirizing English institutions and the obsession with class, part of the enduring legacy of Gilbert and Sullivan was that they reclaimed the theatre as a "respectable" middle-class pursuit (Murphy, 2010). In many ways, then, the R&R could be interpreted as offering a less challenging form of cultural expression because it presented the familiar, benign Gilbert and Sullivan repertoire, which was the same kind of entertainment enjoyed by the middle classes across Britain and Ireland, rather than representing "cutting edge" creativity (Williams, 2017). Its non-threatening familiarity was perhaps one of the reasons why the R&R could become so successful. Furthermore, the choice of repertoire reflects the political tendencies of the Rathmines residents, who generally favoured retaining ties with Britain. The early members of the society tended to be drawn from the respectable middle classes who were not attracted to the music or song of the Celtic

Revival, which was then challenging the status quo by presenting specifically Irish forms of culture as distinct from English culture. From the outset, performances by the R&R took place in city-centre theatres, and its members, as respected members of the upper middle-class, were accepted into the conventional social life of the city. By the 1940s, it was clear that the society drew its membership from across the city rather than being confined to its initial suburban hinterland (Quidnunc, 1945; McCullagh, 1992), although right to the present its rehearsal rooms remain in what is now the inner suburb of Rathmines. The R&R reflected the aspirations of the suburb from which it originated – one that saw itself as being safe and "English" (politically Unionist), rather than Irish.

The example of the Rathmines & Rathgar Musical Society represents a strand of creative expression characteristic of British and Irish suburbs from the late nineteenth century into the 1930s or so (see Jackson, 1991; Harris and Larkham, 1999). At this period, the moneyed classes who sought entertainment and leisure activities in their relatively new suburban locations largely replicated urban social circles and bonds created by socio-economic status and cutting across religious difference. In many cases, these social constraints resulted in limited creative expression within a relatively confined, safe bourgeois context.

Challenging Conformity: Women's Creativity at Dún Emer Guild / Cuala Press

In contrast to the R&R, which reflects the conservative nature of the suburb from which it sprang, the second case turns to a more radical form of creativity, focusing on craftsmanship and applied arts, which challenged contemporary orthodoxy. At the turn of the twentieth century, Ireland was undergoing the so-called Celtic Revival, where a growing sense of national identity became manifest in literature and the applied arts. The Dún Emer Guild was founded in 1902 by English-born Evelyn Gleeson (1855–1944) and two sisters, Elizabeth and Susan Yeats, in the then sleepy distant suburb of Dundrum, County Dublin, about four miles / six kilometres from the city (see Figure 8.1). The achievements of Susan "Lily" Yeats (1866–1949) and Elizabeth "Lolly" Yeats (1868–1940) have largely been overshadowed by those of their brothers, Nobel laureate for literature William and celebrated artist Jack. Gleeson was active in the Irish artistic circle in London and in the women's suffrage movement. In 1900, on the advice of her friend Augustine Henry, Gleeson moved away from the London smog to Ireland to improve her health. Henry also offered her a £500 loan to establish her own craft centre. She discussed these plans with Elizabeth (Lolly) and Susan (Lily) Yeats, both talented craftswomen with a network of influential contacts. The Yeats family had lived in the famous London suburb of Bedford Park, where they were part of the circle of William Morris. Following the decision of their father to abandon a legal career for the more precarious life of an artist, the adult Yeats children found themselves supporting their parents financially. While all four children were talented creatives, all were also conscious of the vital importance of earning a living. Lily worked for Morris & Co. as a seamstress for six years from 1888, contributing among others to the elaborate bed hangings at Kelmscott Manor, which were exhibited at the 1893 Arts and Crafts exhibition. Lolly worked as an art teacher in London and wrote several successful manuals to teach art, while learning the craft of book printing at the Women's Printing Society. In 1902, the three women – Evelyn Gleeson, Elizabeth Yeats, and Susan Yeats – established a craft collective in Dundrum under the medieval guild model favoured by William Morris. Within the guild, each of the women would be involved with a particular craft. Susan (Lily) would manage the embroidery, Elizabeth (Lolly) would look after the printing department, while Evelyn managed the tapestries and rugs, and was responsible for overall finances. Initially, the selection of an outer suburban location was made by Gleeson for health reasons, though the decision-making is not detailed, and there may have been other considerations. In the summer of 1902, Gleeson found a "suitable house" in Dundrum, then, as now, considered an attractive and affluent suburb of Dublin, close to the mountains. Originally a village in its own right, Dundrum had become a popular residential location for the moneyed classes following the arrival of the railway in 1854. The house had previously been named "Runnymede," an English name that could be associated with a suburban dwelling anywhere in the British Isles (the original Runnymede is the water meadow location in Surrey where the Magna Carta was signed), but it was renamed by the guild members in the Gaelic language as "Dún Emer" (literally Fort of Emer). In Irish legend, Emer was the wife of heroic warrior Cú Chulainn and possessed great craft skills. Through the Dún Emer Guild, Dundrum would become a centre of innovation and design, as a hub of the Irish Arts and Crafts movement. These were not just genteel ladies keeping themselves occupied with a "suitable" hobby but were providing gainful employment for themselves, as well as employing and training local girls and women. By 1905, there were thirty women in employment. The Irish Arts and Crafts movement, which was pioneered by the Dún Emer Guild, provides a distinctive visual counterpart to the better known literary Celtic Revival (Gordon Bowe, 1990). The crafts of the

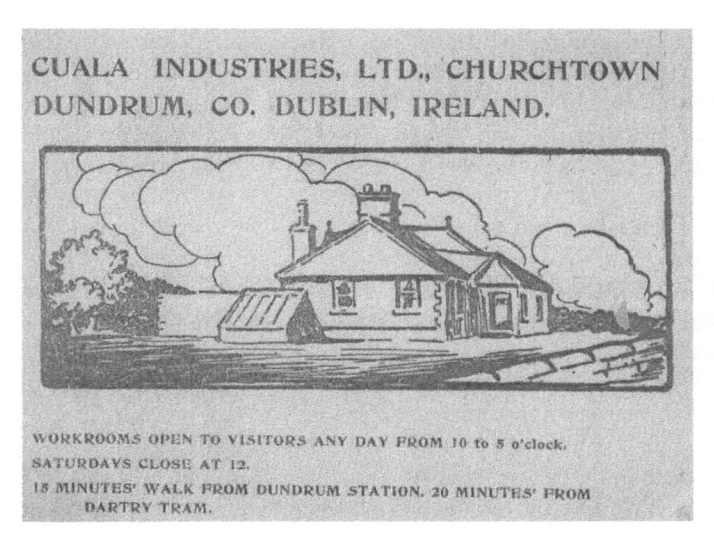

CUALA INDUSTRIES, LTD., CHURCHTOWN DUNDRUM, CO. DUBLIN, IRELAND.

WORKROOMS OPEN TO VISITORS ANY DAY FROM 10 to 5 o'clock. SATURDAYS CLOSE AT 12. 15 MINUTES' WALK FROM DUNDRUM STATION. 20 MINUTES' FROM DARTRY TRAM.

Figure 8.3. Promotional brochure for Cuala Industries, including an image by Elizabeth Yeats. Source: Private collection.

guild were inspired by the native Celtic vernacular; their designs "were a rethinking of tradition … [and] the translation of folk art became a basis for a national style" (Coleman, 2012). But they also had an international reach, reflecting links with William Morris and the broad international Arts and Crafts movement. The emphasis was on using high quality Irish materials to create beautiful, luxury, lasting originally designed objects. Their 1903 manifesto stated: "Everything as far as possible is Irish … The designs are also of the spirit and tradition of the country." In the same year, a circular from Dún Emer Industries usefully introduced this distinctive suburban enterprise, explaining that "a wish to find work for Irish hands in the making of beautiful things was the beginning of Dún Emer." The circular ends with a note that the premises could be found ten minutes' walk from Dundrum station, with visitors welcome on the first Thursday of every month or by appointment.[3] Similarly, a later promotion for the successor Cuala Enterprises notes transport links and suitable hours for visitors (see Figure 8.3). Their suburban location afforded the freedom to break away from the "norms" of the metropolis while also facilitating ease of access for visitors and potential customers.

The Dún Emer Guild displayed its products regularly at a number of exhibitions, including some overseas such as the 1904 St. Louis World's Fair. Their work was acclaimed in national and international press, and they received many prestigious commissions, including one for embroidered vestments and a series of twenty-four banners for Loughrea Cathedral, County Galway. These were designed by George "AE" Russell, Jack B. Yeats, his wife Mary Cottenham Yeats, and Pamela Coleman Smith, an American artist. The guild was thus part of an important artistic movement with international connections, while choosing to be based in this suburban location.

The printing side of the business, staffed entirely by women, published only new material, most of it Irish. The first book published by the Dún Emer Press in 1903 was *In the Seven Woods: Being Poems Chiefly of the Irish Heroic Age* by Lily and Lolly's elder brother, W.B. Yeats. It was printed on paper made in Saggart, County Dublin, and bound in Irish linen. The characteristic colophon (printed in red ink) ran thus:

> Here ends In the Seven Woods, written by William Butler Yeats, printed upon paper made in Ireland, and published by Elizabeth Corbett Yeats at the Dún Emer Press, in the house of Evelyn Gleeson at Dundrum in the county of Dublin, Ireland, finished the sixteenth day of July in the year of the big wind, nineteen hundred and three.[4]

James Joyce alludes to the Dún Emer Press in his magnum opus *Ulysses* in a characteristically sarcastic fashion:

> Five lines of text and ten pages of notes about the folk and the fish gods of Dundrum. Printed by the weird sisters in the year of the big wind. (Joyce, 2000 [1922]: 38)

Eventually, the personal relationship between the sisters and Evelyn Gleeson became strained over financial and administrative matters. In 1904, the operation was reorganized into two parts, the Dún Emer Guild run by Gleeson and Dún Emer Industries under the direction of the Yeats sisters, and in 1908, the groups separated completely. Gleeson retained the Dún Emer name, and the Yeats sisters established Cuala Industries, which ran the Cuala Press and an embroidery workshop, in the nearby suburb of Churchtown,. The Cuala Press, like its forerunner at Dún Emer, concentrated on publishing new works by contemporary writers, often by writers associated with the Irish Literary Revival, rather than established classics. In order to help make ends meet, Cuala Press also printed cards and a broadside series that included illustrations by Jack Yeats. During the forty years that she ran Cuala Press, Elizabeth guided the publication of seventy-seven books and hundreds of pieces of ephemera. The Cuala Press published books by W.B. Yeats (who acted as its editor), John Millington Synge, and (later) Patrick Kavanagh. Although it mainly published Irish writers, Cuala also published work by Ezra Pound, Rabindranath Tagore, and John Masefield. The Cuala Press was unusual in that it was the only Arts and Crafts press to be run and staffed by women.

Meanwhile, the main source of income for the Dún Emer workshop, which made rugs, tapestries, and embroideries, was patronage of the Catholic Church, as in the Honan Chapel tapestries (1917) and vestments for St Patrick's Church, San Francisco (1923), but Gleeson also completed a banner for the Irish Women Workers' Union in 1919. Thus, even though Dún Emer positioned itself outside the mainstream both in a physical and political sense, the suburban products of Dún Emer were adopted by the mainstream in the form of the Roman Catholic Church. Perhaps it is an indication of its success and acceptance by the metropolitan arbiters of taste that Evelyn Gleeson's Dún Emer workshop moved from Dundrum to a more central location in Dublin's Hardwicke Street in 1912, being, in that sense, subsumed by the city. The original house at Dundrum no longer exists, having been overwhelmed by a subsequent suburban wave, which saw it replaced by a housing estate that bears the name Dún Emer.

Neglected Suburbs: 1970s On – Kilbarrack/"Barrytown," and Finglas

The first two examples date from a period in the early twentieth century when suburban life in Dublin was still dominated by the better-off classes, and they exemplified some of the very different forms that suburban creativity could take within this context. In turning to the latter part of the twentieth century, the second two examples explore a time when mass suburbanization had become the norm. In moving from elite to middle- and especially working-class suburbs, both parallels and contrasts can be seen in the nature of, and responses to, suburban creativity.

By the 1970s, the policy of mass suburbanization of the working classes in Ireland had been underway for nearly fifty years, resulting in several "waves" of new development that affected two generations of Dubliners and changed the demographic balance between city and suburbs. While private speculators continued to build suburban homes for the middle classes, many of the residents of the new local-authority-built suburban estates were former slum dwellers who had been reluctant to move from their existing city-centre communities. Poet Paula Meehan (b. 1955), who spent her earliest years in an inner-city tenement on Gardiner Street but grew up in the working-class suburb of

Finglas, wrote about the sense of helplessness as the city's population was shifted to new suburban estates "like tribes of American Indians forcibly settled on reservations" (quoted in Mac Anna, 1991: 23).

Perhaps surprisingly, it was not until the 1970s that the experience of these new working-class suburbanites coalesced and gave rise to significant creative expression. In his introduction to *A Dublin Quartet* by Dermot Bolger, critic Fintan O'Toole observed that "the great tradition of Irish writing is silent on the subject of the suburbs, so you can slip out from under its shadow. No one has ever mythologized this housing estate, this footbridge over the motorway, that video rental shop. It is for the writer, virgin territory" (O'Toole, 1992: 1–2). Now, however, the work of a new generation of poets, playwrights, and authors – born in the late 1950s and raised in suburban Dublin – began to directly address the suburbs and did so in new and different ways. Internationally, the best-known of these is probably Booker prize–winning author Roddy Doyle, while other significant figures include Paul Mercier, Paula Meehan, and Dermot Bolger. Coming of age in the late 1970s and 1980s, these suburban writers "explored elements of Dublin life that many people – politicians, media, the complacent middle classes, older writers – had tried to ignore or sweep under the carpet. They became the guides to a new city, a 'new Dublin' that was developing as fast as its writers" (Mac Anna, 1991: 26). This marginalization sparked particular forms of literary creativity, in some ways echoing Phelps's recognition of the "irrepressible creative economy of the marginalized suburbs of social housing and favelas" (Phelps, 2012: 269).

Dermot Bolger and Raven Arts Press: Literary Creativity at the Margins

In the late 1970s, a young Finglas man, Dermot Bolger, was the leading force behind the establishment of a new publisher, the Raven Arts Press. In common with the Cuala and Dún Emer presses, Raven Arts Press was a ground-breaking publishing house established in suburban Dublin. From small beginnings in 1977, when the local Finglas Writer's Workshop printed a mimeographed magazine, *Suburban Poetry*, Raven Arts grew to become "one of the most important publishers of literature in Ireland during the last part of the twentieth century" (University of Delaware, n.d.).

In the same way that Cuala Press, in its modest suburban setting, provided a conduit to publication for an array of important new talent, Raven Arts Press, from its suburban base in Finglas, also made publication accessible to a host of new writers and poets, becoming Ireland's leading underground and alternative press. Encouraged by the Irish poet

Anthony Cronin, Raven Arts founder and recent school-leaver Dermot Bolger (b. 1959) began publishing his own and others' poetry in book and pamphlet form, as well as fiction, non-fiction, and anthologies designed to introduce new writers to the reading public. In a 1988 essay collection *Invisible Cities: The New Dubliners: A Journey through Unofficial Dublin*, many of the writers explored their roots (Bolger, 1988). "They painted a portrait of 'the New Dublin' and its often ignored or officially invisible suburbs such as Tallaght, Kilbarrack, Howth, Glasnevin, Crumlin, Inchicore and Finglas" (Mac Anna, 1991: 23). The suburban experience that these artists portrayed was seen as being very different and quite removed from the accepted Irish narrative, which continues to perceive itself as being rooted in the countryside, associating "true" Irish identity with an idealized pastoralism in the rural west of Ireland. The notion that urban areas, still less suburbs, were Irish was a strong challenge to the status quo. As O'Toole (1992: 1) wrote, "new places have been born, places without history, without the accumulated notion of Irishness that sustained the State for 70 years. Sex and drugs and rock 'n' roll are more important in the new places than the old Irish totems of Land, Nationality and Catholicism."

The characterization of the suburbs, particularly the non-elite suburbs, as being alien and unproductive – their othering – is something that was clearly identified by the young protagonists of Raven Arts Press. Poet Michael O'Loughlin, who was involved with Bolger in setting up the press, recently observed that "Raven Arts Press immediately received plenty of media attention: we were young, photogenic, loudly opinionated, and to the features writers, a poet from Finglas was as interesting as a dog who could talk. Not much changes" (O'Loughlin, 2017). This condescension from the mainstream media was also reflected in the attitude of the key government funding agency, the Arts Council of Ireland. Bolger has noted the attempt of this establishment organization to circumscribe the work of Raven Arts Press:

> The Arts Council was initially determined that we stick to community arts, which was a sort of way of putting us in a box, however while we published a new generation of Dublin writers we also published books by major European figures like Paul Celan and Pier Paolo Pasolini, because we refused to have any limit set on our imaginations, while we always remained informed by our formative experiences. (Grenham, 2018)

Dermot Bolger's own work, as novelist, poet, and playwright, frequently concerns the experiences of working-class characters who feel alienated from society, depicting "a literary landscape dominated by neon-lit amusement arcades, four-in-one cinema complexes, greasy chippers, garish video bars and dingy basement flats where lives of poverty, violence and squalor are played out to a soundtrack of rock music and tv babble" (Mac Anna, 1991: 22–3). He has observed that nothing in his school education suggested he could write literature about his own suburban experience: "[There were] no books to suggest that my own world was worth writing about, or the world of the people around me (I saw nothing of my life reflected in writing)" (Bolger, 2003: 5); "there was a very strong resistance to writing from the margins" (6). Despite this observation, the new suburban environment appears to have unleashed a surge of creative energies, perhaps encouraged by the juxtaposition of people of varying backgrounds. Bolger has described the cultural hybridization that was generated by the suburban process in Finglas:

> Almost all my neighbours were people like my parents who had come up to Dublin from the country and brought their country habits with them, so the back gardens of my street were rural in many ways with long rows of potato beds and hens being kept, whereas the more modern estates were populated by people who had moved out from the inner city. Therefore you saw a great glimpse into two different backgrounds co-existing side by side with the children of each creating their own new culture from it, be it in the poems of Paula Meehan, Michael O'Loughlin or Rachel Hegarty or the songs of Christy Dignam or dozens of bands creating their own voice in a new space. (Grenham, 2018)

New voices emerged as the early stage of community formation in the suburb lent itself to new forms of creative energy. Furthermore, opportunities for self-expression were enhanced by improved educational access from 1967, when all children in Ireland became entitled to free second-level education. Michael O'Loughlin recalled: "In retrospect, Raven Arts Press was giving a voice to the first generation of Irish citizens to have access to higher education as a consequence of the introduction of free secondary schools in the 1960s, a huge release of battened-down energies" (O'Loughlin, 2017). The availability of free secondary education would also have unintended consequences, as demonstrated in our final case study of Kilbarrack.

Greendale Community School and Roddy Doyle's "Barrytown"/ Kilbarrack: A Creative Hub Remaking the Suburban Image

While the Raven Arts Press acted as an important vehicle to transmit the works of a new generation of suburban poets and writers to a wider audience, in the nearby north Dublin suburb of Kilbarrack, a

Figure 8.4. Children at play in Kilbarrack ("Barrytown"), 1971. Source: Dublin City Council Photographic Collection, courtesy of Dublin City Library and Archive.

remarkable hub of creativity emerged in the local secondary school. Greendale Community School was established in 1974 to cater to hundreds of children in a rapidly growing new suburb, which was dominated by local-authority housing designed for the working classes (see Figure 8.4). Described as "one of the most striking examples in Irish culture of the effects of a creative environment" (O'Toole, 2007), Greendale featured among its teaching staff Roddy Doyle and two of his college friends, Paul Mercier and John Sutton, as well as Catherine Dunne, who would also later become a successful novelist. Doyle became a celebrated novelist, screenwriter, and dramatist, winning the Booker Prize in 1993. His work, while humorous, challenged perceptions of suburban life in Ireland and legitimated the experiences of the working-class suburbs. Since 2009, he has also spearheaded Fighting Words, a charity that promotes resilience and creativity through the creative practice of writing and storytelling (Fighting Words, 2018). Playwright, film and theatre director, and screen writer Paul Mercier was the founder (with cultural entrepreneur John Sutton) and artistic director of the Passion Machine Theatre Company in 1984. Passion Machine staged only original Irish work. They presented theatre depicting contemporary everyday life and succeeded in attracting many audience members who had never previously set foot in a theatre. Mercier's work is famous for its gritty poetic realism and examination of ordinary, contemporary Irish life.

While these individuals associated with Greendale Community School undoubtedly had innate talent and creativity, their work was shaped by their environment:

> Greendale certainly influenced the kind of writers they became. The language, the humour, and the deep sympathy that marked Doyle's *Barrytown* trilogy and, in particular, *Paddy Clarke Ha Ha Ha*, owed its depth in some measure to daily contact with the children in the school. Equally, Mercier's extraordinary confidence with the worlds and words of working-class suburban youth in his Passion Machine plays reflected a real involvement in their lives. (O'Toole, 2007)

In this way, the attributes of the locality acted as a catalyst for individual creativity, echoing Drake's (2003) findings for his craft and design interviewees, who found inspiration in their surroundings. In addition to taking creative inspiration from their suburban surroundings, the writers and creatives working in Kilbarrack and Finglas also consciously created new audiences. Their work was of, and for the suburbs. It was an important step in challenging traditional perceptions of Irish cultural identity. Roddy Doyle and Dermot Bolger were among the first modern Irish writers to write about their experiences of the suburbs where they had grown up and in which they worked. In his portrayal of Kilbarrack/Barrytown, Roddy Doyle "wrote about its working-class inhabitants not as exotic anthropological specimens but as people with rich lives" (O'Toole, 2016). This approach challenged the denigrating tendencies of the mainstream in the same way that Raven Arts contended with patronizing perceptions of what it meant to be a poet from Finglas. Popular perceptions of the twentieth-century northside suburbs were, until now, largely shaped by a media based – along with the state's largest university – in prestigious Dublin 4, on the south side of the city. Unfamiliarity with these areas tended to breed indifference and even contempt.

Despite their comic warmth, Doyle's trilogy of novels (subsequently made into movies) set in "Barrytown" (the fictionalized Kilbarrack) – *The Commitments* (1987), *The Snapper* (1990), and *The Van* (1991) – also explore some of the bleak realities of life in this poorer suburb, including unemployment, poverty, alcoholism, and sexual violence. As Liam Harte suggests, Doyle's work shines a light into more marginalized, less palatable territories of Irish society: "Not only were such economically blighted suburbs airbrushed from tourist-board images of Dublin, they also remained beyond the pale of literary representation" (Harte, 2014: 27).

The gritty realism with which these suburban writers (Doyle, Bolger, Mercier, and others) expressed themselves was shocking to their middle-class audiences, while striking a chord with the new working-class audiences about whom they wrote. They challenged the status quo by documenting socially and physically marginalized suburban experiences. In many respects, their work derives its meaning from its suburban/outcast nature, often being consciously created in opposition to the metropolitan core. Greendale Community School in the suburb of Kilbarrack had a critical mass of individuals whose work was shaped by being in this location, while they, in turn, shaped popular perceptions of the area through their various literary, theatrical, and cinematic expressions. Their work – like much of what was published by Raven Arts Press – was a challenge to the reader and/or viewer to "move beyond the stereotypical images of Ireland that belong to a moribund tradition" (Cosgrove, 1996: 232–3) and ignore the Ireland of the Irish Tourist Board with its glamorized, idealized, romanticized Ireland reliant on pastoral imagery.

Moving further, this expression of creativity can be positioned in terms of power dynamics. Persson (2006: 62) has argued that Roddy Doyle's novels portray the "third space," strategies of the radically disempowered, which might be termed "everyday resistance." In this way, they "offer a resistance as well as an alternative to official Ireland." The characters insist that their experiences be taken seriously, refusing to accept the top-down treatment and attitudes instigated and perpetuated by the state (McGlynn, 2008). The writers of and from Dublin's working-class suburbs similarly challenged perceptions, forcing the metropolitan core to recognize and value these previously ignored, invisible spaces. By refusing to accept the dominant narrative of the centre, these writers, it could be argued, have been doubly resistant to dominant power relations. Ultimately, Roddy Doyle's achievement in receiving the 1993 Booker Prize for *Paddy Clark Ha Ha Ha*, his brilliant depiction of the effects of suburbanization, heralds the success of this generation of writers in making the suburbs visible.

Conclusion

All the case studies analysed in this chapter have demonstrated the diversity and variety of suburban creativity in an Irish context. Clearly, the suburban experience is far richer and more nuanced than has been generally acknowledged, reflecting a variety of elements of creativity that operate at different levels and have been received and understood in different ways by the urban elite. In the first example discussed here, the R&R Musical Society was readily accepted and absorbed into the mainstream of urban life, largely because its founding members were – broadly speaking – part of that elite. It offered a non-threatening outlet for creative expression to middle-class suburbanites, largely choosing to perform established popular works by Gilbert and Sullivan. By contrast, the women who established their Arts and Crafts–inspired guild in the neighbouring suburb of Dundrum can be seen as representing a more radical strain of creativity because of their Celtic Revival focus as well as their female workforce and guild organization. Dún Emer Guild and Cuala Press, then, challenged the establishment but were, ultimately, absorbed into the mainstream. In a sense, their experience reflects what Mark Gibson has described, whereby "the creative resources extracted from suburbia are not always re-invested there" and "the center often capitalizes the value which is mined from the periphery" (Gibson, 2011: 536).

Our final two cases, looking at literary expressions of suburban life in the more marginalized northside suburbs of Dublin, contribute a further layer to our understanding of how creativity can operate and flourish in a suburban setting. Whereas Rathmines/Rathgar and Dundrum/Churchtown were all well-off suburbs, the locations from which this last group of poets, playwrights, and authors derived their meaning were very different. These were largely the marginalized local-authority housing schemes dominated by working-class, often welfare-dependent families. When the suburban creatives living or working there began to write about such areas, it was a considerable challenge to urban elites. For the first time, these ignored, invisible, and forgotten places were being treated as having value, their residents being recognized as having stories worthy of expression. Critics who emphasized the "gritty realism" of these works reflected the discomfort of the elite when confronted with the harsh realities of life in parts of the city they had previously chosen to ignore. These creative expressions grew out of their environment and, in so doing, began to challenge dominant power relations in new and often uncomfortable ways.

In attempting to understand the experience more broadly, can it be suggested that this particular literary flowering came into being only when the suburban environment had attained a certain critical mass and when the voices of the residents could therefore become loud enough to gain expression? Certainly, there is an element of serendipity, such as the fact that a critical mass of creative people worked together in Greendale Community School at the same time, whose experiences of interacting with their pupils and locale would light the touch-paper of imagination that burned so brightly for a period.

These cases raise a number of further questions. Is there a particular point in the suburban life-cycle when creativity is most likely to come to the fore? For each of these examples, the suburb in question was at the leading edge of the city or in its early stages of development. It was still a raw, emergent entity. In the case of the women of Dún Emer and Cuala, they were free from the constraints of urban life by choosing a semi-rural setting, whereas Rathmines and Rathgar had attained a critical mass of population that facilitated the formation of associations for leisure activities. The Finglas of Raven Arts Press was a raw new suburb accreting around a tiny historic village core, with a rapid influx of population from both rural and urban backgrounds creating a new hybrid identity. Kilbarrack was still under construction, while Greendale Community School was a new facility with young staff, a vibrant place teeming with young children and families, despite the degree of deprivation experienced there. Is it a coincidence that these examples arise at these relatively early stages in the life-cycle of a suburb? Perhaps it is not so remarkable, in that, as the suburb matures, it gradually comes within the orbit of the city and is absorbed by it, so that the creative endeavour undertaken there is automatically appropriated by the metropolitan core. As suburbs evolve and grow, their composition changes, and the nature of the creative process may also change as they become more established. One might then ask, are newer suburbs creatively "edgier"?

It seems, from these cases, that the response from the core to suburban creativity also depends on the existing relationship the metropolis has with the suburb and, indeed, the class and power relations or networks that exist between core and periphery. For the R&R Musical Society, the response was a ready acceptance, as its members were already part of the same world, whereas the radical nature of the endeavours of the Yeats sisters and Evelyn Gleeson led to greater suspicion and unease – as reflected in Joyce's characterization of "the weird sisters." Nevertheless, the women of Cuala and Dún Emer did gain a certain level of acceptance for their work, probably because they were of a class and background that made it easier for them to interact with the elites. The publishers at Raven Arts Press had to push back against the perception from the Arts Council that they were "merely" a community-based group in order to earn the right to be treated as more than the human equivalent of performing dogs, a novelty to be regarded with condescension. Those working creatively in Finglas, Kilbarrack, and other "invisible suburbs" found it far more difficult to receive due recognition from the metropolitan elite, who found their work disturbing. Their portrayals of a very different suburban world with its harsh deprivation and ready profanities were challenging to the arbiters of taste, who found it difficult to accept the raw unvarnished reality of life in these forgotten locations. It was made doubly difficult when the work of Roddy Doyle became internationally popular through the screen adaptations of several novels, depicting the impoverished, run-down city in a way that ran counter to the millions being spent on tourism agency advertisements.

These contrasting examples reveal a variety of cultural and creative production taking place beyond the city but intimately bound up in and engaging with the urban core in different ways (both physical – the location of performance space or clients; and less tangible – the challenge to the Dublin imaginary and the interaction with urban critics). In various ways, the city affected all of these suburban activities, while, in turn, the suburban creatives gradually reshaped the aesthetics of the core. Rather than conceptualizing the dominant metropolitan core sifting through the products of the periphery, absorbing and appropriating those of which it approves, while rejecting the rest, can this notion not be turned on its head? In the case of the Raven Arts Press and the Greendale writers, the suburban creatives ultimately reshaped the sensibilities of the core. Despite strong resistance to the portrayals of these marginalized suburbs, ultimately the core was forced to accept and adopt this work as forming part of the image of the city. It can thus be argued that the relationship between core and periphery is more nuanced than a purely dominant-subservient binary.

NOTES

1 I would like to acknowledge the assistance of Jonathan Cherry, Finnian Ó Cionnaith, Billy Shortall, and the staff of the National Library of Ireland and Dublin City Library and Archive in the preparation of this chapter.
2 These abbreviations refer to the two universities, University College Dublin (UCD) and Trinity College Dublin (TCD).
3 Both the manifesto and the circular are collected in the papers of Evelyn Gleeson and the Dún Emer Guild, Trinity College Dublin, Archive Reference: IE TCD MS/10676.
4 Elizabeth Yeats used an Albion printer, built in 1853, and the typeface used was an eighteenth-century Caslon 14 point. A striking feature of the seventy-seven books produced by Dún Emer and Cuala was the way in which Elizabeth identified the date of printing with an event or an anniversary such as "the big wind" or "Eve of Lady Day in Harvest in the year 1906."

REFERENCES

Bain, A. 2016. "Suburban Creativity and Innovation." In R. Shearmur, C. Carrincazeaux, and D. Doloreux, eds., *Handbook on the Geography of Innovation*. Cheltenham, UK: Edward Elgar Publishing, 266–76.

Bolger, D., ed. 1988. *Invisible Cities: The New Dubliners: A Journey through Unofficial Dublin*. Dublin: Raven Arts Press.

Bolger, D. 2003. "Interview with Dermot Bolger by Maria Kurdi." *Études Irlandaises*, 28(1): 7–22. https://www.persee.fr/doc/irlan_0183-973x_2003 _num_28_1_1645

Brady, J. 2014. *Dublin 1930–1950: The Emergence of the Modern City*. Dublin: Four Courts Press.

Brady, J. 2016. *Dublin 1950–1970: Houses, Flats and High-Rise*. Dublin: Four Courts Press.

Coleman, Z. 2012. "Susan and Elizabeth, the Yeats Sisters, from the Dun Emer Guild to Cuala Industries." Women's Museum of Ireland. https://www .womensmuseumofireland.ie/exhibits/yeats-sisters

Collis, C., S. Freebody, and T. Flew. 2013. "Seeing the Outer Suburbs: Addressing the Urban Bias in Creative Place Thinking." *Regional Studies*, 47(2):148–60. https://doi.org/10.1080/00343404.2011.630315

Comunian, R., C. Chapain, and N. Clifton. 2010. "Location, Location, Location: Exploring the Complex Relationship between Creative Industries and Place." *Creative Industries Journal*, 3(1): 5–10. https://doi.org/10.1386/cij.3.1.5_2

Cosgrove, B. 1996. "Roddy Doyle's Backward Look: Tradition and Modernity in 'Paddy Clarke Ha Ha Ha.'" *Studies: An Irish Quarterly Review*, 85(339): 231–42. https://www.jstor.org/stable/30091211

Dawson, W. 1971 [1912]. "My Dublin Year." *Studies: An Irish Quarterly Review*, 60(239/240): 243–58.

Drake, G. 2003. "'This Place Gives Me Space': Place and Creativity in the Creative Industries." *Geoforum*, 34(4), 511–24. https://doi.org/10.1016 /S0016-7185(03)00029-0

Dungan, M. 2013. *If You Want to Know Who We Are: The Rathmines & Rathgar Musical Society 1913–2013*. Dublin: Gill & Macmillan.

Felton, E., and C. Collis. 2012. "Creativity and the Australian Suburbs: The Appeal of Suburban Localities for the Creative Industries Workforce." *Journal of Australian Studies*, 36(2): 177–90. https://doi.org/10.1080/14443058 .2012.676560

Fighting Words. 2018. "About Us." fightingwords.ie (website). https://www .fightingwords.ie/about

Gibson, C., and C. Brennan-Horley. 2006. "Goodbye Pram City: Beyond Inner/Outer Zone Binaries in Creative City Research." *Urban Policy and Research*, 24(4): 455–71. https://doi.org/10.1080/08111140601035275

Gibson, M. 2011. "The Schillers of the Suburbs – Creativity and Mediated Sociality." *International Journal of Cultural Policy*, 17(5): 523–37. https://doi.org/10.1080/10286632.2010.531717

Gordon Bowe, N. 1990. "The Irish Arts and Crafts Movement (1886–1925)." In *Irish Arts Review Yearbook (1990/1991)*. Dublin: Irish Arts Review, 172–85. https://www.jstor.org/stable/20492642

Gregory, J.J., and C.M. Rogerson. 2018. "Suburban Creativity: The Geography of Creative Industries in Johannesburg." *Bulletin of Geography, Socio-economic Series*, 39(39), 31–52. https://doi.org/10.2478/bog-2018-0003

Grenham, S. 2018. "Writer's Block with Dermot Bolger." *The Gloss Magazine*, 7 December 2018.

Harris, R., and P.J. Larkham. 1999. "Suburban Foundation, Form and Function." In R. Harris and P.J. Larkham, eds., *Changing Suburbs, Foundation, Form and Function*. London: E & FN Spon, 1–31.

Harte, L. 2014. *Reading the Contemporary Irish Novel, 1987–2007*. Chichester. UK: Wiley-Blackwell.

Jackson, A.A. 1991. *Semi-Detached London: Suburban Development, Life and Transport, 1900–1939*. London: Wild Swan.

Joyce, J. 2000 [1922]. *Ulysses*. London: Penguin Books.

Mac Anna, F. 1991. "The Dublin Renaissance: An Essay on Modern Dublin and Dublin Writers." *The Irish Review*, 10(Spring):14–30. https://doi.org/10.2307/29735579

McCullagh, R. 1992. "Seventy-Five Years of the Rathmines and Rathgar Musical Society." *Dublin Historical Record*, 45(2): 109–25.

McDonald, F. 2018. "Suburban Mindset – An Irishman's Diary on Galway's Search for a European Vision." *Irish Times*, 25 July 2018. https://www.irishtimes.com/opinion/suburban-mindset-an-irishman-s-diary-on-galway-s-search-for-a-european-vision-1.3576361

McGlynn, M.M. 2008. "Barrytown Irish: Location, Language and Class in Roddy Doyle's Early Novels." In M. McGlynn, *Narratives of Class in New Irish and Scottish Literature: From Joyce to Kelman, Doyle, Galloway, and McNamee*. New York: Palgrave MacMillan, 77–130.

McManus, R. 2002. *Dublin 1910–1940: Shaping the City and Suburbs*. Dublin: Four Courts Press.

McManus, R. 2018. "Brave New Worlds? 150 Years of Irish Suburban Evolution." In E. Smith and S. Workman, eds., *Imagining Irish Suburbia in Literature and Culture*. London: Palgrave Macmillan, 9–38.

McManus, R. 2019. "Suburbanization in Europe: A Focus on Dublin." In B. Hanlon and T.J. Vicino, eds., *The Routledge Companion to the Suburbs*. London: Routledge, 87–99.

McManus, R., and P.J. Ethington. 2007. "Suburbs in Transition: New Approaches to Suburban History." *Urban History*, 34(2), 317–37. https://doi.org/10.1017/S096392680700466X

Murphy, C. 2010. "The Model of a Very English Irishman." *Irish Independent*, 23 October 2010. https://www.independent.ie/incoming/the-model-of-a -very-english-irishman-26692323.html

O'Loughlin, M. 2017. "This Finglas Arts Press Became the Spiritual Home of a New Generation of Writers." RTÉ, 2 November 2017. https://www.rte.ie /culture/2017/1029/916093-this-finglas-nest-helped-spawn-a-generation -of-new-writers/

O'Toole, F. 1992. "Introduction." In D. Bolger, *A Dublin Quartet*. London: Penguin.

O'Toole, F. 2007. "Greendale RIP: The End of an Academy of Irish Writing." *Irish Times*, 12 May 2007. https://www.irishtimes.com/news/greendale -rip-the-end-of-an-academy-of-irish-writing-1.1205654

O'Toole, F. 2016. "Modern Ireland in 100 Artworks: 1993 – *Paddy Clarke Ha Ha Ha*, by Roddy Doyle." *Irish Times*, 14 May 2016. https://www.irishtimes .com/culture/modern-ireland-in-100-artworks-1993-paddy-clarke-ha-ha -ha-by-roddy-doyle-1.2645577

Persson, A. 2006. "Between Displacement and Renewal: The Third Space in Roddy Doyle's Novels." *Nordic Irish Studies*, 5, 59–71. https://www.jstor .org/stable/30001543

Phelps, N.A. 2012. "The Sub-Creative Economy of the Suburbs in Question." *International Journal of Cultural Studies*, 15(3): 259–71. https://doi.org/10.1177 /1367877911433748

Quidnunc. 1945. "An Irishman's Diary." *Irish Times*, 2 November 1945, 3.

University of Delaware. n.d. "Raven Arts Press Archive: Bibliographic and Historical Notes." University of Delaware Special Collections. https:// library.udel.edu/special/findaids/view?docId=ead/mss0467 .xml;tab=notes

Williams, C. 2017. "Comic Opera: English Society in Gilbert and Sullivan." In R. Gordon and O. Jubin, eds., *The Oxford Handbook of the British Musical*. Oxford: Oxford University Press, 91–116.

9 Recreating Locality: Community and Identity in Budapest Suburbs, 1995–2020

JÁNOS B. KOCSIS

Introduction

The outer, less dense areas of Budapest, Hungary, are often depicted as vast, dull expanses of endless rows of family houses of poor and mixed aesthetics, with a general lack of services and amenities, from which people commute by car to central areas to work or study, do their shopping, socialize, and be entertained. The fringes are unilaterally connected to the core in ways that environmentally and socially menace the liveability and sustainability of the city; the tendency to immense growth of these areas in the last thirty years is thus to be forestalled, and access to such outer areas should be constrained in order to lure residents back to the city proper.

The statement above, however, seems to be inadequate in terms of construing why people of various levels of social standing and values are nevertheless attracted and then "glued" to the urban periphery; additionally, the patterns of socio-economic connections and dynamics are more complex and reciprocal than this picture would suggest. It lacks an analysis of the effects that migrants have on local communities and vice versa. Either traditionally or recently, several such locations appear to have been emitting strong vigorous signals pointing to a vivid, appealing community and vibrant, developing neighbourhoods.

Transformation of the urban fringes has been well documented in recent international publications about post-socialist cities (Sailer-Fliege, 1999; Tsenkova and Nedović-Budić, 2006; Stanilov and Sýkora, 2014a), and particularly for Budapest from a macro perspective (Tosics, 2013; Kovács and Tosics, 2014; Laki, 2018; Kovács et al., 2019). This chapter, however, focuses on the macro level and aims at describing and analysing how an influx of new social strata and general changes in lifestyles and attitudes both begets and contributes to the emergence of creativity and the re-creation of locality and local community. In turn, the traits of locality, such as the composition of local community, the traditions, historical factors, and built and natural environment, serve as a source of inspiration for newcomers and earlier inhabitants alike, triggering new activities and the revival of old ones based on the interplay of the locality, traditions, and novel techniques and approaches. Different people play different roles in suburbs that are creative areas; thus, the chapter pays attention to detecting and delineating the characteristics of the most prominent actors, placing special emphasis on the role of women in creating locality and creative milieu.

The officially defined agglomeration belt comprises thirty-six towns and forty-three villages, with a total population of over 872,000 people scattered across over 7,000 square kilometres around the Hungarian capital of over 1.7 million people (Hungarian Central Statistical Office, 2019). Despite proximity to the central city and the generally high rate of daily commuting among the residents, even abutting settlements in the belt differ significantly from one another in terms of social composition, historical givens, attitudes, economic structure, and recent development paths. This variance hamstrings a viable, valid, and reliable perusal of the entire area in the form of generalizable statements. A handful of abutting towns have instead been selected for analysis where socio-economic change related to the emergence of creativity is most striking and prominent (Figure 9.1), allowing for deeper scrutiny, understanding, and comparison of them from a micro perspective – which is admittedly of limited validity on the macro scale. Due to its prominence, the area has been widely and deeply analysed in the Hungarian scientific literature from various perspectives by different authors (Csanádi and Csizmady, 2002; Csanádi et al., 2010; Dövényi and Kovács, 1999; Izsák, 2001; Koós, 2004), with a few studies concentrating on transforming local societies (Csurgó, 2013; Kocsis, 2015; Kondor, 2016).

Besides using the results of other researchers, this chapter is mainly based on primary resources collected between 1997 and 2020 in the area. Over 190 in-depth interviews with residents and local decision-makers were carried out in four major waves between 1997 and 2020. Two focus group meetings were held in 2017 in Törökbálint with residents and stakeholders, and two large-scale representative surveys were implemented – one in 2012 (representative sample of 1,204 in

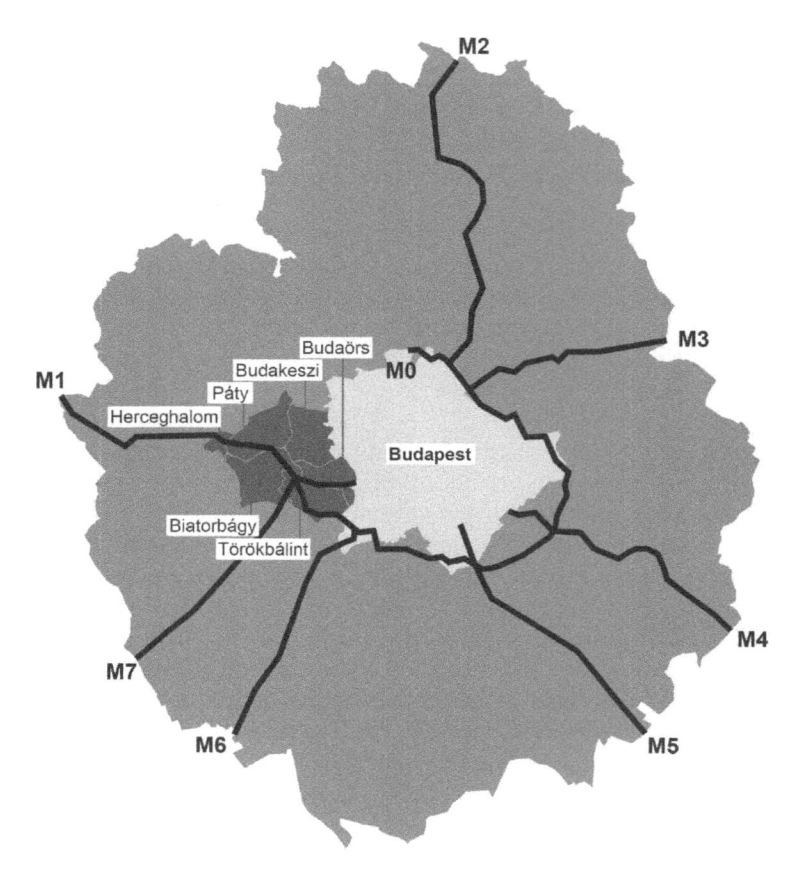

Figure 9.1. Position of the analysed region in the Budapest inner metropolitan area. Source: Map by Virág Varga.

Biatorbágy, Budaörs, Budakeszi, Herceghalom, Páty, and Törökbálint) and the other in 2018 (representative sample of 1,009 in Törökbálint; Kocsis, 2013; Kocsis and Tomay, 2018).

Analysis of the phenomena related to suburban creativity necessitates first a description of the main particularities of suburbanization and urban sprawl around Budapest in comparison to general urbanization tendencies. In a next section, different groups of suburban societies, as well as major actors and agents of suburban creativity, are then scrutinized, and the effect of their activities on the local community, economy, and townscape is analysed within the context of the socio-economic transformation of the entire Budapest agglomeration.

Suburbs and Creativity

The influx of better-off, influential, often creative people to specific neighbourhoods and towns, together with the spread of urban lifestyles and attitudes, is contributing to changing the characteristics of local life, the economy, power structures, and politics – a process that is often collated under the umbrella term "gentrification of the suburbs" – the effects of which may be clearly traced in these areas. Such developments supplement, or even reverse, the traditional functions and links between "creative," "productive," dense core areas and predominantly residential urban and semi-rural peripheries, previously perceived as exclusively dependent, or even parasitic, on the core.

In this chapter, the term "suburbanization" is used in a narrow sense as a social phenomenon whereby well-defined social strata (that is, better-off, affluent ones) move from central, dense urban areas to the periphery seeking an appropriate residence due to various social and historical factors (Fishman, 1989), as opposed to the more general formulation of the term (Stanilov and Sýkora, 2014c: 275). Suburbanization thus forms a distinct part of the more general process of urban sprawl: the spatial expansion of urbanized areas towards the peripheries, which has been experienced on a large scale in Europe over the last 150 years (Couch et al., 2007). A further process of sprawl and transformation in the functions and structure of external areas – namely, polycentric urban development and the formation of metropolitan areas – has recently fundamentally altered the socio-economic layout and dynamics of peri-urban areas (Clark, 2003), often inverting the traditional cast of productive, creative core versus dull, residential, dependent fringes to active, creative, colourful, and diverse suburbs versus homogeneous, monofunctional inner areas, frequently heavily dependent on the suburbs (Marshall, 2001).

Besides the various factors involved in gentrification that are outside the scope of this analysis (Hamnet, 2000), the revival of urban nuclei has of late been further reinforced by changes in the preferences of Millennials for urban values (Speck, 2013; Fishman, 2005), which has affected most prominently the central core areas of larger cities, but also numerous former subcentres, and has led in many cases to the emergence of completely new ones (Hanlon et al., 2010).

The success of urban economies lies in the spatial denseness of various, interlinked economic activities, a diverse workforce, powerful communication links, and a developed infrastructure with an intense flow of knowledge among various economic actors and individuals based on strong social networks (Jacobs, 1969; Florida, 2002). Creativity and economic productivity are usually associated with dense urban centres (Glaeser, 2011; Sassen, 2009; Porter, 1995; Hall, 1998), although the situation is far from unequivocal (Keil, 2018). In accepting the definition of "gentrification" as the spatial concentration of a higher status social group occupied in white-collar or creative jobs in a specific geographical location (Tomay, 2019), we can talk about rural gentrification in select communities in general (Phillips, 1993) or about the gentrification of the suburbs (Markley, 2018; Hanlon and Airgood-Obricki, 2018), where the physical proximity of creative people (Florida, 2002) may lead to a high level of creativity in less dense locations, especially those close to larger economic centres (Florida, 2018).

The abundance and proximity of people categorized as belonging to the creative class, however, does not necessarily lead to the actual emergence of a creative milieu and creative activity. Besides the aforementioned diversity associated with workforce, ease of access, and developed infrastructure, strong social links and communication across different social strata are prerequisites, which require a suitable physical arrangement of public spaces and places (Kocsis and Dúll, 2016). The financial effects of the proliferation and diversification of economic activities that manifest in the development or renovation of suburban town centres, with an emphasis on services, entertainment, and places for socialization and walking (Duany et al., 2010), may lead to the emergence of the concentration of agents of community-building, such as carefully designed pavements, streets, squares, parks, playgrounds, and cafés in a diverse, more vibrant environment than is traditionally associated with dense urban centres (Jacobs, 1961; Mumford, 1961).

The Post-Socialist Periphery

Suburbanization in Hungary, especially in the case of Budapest, has been thoroughly scrutinized in Hungarian literature since the mid-1990s from various perspectives. The main focus has been on the causes, large-scale dynamics, and socio-demographic patterns thereof (Csanádi and Csizmady, 2002; Ladányi and Szelényi, 1999; Kovács and Tosics, 2014; Szirmai et al., 2011; Kovács et al., 2019), but has also taken into account its effects on core areas and regional governance (Tosics,

2013; Egedy et al., 2017; Ricz et al., 2009; Stanilov and Sýkora, 2014b) and recent socio-economic transformation creating a wider, polycentric metropolitan region of some 2.9 to 3.1 million people within a diameter of 160 to 180 kilometres (Szabó et al., 2014; Salamin et al., 2008; Salamin, 2004; Salamin et al., 2014; Salamin et al., 2009; Hungarian Central Statistical Office, 2018).

Despite many similarities with the rest of the continent, urban sprawl has had some particularities in Eastern Central Europe. Urbanization in general started later, and for a long period occurred more slowly than in Western Europe (Enyedi, 1996). The expansion of the outskirts started due to the influx of low-status migrants from the rural countryside, and until the late 1980s, this pattern remained by far the most dominant process, whereas the outflow of the middle strata – suburbanization – played an inferior role (Stanilov and Sýkora, 2014c; Kovács and Tosics, 2014; Kocsis, 2009). Residential areas in the urban periphery remained rural in character, with significant agricultural production in the surrounding gardens (Enyedi, 1996) due to the lack of investment in the "non-productive sectors" (Kocsis, 2012), resulting in "underurbanization" (Szelényi, 1996; Kocsis, 2009). The strong "going out" culture of urban life in the interwar period was replaced in general by a predominant "staying home" lifestyle (Konrád and Szelényi, 1969), and urban societies became "proletarianized" (Enyedi, 1996). Patterns of social segregation differed as higher status residential areas appeared on a large scale, whereas lower status social groups clustered only on a micro scale, scattered across the whole urban area (Csanádi and Ladányi, 1992).

Socialist urban policies and practices had long-standing consequences for the agglomeration belt. The conditions of earlier nuclei outside the very centre were exacerbated, and they stopped functioning as subcentres. Together with the massive influx of people from the countryside, the lack of development led to the proliferation of low-status, homogeneous, characterless residential peripheral areas, with only basic amenities and services, both within and outside the official boundaries of the city.

Notwithstanding, suburbanization occurred in different forms and to varying extents as middle and higher strata individuals moved outwards throughout the entire epoch, first to distinguished housing estates and then to condominiums in prestigious areas within the city limits (Kocsis, 2009). More peripheral areas beyond the city limits, on the other hand, bore the impact of "provisional" suburbanization, with the mass appearance in some more tempting locations of clusters of

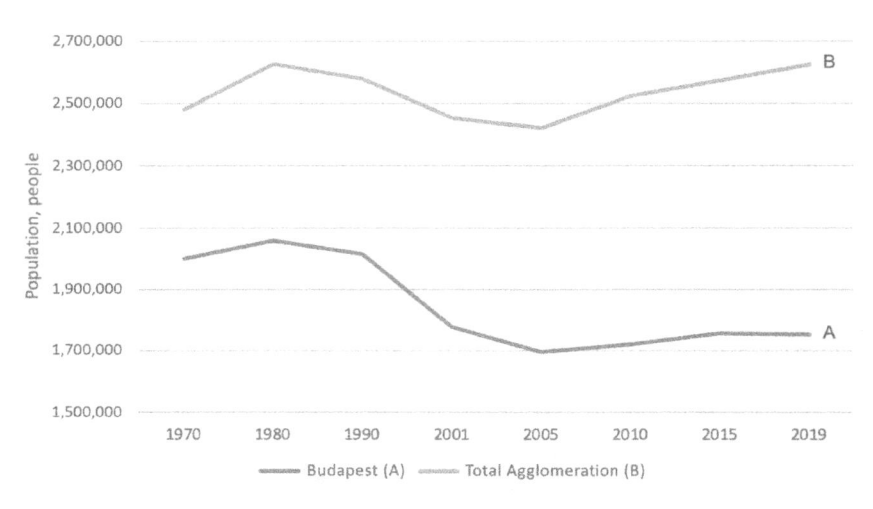

Figure 9.2. Population change in the Budapest agglomeration between 1970 and 2019. (Note that data points for 1970 through 1990 are plotted more closely together than are data points after 2000.) Data source: Hungarian Central Statistical Office.

garden plots and weekend cottages (Stanilov and Sýkora, 2014c; Kovács and Tosics, 2014).

The years surrounding the collapse of the socialist system (1989–90) created features strikingly different from the previous period. Better-off people began rapidly flowing from the inner areas to the predominantly rural, obsolete periphery in large numbers, the dynamics of which deservedly imply the phenomenon "suburban revolution" (Keil, 2013). The population of Budapest suddenly declined, whereas the outer areas started rapidly growing (Figure 9.2).

Numerous settlements actively encouraged suburbanization in the hope of rapid development and catching up. Availability of their own resources and funds, and their greater liberty later allowed the local municipalities, especially the richer ones, to enact local policies and developments that fostered the expression of local creativity and social inclusion. The former socially rather homogeneous agglomeration belt bifurcated into predominantly better-off and poorer segments, reflecting the presumed prestige of areas, proximity to already high-status neighbourhoods, and access to green hills and the like (see Figure 9.3). Some areas, especially in the western and northern sectors, thus experienced the massive arrival of wealthy families,

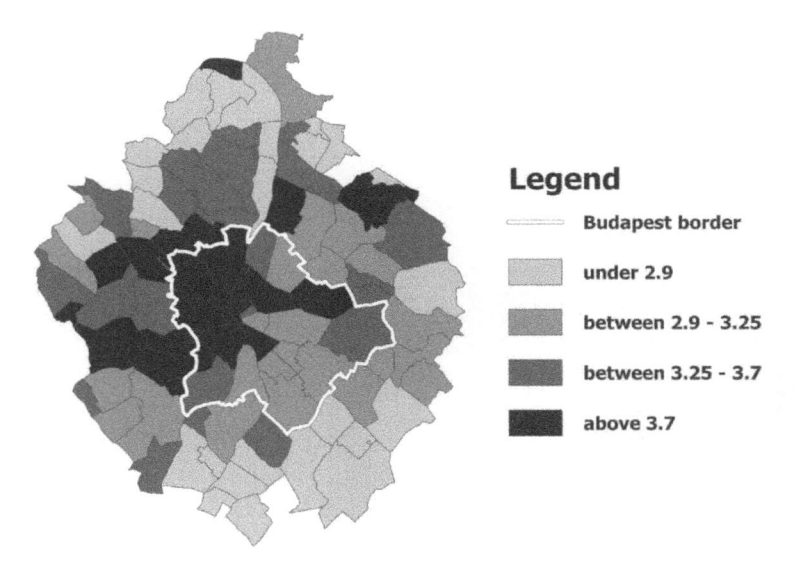

Legend

- - - - - Budapest border

under 2.9

between 2.9 - 3.25

between 3.25 - 3.7

above 3.7

Figure 9.3. Tax base per capita in million Hungarian Forints in the Budapest agglomeration, 2018. Source: Map by Ádám Bauer based on data from the Hungarian Central Statistical Office.

usually with children, from the inner areas, while others, especially in the southeast, faced the arrival of the less well-to-do, either from the city or from the countryside.

The outer ring of the metropolitan region comprises mid-sized towns and cities, often with traditionally developed services and job opportunities (Varga et al., 2020), but the area adjacent to Budapest has lacked such subcentres as a direct consequence of socialist territorial policies. Development, first of commercial, then logistical office and other business activities, started in the late 1990s in favourable locations and culminated in the emergence of a centre adjacent to the triangle formed by motorways southwest of Budapest proper, with the flagship towns of Biatorbágy, Budaörs, and Törökbálint (see Figures 9.1 and 9.4).

The Substratum: Native Residents in the Southwest

Despite the rapid and enormous changes brought about by suburbanization, local communities still display characteristics deeply rooted in their rural past. Historical events and processes, as well as geographical

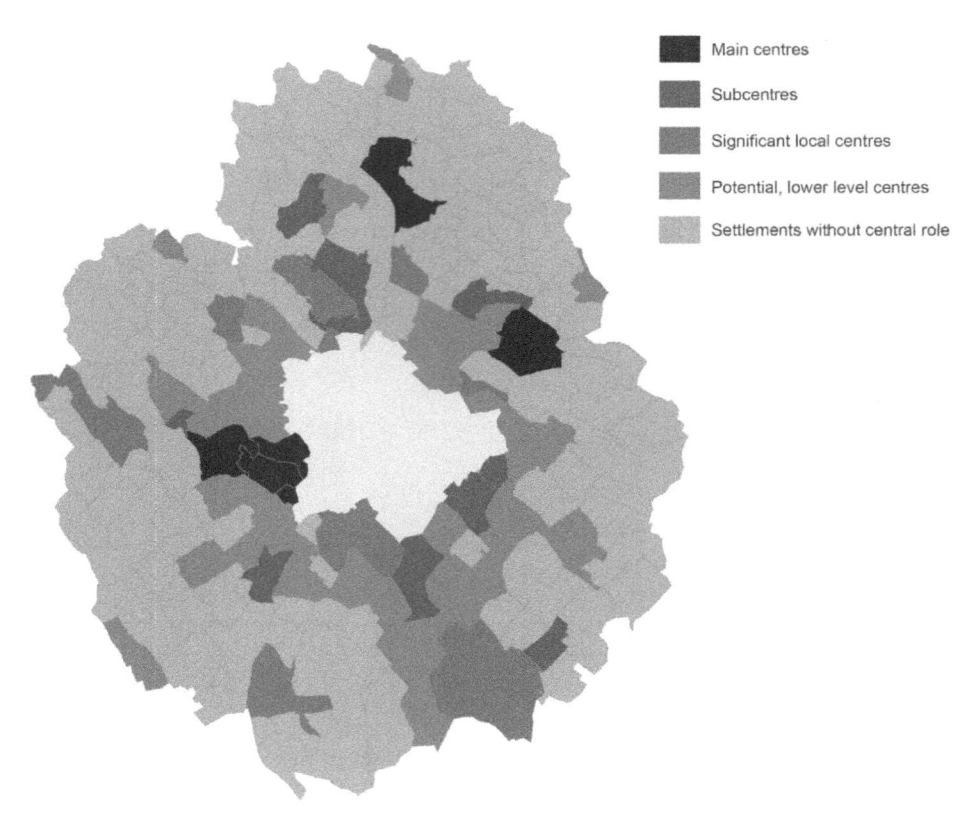

Main centres

Subcentres

Significant local centres

Potential, lower level centres

Settlements without central role

Figure 9.4. Centres and subcentres in the inner metropolitan belt around Budapest, 2020. Source: Based on calculations and drawn by Virág Varga.

features, have left enduring marks on local communities and their activities in the present. Most towns in the southwestern fringe are perceived as "Swabian"[1] by locals, newcomers, and outsiders, and German traditions are actively and proudly kept alive, even though the overwhelming majority of inhabitants are not of German background; this heritage is an attractive factor for newcomers.

The original predominantly German, but also Hungarian, Serbian, and Slovakian inhabitants settled down in separate, internally homogeneous, small rural communities in the valleys, small basins, and hillslopes of this varied terrain, where first stock-raising, then gardening and viticulture were the dominant economic activities. The

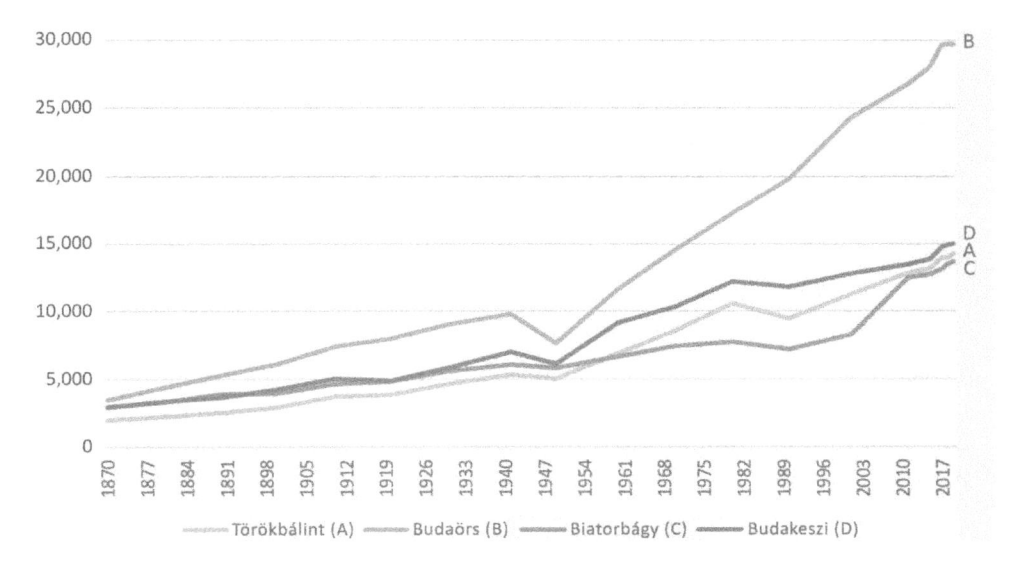

Figure 9.5. Population of major towns in the southwestern agglomeration belt of Budapest, 1870–2019. Source: Népesség.com (2020).

city started to exert a discernible influence on the area during the second half of the nineteenth century, and the population began to grow rapidly due to natural demographic growth and migration (Figure 9.5). Adjacent to the popular leisure areas of the city, restaurants and taverns for weekend visitors were set up in Budakeszi,[2] whereas Budaörs,[3] Törökbálint,[4] Bia,[5] and Torbágy[6] attracted health services and weekend gardeners. Despite these changes in population size, economic activities, and services, the towns remained predominantly rural in character.

The end of the Second World War brought about drastic changes. In 1946, most of the ethnic German minority population was deported. In exchange, ethnic Hungarians evicted from former Hungarian territories were settled there in large numbers. The remaining Swabians, in fear of future eviction, attempted to hide their nationality, and the next generation barely spoke German or displayed any visible signs of their ethnicity. The incoming Hungarians had the same negative psychological attitude as that of ethnic refugees – a desire not to stand out but rather to blend in – and thus local identity quickly started to become diluted. From the 1950s, industrialization attracted a large number of workers who migrated in from the countryside. In addition, vast amounts of the outer, usually hilly areas of the towns were turned into garden plots, exponentially expanding the tradition from before the Second World War, resulting in thousands of parcels of land with no infrastructure, narrow unpaved roads, and poor accessibility emerging as a form of "temporary suburbanization" during socialism (Kocsis, 2009).

The development of industrial and other economic activities, services, amenities, and cultural functions was concentrated in Budapest proper, where the supply of food and other goods was of a much higher standard than even some kilometres farther out, while cultural and educational institutions were situated in the inner city. As a consequence, the Budapest agglomeration turned into a highly monocentric area in which creative activities were centralized in the very core areas, and the fringes remained dull, grey, residential, and underserviced. Specialized services, restaurants, cafés, parks, and other such "luxuries" were seen as superfluous in a rural or, at best, semi-rural environment. Socialist economic and territorial policies exacerbated the loss of local identity and community spirit. Apart from the official socialist policy of purposefully disrupting social networks, opportunities for socializing under these "post-rural–pre-urban" conditions were scant, further contributing to the weakening of social connections.

To some extent, Budaörs and Budakeszi represented unusual exceptions with their significant housing estates built in the 1970s for the then preferred social categories. Mid-level cadres working in the administrations of various bureaucracies and state-owned companies were concentrated in Budaörs, while Budakeszi housed many from the army. These projects reflect the positive perception of the western areas of the capital and the bizarre particularity of socialist housing policies that favoured the middle strata during the centralized allocation of housing (Kocsis, 2012). Thus, these areas became long-standing social islands in otherwise demographically different, generally lower status, and more poorly serviced neighbourhoods. These housing estates prognosticated in some ways the potential of the entire area.

The preceding description is not to say that more educated, better-off people – potentially creative ones – were not present in this territory but rather that their creativity had to be expressed in the core areas of Budapest where the requisite infrastructure was available. As the former area gradually became more desirable due to its proximity to the city and its prestigious areas, increasing numbers of intellectuals spontaneously filtered there, often as pioneering suburbanites looking for relatively cheap but cozy homes with gardens. They joined the local intelligentsia, who were either the remnants of the traditional local elite, the clergy, teachers at local schools, or other upwardly mobile local residents. The existence and sheer number of the former suddenly became apparent in the years after the transition of 1989, when opportunity and place became available.

The Superstratum: The Suburbanites

The transition to a market economy suddenly opened up the gates to suburbanization and other momentous migration processes that resulted from the new political and economic structures, as well as to the deep recession that followed 1989 – a complex phenomenon that introduced new tendencies or, more probably, revealed and boosted pre-existing ones (Csanádi et al., 2010; Kovács and Tosics, 2014). Municipalities with their newly awarded autonomy started constructing local infrastructure in a hurry, and the more advantageous ones soon caught up to the average standard in Budapest. Such developments needed huge additional investment from the local municipalities. Two major paths seemed viable: encouraging suburbanization by selling land for housing construction and strengthening the local economy, especially through larger investments. Both approaches proved adequate for the analysed towns due to their geographic and historical situation.

In the early 1990s, local elites initially construed suburbanization in three main ways, clearly reflecting the state of local communities. Those with weaker communities, either due to inherited weakness or to massive migration that had already transformed local neighbourhoods (such as Budaörs), actively triggered or did nothing to stop the massive influx of better-off people from the city. Settlements with stronger traditional local social networks had two major strategies to somehow lessen the extent and effect of the transformation. Some attempted to desist by at first declining to allocate lands for development – Biatorbágy and Budakeszi – while towns in the next category sought to retain the intactness of their community by allocating land not directly adjacent to traditional neighbourhoods – Törökbálint. Not being able to resist the huge pressure, however, both these latter communities had to abandon their stance by the end of the 1990s.

Besides the aspirations and attitudes of the local elites, the supply side of suburbanization was additionally stimulated by the abrupt large-scale availability of cheap construction sites resulting from the

Figure 9.6. Neighbourhood types in Törökbálint. Source: Map by Virág Varga.

reinstatement of agricultural lands. Furthermore, the completion in 1995 of the southwestern section of the M0 circular motorway that cut through the area (see Figure 9.6) coincided with all the aforesaid developments, rendering the area attractive for commercial, logistics-related, and other business investment.

The demand side of suburbanization swiftly multiplied as well. City dwellers were pushed out of central areas by the decay and blight caused by a long period of neglect and deteriorating social conditions. Growing social polarization provided the necessary funds for the better-off to leave for more attractive areas and to segregate.

With improvements in the road network and rising car ownership, more distant areas were discovered by increasing numbers of families in the early 1990s, and young couples, usually with children, sought out appropriate living environments at an acceptable price in green, quiet environments close to familiar neighbourhoods. This phenomenon was quite similar to the urban flight that occurred in the West in the 1960s (Fishman, 1989) but has some distinct features.

Suburbanites gradually moved farther out, while staying within access of well-known areas and their services. Outward migration patterns followed radial axes, initially often involving one leap at a time, such as the movement from inner areas to inner quasi-suburbs during socialism, and thence further out following 1990. Only much later – when having a residence in some of the suburbs became an expression of social status – did people from more distant areas in Budapest start migrating to prestigious neighbourhoods in towns like Biatorbágy.

Our research in Törökbálint showed that the majority of migrants in the last thirty or so years came from nearby suburbs within the Budapest city limits – that is, from lower density, higher status areas or from prestigious housing estates that were developed as quasi-suburbs during the later decades of the socialist era (Kocsis, 2008). In general, approximately half of all newcomers had a university degree (Kocsis, 2013), while in some neighbourhoods the ratio exceeds one-third of all the population aged eighteen years or older (Kocsis and Tomay, 2018).

The just-described radial nature of migration quickly led to the present situation, wherein the social composition of suburban areas outside the city limits mirrors that of the adjacent area within (see Figure 9.4). The expectations, lifestyles, and material status of the new, wealthier residents differ significantly from those of the locals, although neither may be categorized as a homogeneous group. Tensions between the original inhabitants and newcomers started to rearrange local power structures in numerous settlements following the late 1990s, and in many cases new elites have now partially or totally replaced older ones (Kondor, 2016; Kocsis, 2015; Csurgó, 2013). Additionally, different categories of suburbanites chose different neighbourhoods in the existing – or emerging – local fabric in relation to their preferences and as an expression of their aspirations, thereby actively transforming their physical and social characteristics.

Neighbourhoods and Social Groups

Suburban towns in the region include a variety of neighbourhoods and social groups in close proximity. A study of two suburbanizing villages in the western fringes described two major social groups from the recent waves of migration who now reside in disjunct neighbourhoods: namely, "oversleepers" with urban values and no or low place attachment; and "village-saviors," who seek to discover and preserve the authentic, local rural culture and have a strong sense of, and attachment to the local community (Kiss, 2007). Another recent extensive study in five nearby suburban villages identified three major groups according to the perceived function of homes: physical status symbols in the form of luxury residences; family nests with a detached house; and authentic dwellings in the form of renovated peasant houses (Csurgó, 2013).

Our analysis distinguished four major neighbourhood types, roughly corresponding to four social groups and complementing the above-described picture: "prestige seekers," "economic refugees," "home settlers," and "utopia hunters" (Kocsis and Tomay, 2018). The neigh-bourhoods themselves are fragmented; their situation is rapidly chang-ing, and people's choice of residential location only partly reflects their aspirations, permitting only cautious generalization. That said, valid and reliable conclusions may be drawn and comprehensible segrega-tion patterns may be identified, as in the case of Törökbálint (Figure 9.6). Two social categories are of special interest with regard to suburban creativity: home settlers and utopia hunters.

Developments of new estates on formerly outlying agricultural lands gradually appeared in the second half of the 1990s. Various private for-profit investors bought up formerly collectivized agricultural lands from their reinstated owners in favourable locations and convinced local municipalities to reclassify them as residential areas. The estates built thereon were aesthetically, architecturally, and socially in sharp contrast to adjacent neighbourhoods (Figure 9.7). The architectural styles of the buildings are chaotic, reflecting the desires of the owner, and range from Transylvanian and Alpine to Provençal, from Bauhaus and Art Deco to modern Finnish and New Englandesque, in contrast to the uniformity of other areas. Social tensions were sometimes harshly expressed, with roadblocks erected at some entry points to keep away the traffic of newcomers from traditional areas (Figure 9.8). Social dif-ferences have recently decreased significantly as numerous wealthier inhabitants have moved from traditional areas to the new estates, and the influx of suburbanites into traditional areas has recently eliminated most of the major socio-demographic difference; nonetheless, the inhab-itants of the new estates are generally perceived as less integrated into the wider community and as forming their own separate realm with their own lifestyles and values.

Predominantly, but not exclusively, the new estates have been popu-lated by a distinct type of "archetypal" suburbanite who has found a location that reflects their existing, or desired, social status, in parallel to securing a quiet, green, and cozy environment with little regard for the exact location and its pre-existing community. These *prestige seekers*

Figure 9.7. Newly developed area in Biatorbágy. Source: Photo by János B. Kocsis.

Figure 9.8. Roadblock in Törökbálint, looking towards a neighbourhood, early 2000s. Source: Photo by János B. Kocsis.

Figure 9.9. Outer area in Törökbálint. Source: Photo by János B. Kocsis.

looked around for the appropriate social composition and reputation that these neighbourhoods started to offer around the turn of the millennium. Actual layouts reflect the financial status of the inhabitants, with detached residences for the wealthier families and denser forms, such as semi-detached and condominiums, for less affluent ones. Thus, complex arrays of mid-strata condominiums may be found at the ends of towns next to agricultural fields, while expensive, low-density neighbourhoods prevail closer to the city.

At the very other end of the spectrum, the vast outer areas are composed partly of remnants of small estates for agricultural workers, but nowadays expanses of haphazard garden plots with small sheds dominate (Figure 9.9). Owing to their remote location and lack of facilities, a relatively small number of people have chosen to settle in these areas, and those who do often feel segregated and left out of the local community. Some areas in closer proximity to traditional areas, however, seem to have become more integrated parts of the respective towns and have developed significantly into more or less proper residential neighbourhoods in recent decades.

Deep economic hardship around the time of the collapse of the socialist system, along with soaring living costs in the inner areas, impelled a large number of poor people to move to outer areas where their expenditure could be better controlled and cut back

(Csanádi and Csizmady, 2002). *Economic refugees* arrived mostly to less favourable, more remote settlements, but former garden plot areas in more advantageous areas were also affected.

Vast areas of neighbourhoods built during the socialist period are dominated by detached houses on small lots in dense rows, usually with both front and back gardens. Predominantly residential in character, some smaller scale facilities have recently sprung up in different locations, especially various grocery shops, cosmeticians, restaurants, car mechanics, and the like. Neighbourhoods were usually built up in waves, with people of a similar age cohort moving in; with the original settlers dwindling, these areas opened up for newcomers to buy up houses. The densification of the population without new land development can thus be traced in many areas, as homes formerly lived in by one or two elderly persons are now inhabited by families of four or five. The massive wave of newcomers in recent years has led in some more preferred areas to a social composition rather similar (in terms of the proportion of educated individuals, material status, and social capital) to that of the residents of new estates, but with striking differences in terms of their embeddedness in the local community.

Home settlers may be portrayed as another group of "archetypical suburbanites" who wanted to secure an appropriate environment for their families, usually including small children. Members of this group aimed to find a strong social network with like-minded local residents. They sought out detached family houses with a garden in a decent, green, quiet, and secure environment. In the early years of suburbanization, such pioneers were quite adventurous and cared little about amenities, infrastructure, or the accessibility of schools or other services. Later on, targeted, large-scale developments became one of the predominant forms of housing, whereas other individuals opted to refurbish existing houses in neighbourhoods in former socialist areas.

Traditional rural cores used to be dominated by dense rows of traditional peasant homes with large gardens behind them. With their original inhabitants expelled, and properties nationalized in the late 1940s, they quickly declined in status and often became obsolete and blighted (Figure 9.10). Most of these neighbourhoods even nowadays have the lowest proportion of graduates and the highest number of original inhabitants. More aesthetically appealing areas, however, are now preferred by utopia hunters and certain other segments of newcomers, who rapidly transform and upscale them.

The group of *utopia hunters* had an idealized and romanticized concept of a rural type of living and community, and aimed to move to traditional quarters, preferably to old homes that they could renovate

Figure 9.10. Low-status traditional core area in Törökbálint, early 2000s.
Source: Photo János B. Kocsis.

in a make-believe style, if they could afford to do so. Those with limited financial resources often chose houses built in the socialist epoch or even earlier. The former mostly belong to the most educated echelon of society, who have highly urbanized patterns of living and working. They intentionally form connections to neighbours, and those in central areas have made strong efforts to mingle with locals and form strong networks with similar people. The reputation and perception of pre-existing communities played an important role in the selection of locality, but good accessibility to the city and services was an important prerequisite, as well as the presence of sophisticated telecommunication networks.

The spatial separation of these groups is more a signal of concentration via processes of cultural succession than exclusivity. Whereas the above spatial units form clearly distinguishable neighbourhoods, the categories of social groups are more like useful representative labels within a multidimensional range, while individual attitudes are changeable according to personal histories. The distinction between prestige seekers and home settlers is especially permeable during the individual life course, such as when either parent (or both) in a family become embedded in local social networks through connections built up while taking children to local kindergartens and schools – or when one of the parents gives up tedious and laborious commuting to the centre and starts looking for work, entertainment, services, and social connections locally.

Members of all the groups but one can usually be categorized within Richard Florida's creative class of better-off, well-educated people. When it comes to the suburbs, the main distinction, however, is that prestige seekers with jobs and social networks tied to the city express their creativity in the inner areas, whereas home settlers and utopia hunters pursue locally linked goals, either in the form of an appropriate home-based social network and environment or an idyllic, rural, authentic, unspoiled habitat with a slow life and real social connections.

The effects of the migration of utopia hunters on local communities have been extensively discussed as rural gentrification internationally (Phillips, 1993) and in the Hungarian context, especially in the context of fashionable rural areas, focusing on socio-economic (Csurgó, 2013; Tomay, 2019; Nemes et al., 2019; Tomay and Nemes, 2020) and architectural (Tamáska, 2011) aspects, where this feature is more prominent and palpable. In the suburbs, home settlers and prestige seekers dominate, but the effects of utopia hunters searching for an idyllic rural lifestyle are also detectable.

Expressions of Creativity and Related Actors

Creativity in the suburbs appears in three, interconnected dimensions in which different groups of migrants, as well as traditional residents, play their part. Local economies are transformed and the local communities reinterpreted and re-created – both factors affecting the physical form of the towns – whereas the urban fabric and the townscape affect both the local economies and communities.

The role of higher status urban migrants in rural gentrification has been widely analysed in the international (Phillips, 1993) and Hungarian context (Csurgó, 2013; Tomay, 2019; Kovách and Kristóf, 2005), along with the role of changing attitudes towards the urban way of life (Hack-Handa and Kocsis, 2018). Similar patterns may be detected in the transforming agglomeration towns, where newcomers act as creative or intermediate actors, as they have more entrepreneurial drive, self-confidence, courage, prowess, a broader perspective and deeper knowledge of markets, along with more social as well as financial capital.

Creative actors may mostly be found among the utopia hunters, but home-settler families also play important roles. Looking for authenticity and wanting to rely on local goods and services and specialties, they discover old, Swabian, and other local traditions, methods, and products and re-create them in ways that match present requirements and market demands. Diverse agricultural products thus appear in local markets, especially various sorts of cheeses, jams, preserves, meat, and alcoholic products, while handmade artefacts are also present. Most of the new producers previously had no connection to horticulture and food production, apart from being enthusiastic consumers.

Intermediate actors thus play a key role in creating and shaping the local production of various goods and channelling them to wider markets. Small-scale farmers' markets, as well as broadly publicized local fairs and festivals, are obvious as marketplaces, but the networks of specialty shops in large urban centres as well as internet-based sales are also essential. Additionally, some newcomers have an explicitly negative attitude to supermarkets and other forms of large-scale retail and prefer local, independent small stores for their everyday shopping and seek out local products.

Viticulture and winemaking were the most important economic activities in the region in early modernity, but an epidemic of phylloxera, the collectivization of land during socialism, and urban sprawl have almost eradicated the vineyards and the skills required to make good wines. The prominence of wine production, centred on Etyek, a large village next to Biatorbágy, has been reinstated mostly by urban migrants, often winemakers, but usually in the role of intermediate actors who discover and reveal local producers, transform their products into marketable goods, and facilitate their path to market.

The action typically taken by home settlers shapes the local economy too. Instead of authenticity, they tend to look for urban functions in a suburban environment, as offered by the suburbs in Western Europe and Northern America (Fishman, 2005; Keil, 2018), although the newly suburbanized towns around Budapest may not satisfy these requirements yet. They create demand for local shops as well, but to a lesser extent, since home settlers are not reluctant to shop in supermarkets and discount chains. Demand for other local urban-like services, such as restaurants, pubs, entertainment, cultural and sports facilities, and opportunities for self-expression, however, is thus rising, and newcomers are active in supplying them.

Suburbanites are creative at finding new functions for pre-existing structures and using them in more innovative ways, which also triggers locals to act likewise. New small and medium-sized enterprises are thus set up that attract local and commuting employees.

Most of the suburbanites work long hours, commute, and raise children and thus have little time for local affairs. Plenty of young professional spouses, especially women, however, are trying to find more flexible, often part-time jobs in the vicinity to allow them to spend more time with children and family. Opportunities for doing so have arisen due to the recent polycentric development of the economy around the metropolitan core, which has brought into being numerous rather well-paid jobs, as well as sprouted local services and functions. In turn, such women, having more time for various activities and social networks, have become more embedded in local communities. Abounding with ideas and initiatives, they are critical elements of local creativity.

A detailed analysis of the role of women has revealed deeper patterns (Csurgó, 2013), similar to those explored elsewhere (Halliday and Little, 2001). Especially among utopia-hunter families, the turn towards traditional values in general is also manifest in their engagement in more traditional family roles and the division of labour, whereby men act "outwardly" and women "inwardly"; that is, men work at jobs, usually in the city, while women manage the inner life of the family and maintain social connections. In opposition to the traditional model, however, this division does not necessarily entail an unbalanced hierarchy between men and women: women participate equally in decision-making in the family and do not perceive giving up their jobs as detrimental but rather think it represents an opportunity to participate more freely in their chosen community and to express themselves. The women thus embody the aspirations of the entire family in relation to having a quiet, idyllic life. This symbolic force gives female roles exceptionally high prestige within families (Csurgó, 2013). Similar tendencies may be observed among other, less tradition-oriented families, as many women choose, or create, more local, even part-time jobs. Family life becomes of focal importance, but in highly automated households – and in those able to afford domestic helpers – members have time to turn their attention to the social side of family, as well as to the local community.

Agents of Locality and Creativity

Face-to-face interactions and intended and chance encounters among people of various kinds play a vital role in sparking creativity. Fostering the high density of such occurrences is one of the most important functions cities have in social and economic life. To be creative, suburban communities therefore need analogous agents that enable and ease the development and maintenance of social networks.

Educational institutions act as a means of social integration, as well as facilitators of local creativity and identity. Newcomer families often expected to take their children back to the city, but kindergartens and primary schools were soon full to overflowing with the offspring of freshly arrived families. The hassle of taking young children to the city was soon abandoned in favour of the convenience of local schooling. Municipalities, in addition, typically take pride in the quality of local schools and have put huge efforts into modernizing and improving them.

Besides their integrative function, educational institutions, especially secondary schools, serve as large assemblies for local intelligentsia. School teachers typically have deep knowledge about local affairs and extensive personal ties. As is usual in smaller communities, municipalities employ many people interested in local affairs, and the bulk of members of civil society groups are also closely and personally connected. Successful secondary schools offer additional, more independent positions to local intellectuals, thereby strengthening the identity of the community, as well as the demand for more sophisticated services and cultural activities.

Cafés, restaurants, social and cultural clubs and events, gyms and other sports facilities act as more "urban-like" services. Hairdressers and beauticians in particular perform a dual function, providing places to meet and socialize, especially for those who spend more time at home. Most of these services are relatively new and are managed and run by more entrepreneurial, creative residents who are rooted more locally, in relation to which women have an outstanding role. Weekly farmers' markets play a minor, but important and increasing role in local life. Playgrounds and streets have integratory functions as well, the former serving as locations for the socialization of parents of younger children and as places for building up and strengthening social links among older children from various educational institutions. In spite of the infamy awarded to suburban streets in literature (Jacobs, 1961), some streets have a strongly integrative function. Most of the pupils of local schools walk to school and thus meet on the way there and back. Cul-de-sacs and other quiet streets offer the place and opportunity for children and parents to meet, talk, and play. The layout of suburban areas is dense, and this proximity allows for the formation of social contacts, especially in neighbourhoods with parents of a similar age, attitude, and roughly similar social status.

(Re)Discovery of Locality and Identity

Most of the creative activities are centred on local community and on an identity that had been deliberately weakened during socialism for ideological and practical reasons. The situation here was further exacerbated by the expulsion of local exemplar elites and intellectuals during the post–Second World War turmoil. The strata that was able to bridge the gap between the past and the present vanished as large proportions of the original inhabitants were expelled – either for being "German" or deemed too bourgeois or aristocratic, or both. Those who remained tried to hide their identity in fear of further repercussions. The continuity of local identity was therefore severely damaged, to the degree that many argue that distinct local identity no longer exists.

Social processes and transformation since the fall of the socialist system, however, have contributed to and precipitated the emergence of a new identity. Although still visibly segregated, the fragmented, variegated towns are slowly being reintegrated in a variety of ways. Barriers seem to be diminishing, both psychologically and physically, between neighbourhoods. Traditional locals move to new, high-status estates, and newcomers intermingle with the original inhabitants at community events. Residents use services in neighbourhoods sharply contrasting with their own that offer the opportunity to socialize.

Whereas a Swabian ancestry used to be something to hide, it has now turned into a source of pride and self-esteem. Quite a few residents speak German, and the presence of a strong Swabian community is still tangible, as their strong, well-forged network determines the local elites. Furthermore, numerous incomers consider the Swabian characteristics and past of areas to be favourable, which plays a decisive factor in the selection of a town to move to.

The new identity is centred on a garden city ideal – that is, living in a calm, lower density neighbourhood in proximity to a green environment – where families may sustain a healthy, decent way of life, and from where the highly developed services and high-salary workplaces of the city are still easily accessible. The towns are not associated with the cacophony of urban neighbourhoods, and streets are well kept and tidy. There is no real nostalgia for a "backward" rural past and way of living, even among those who seem to favour a utopianized rural way of life. As an interviewee observed, some towns now have an infectious positive and strong identity.

The strength of the community is reflected in the large number and variety of civil society organizations, from churches, ethnic choirs, and sports clubs to town beautification associations and the like. Local social life is intertwined with strong and expansive social bonds that abruptly come to light when interests are severely threatened. A stakeholder likened the situation to a swamp that expresses faceless, obstinate, wearisome resistance to initiatives that are not unequivocally welcomed, but which rarely becomes openly hostile.

The arrival of new social strata is becoming apparent in local politics, as newcomers are bringing in new ideas, goals, preferences, techniques, and alliances. In suddenly suburbanizing communities, decision-makers have often been flabbergasted when some of their initiatives were abruptly fiercely opposed, and even successfully frustrated, by previously unnoticed yet powerful groups of newcomers. Some towns have experienced a silent, slow but absolute takeover, while others have seen long struggles, whereas the progressive amalgamation of traditional and new elites may also be traced.

Energizing Local Economy

The towns are the foundation of the strongest economic centre in the agglomeration besides the city proper, and the creativity of their residents contributes to a large extent to the dynamics of the sub-centre in general and to the transformation of local economies in particular.

Proximity to the city acts as one of the most determining factors in terms of the actual socio-economic characteristics and dynamics of the region. It allows people to easily commute to work, but hinders the expansion of services, especially more sophisticated ones. The high level of demand and the concurrent economies of scale within the central areas of the city mean that specialized services can flourish in a way that few initiatives in outer areas can, despite the high average purchasing power of residents. Due to the minimal presence of even basic services during socialist times, people have traditionally been used to going out and to shopping for non-standard goods and engaging in cultural activity in the city. As a result, the neighbourhoods are almost exclusively residential and lack urbanism.

The effects of polycentric urban development have been accentuated in this area since the second half of the 1990s. International retail chains were the first to act, due to the high and increasing purchasing power of residents in the wider catchment area. From the 2000s onwards, other economic functions appeared in the region – first logistics-related, then office and other business and industrial ones. The area now has more demand for employment than the three towns can offer, although mostly in moderately well-paid jobs. Operators of more sophisticated activities, however, have thus far been reluctant to move to the area, although some change in their attitude is detectable, and some pioneering enterprises have relocated. In consequence, high-status people continue commuting into the city, whereas labourers have started to commute out from the city to fulfil demand.

Most of the economic developments have taken place on the empty fringes of the towns. The pace of transformation is felt less in the traditional town centres, where many shops and services are still trying to satisfy the demand of a bygone past and target a lower status clientele. Shopkeepers who set up their ventures twenty or thirty years ago under totally different circumstances have less expertise in the markets, and are rather path-dependent and unwilling to change. The slow replacement of shops and services is occurring to a varying extent, but the total gentrification of the retail outlets and services of the town centres is yet to come. New restaurants, cafés, shops, and services, on the other hand, have appeared scattered throughout residential areas, reflecting local demand, or near the new retail centres, building on patrons arriving by car. Due to the lack of appropriate premises, garages and other facilities are often turned into commercial properties, such as beauty parlours, even in the most prestigious neighbourhoods.

To reflect their reinforced identity, all towns have come up with and partly initiated complex policies and expensive programs to revitalize, beautify, and transform their centres to reflect their prosperity, identity, and community.

Of Small-Town Centres, Town Halls, Landmarks, and Pavements

Until now, town centres have not really acted as the foci of local social and economic activities. Traditional core areas, having become neglected and even blighted, were not at the forefront of the local agenda. After several ambitious plans and failed attempts, the steady and significant economic growth of the country and of the area, in particular in the 2010s, created the financial background for development, and major progress has been achieved, while the plans have become more realistic and reasonable. The overall vision and detailed outlines regarding design, function, and scale, nonetheless, are still hazy, changeable, and severely disputed within the communities.

Town centres in the core neighbourhoods exist under varied conditions. Some towns have a rather well-developed pattern of urbanized squares with pedestrian areas, shops, schools, and churches, while others are at a more preliminary stage, hosting basically anchor developments such as community centres and town halls, as in Törökbálint (Figure 9.11). Cafés, restaurants, other forms of entertainment, and specialized services are less common, which is where the calling power of the city is most perceptible. Demand for more sophisticated amenities presently seems insufficient to sustain a grid of services and facilities,

Figure 9.11. Community Centre with the new Town Hall in the background in Törökbálint. Source: Photo by János B. Kocsis.

although exceptional examples of successful solitary initiatives may readily be found.

Lack of sufficient demand is usually identified as the major cause of the underperformance of town centres, while densification and erection of condominiums arises as a viable remedy. There is an abundance of prospective young residents, with a more general susceptibility towards urban values that has been detected in research (Hack-Handa and Kocsis, 2018), who would be frequent patrons of such facilities. Others fear that the former approach would result in more uncontrolled population growth, and the small-town charm of the areas, the major magnet, would be lost.

One's perception of the substance of a town determines whether one regards the development of the town centre and main square as a superfluous, exorbitant, and pointless gamble or a salient investment in the future. Residents who regard the town as a mere residential annex to the city do not see either the economic or the societal rationale for such development, whereas others who consider it a unique community with a particular identity usually argue that a town centre could be a

Figure 9.12. Community Centre, partly a former train station, in central Biatorbágy. Source: Photo by János B. Kocsis.

hallmark, an expression of locality, and the scene of local social life. It could represent an opportunity and a place for residents to meet, socialize, go out, and take care of their affairs, and do some shopping while walking, seeing, and being seen.

Town halls and community centres, with the intention of creating new symbols of locality and prosperity, and as major elements of new centres, were among the first to be developed, besides schools and kindergartens. The development of community centres, usually involving new construction, like in Törökbálint (see Figure 9.11), or creative renovation of a symbolic building, like in Biatorbágy (see Figure 9.12), was usually completed well ahead of the development of town halls. They usually contain a theatre, other smaller rooms for a variety of activities, and a café or buffet. Generous funding enables the organization of elaborate programs and activities attractive to the entire community, often drawing people from neighbouring towns. Smaller scale events, like sports training, dance clubs, and the like appeal to almost all circles, but theatrical performances are frequented mostly by individuals from the traditional parts of the towns. Despite all their efforts and creativity,

Figure 9.13. Viaduct now defunct and serving as a popular landmark and leisure destination in Biatorbágy. Source: Photo by János B. Kocsis.

they are neither able nor willing to compete with the major institutions in Budapest. For high culture, the well-off go to the city.

Landmarks and monuments, and the lack thereof, as symbols of local identity and place attachment, appear to be relegated to the background. The viaduct of Biatorbágy (Figure 9.13), for instance, a strong and characteristic visual sign, forms an essential part of the local identity. Communities with a stronger local identity tend to have more powerful and widely recognized landmarks and other physical symbols of locality, and the intent behind the development of town centres, major institutions, and schools is partially due to recognition of the need for strong symbols.

The aesthetics of roundabouts appear to be another important element of concern for locals, and were often mentioned in the interviews. The central area of a roundabout offers an excellent site for anything symbolic, from a simple flowerbed to more elaborate statues and the like. They may be combined with other landmarks for greater effect, as in the case of the viaduct in Biatorbágy (see Figure 9.13).

This newly developed local identity poses the question of the connection to local traditions and the tension between new developments and built heritage. Historic homes, mainly peasant houses and some mansions, are often in neglected, dire condition, especially in central areas of towns, as many are not attractive to utopia hunters either. Some have been renovated, but most are facing demolition, which would lead to a complete change in the appearance of the central neighbourhoods. The protection of traditional buildings and relics of bygone implements, such as the *"hintuska kő"* – a stone on which women could adjust the heavy baskets on their backs, some of which are still to be found in central areas – is of concern for many in relation to the rapid transformation of townscapes. As the new identity of these locations gains ground, the issue of imaginatively preserving the remains of the past, mental or physical, while creating new forms and symbols, comes to the forefront.

Conclusion: New Localities of Creativity

The creative milieu in suburban towns arises from the fruitful encounter of migrating creative people who are inspired by the challenging possibilities and stimulating traditions of their chosen new residence, their desire to live in a community with strong, distinct identity, and the parallel aspiration of original inhabitants for more prosperity, modernization, and reconnecting with their pre-socialist past and heritage. Polycentric socio-economic development in the metropolitan area is supporting and nurturing such efforts with financial assets, securing local employment and demand for local products and services.

Quest for authenticity is a determining factor that is underpinned and reinforced here by the availability of deeply rooted traditions, previously neglected or deemed obsolete, which serve as inspiration for new services, artisanal activities, craftsmanship, and specialty products. Creativity is brought about, triggered, and strengthened by a productive interaction between the arrival of creative classes, proximity of business, diversity, and underlying, veiled local heritage, traditions, assets, and local ambitions.

NOTES

1 Germans are almost exclusively (self)defined in present-day Hungary, regardless of their actual ancestry.
2 *Wudigeß* in German.
3 *Wudersch* in German.
4 *Großturwall* in German.
5 *Wiehall* in German.
6 *Kleinturwall* in German.

REFERENCES

Clark, W.A. 2003. "Monocentric to Polycentric: New Urban Forms and Old Paradigms." In G. Bridge and S. Watson, eds., *A Companion to the City*. Malden: Blackwell, 141–54.

Couch, C., L. Leontidou, and K.-O. Arnstberg. 2007. "Introduction: Definitions, Theories and Methods of Comparative Analysis." In C. Couch, L. Leontidou, and G. Petschel-Held, eds., *Urban Sprawl in Europe: Landscapes, Land-Use Change and Policy*. Oxford: Blackwell, 3–38.

Csanádi, G., and A. Csizmady. 2002. "Szuburbanizáció és társadalom." *Tér és Társadalom*, 16(3): 27–55. https://doi.org/10.17649/TET.16.3.1978

Csanádi, G., A. Csizmady, J.B. Kocsis, L. Kőszeghy, and K. Tomay. 2010. *Város Tervező Társadalom*. Budapest: Sík.

Csanádi, G., and J. Ladányi. 1992. *Budapest térbeni-társadalmi szerkezetének változásai*. Budapest: Akadémiai Kiadó.

Csurgó, B. 2013. *Vidéken lakni és vidéken élni*. Budapest: Argumentum.

Dövényi, Z., and Z. Kovács. 1999. "A szuburbanizáció térbeni-társadalmi jellemzői Budapest környékén." *Földrajzi Értesítő*, 48(1–2): 33–57.

Duany, A., E. Planter-Zyberk, and J. Speck. 2010. *Suburban Nation: The Rise of Sprawl and the Decline of the American Dream*. 10th ed. New York: North Point Press.

Egedy, T., Z. Kovács, and C.A. Kondor. 2017. "Metropolitan Region Building and Territorial Development in Budapest." *International Planning Studies*, 22(1):14–29. https://doi.org/10.1080/13563475.2016.1219652

Enyedi, G. 1996. "Urbanization under Socialism." In G. Andrusz, M. Harloe, and I. Szelényi, eds., *Cities after Socialism: Urban and Regional Change and Conflict in Post-Socialist Societies*. Oxford: Blackwell, 100–18.

Fishman, R. 1989. *Bourgeois Utopias: The Rise and Fall of Suburbia*. New York: Basic Books.

Fishman, R. 2005. "Longer View: The Fifth Migration." *Journal of the American Planning Association*, 71(4): 357–66. https://doi.org/10.1080/01944360508976706

Florida, R. 2002. *The Rise of the Creative Class: And How It's Transforming Work, Leisure, Community and Everyday Life*. New York: Basic Books.

Florida, R. 2018. "Foreword." In M. Cornett and J. White, eds., *The Next American City: The Big Promise of Midsize Metros*. New York: G.P. Putnam's Sons, 8–15.

Glaeser, E. 2011. *Triumph of the City: How Our Greatest Invention Makes Us Richer, Smarter, Greener, Healthier, and Happier*. New York: Penguin.

Hack-Handa, J., and J.B. Kocsis. 2018. "A lakóhelypreferenciák változásai és az Y generáció." In L. Józsa, E. Korcsmáros, and E. Seres Huszárik, eds., *A hatékony marketing*. Komárom: Selye János Egyetem, 406–15.

Hall, P. 1998. *Cities in Civilization*. London: Weidenfeld & Nicolson.

Halliday, J., and J. Little. 2001. "Amongst Women: Exploring the Reality of Rural Childcare." *Sociologia Ruralis*, 41(4): 423–37. https://doi.org/10.1111/1467-9523.00192

Hamnet, C. 2000. "Gentrification, Postindustrialism, and Industrial and Occupational Restructuring in Global Cities." In G. Bridge and S. Watson, eds., *A Companion to the City*. Malden: Blackwell, 331–41.

Hanlon, B., and W. Airgood-Obricki. 2018. "Suburban Revalorization: Residential Infill and Rehabilitation in Baltimore County's Older Suburbs." *Environment and Planning A: Economy and Space*, 50(4): 895–921. https://doi.org/10.1177/0308518X18763607

Hanlon, B., J.R. Short, and T.J. Vicino. 2010. *Cities and Suburbs: New Metropolitan Realities in the US*. New York: Routledge.

Hungarian Central Statistical Office. 2018. *Munkaerőpiaci körzetek Magyarországon*. Budapest: Hungarian Central Statistical Office.

Hungarian Central Statistical Office. 2019. *Területi atlasz – Egyéb területi lehatárolások*. www.ksh.hu/teruletiatlasz_egyeb_teruletilehatarolasok

Izsák, É. 2001. "Szuburbanizáció és gazdasági fejlődés: Budaörs, a legsikeresebb magyar város." In M. Keune and J. Nemes Nagy, eds., *Helyi fejlődés, intézmények éskonfliktusok a magyarországi átmenetben*. Budapest: ELTE, 35–54.

Jacobs, J. 1961. *The Death and Life of Great American Cities*. New York: Random House.

Jacobs, J. 1969. *The Economy of Cities*. New York: Random House.

Keil, R., ed. 2013. *Suburban Constellations: Governance, Land and Infrastructure in the 21st Century*. Berlin: Jovis.

Keil, R. 2018. *Suburban Planet*. Cambridge: Polity Press.

Kiss, R. 2007. "Falusi értékek, városi igények. A szuburbanizációs folyamatok hatása a helyi társadalom átalakulására Budajenőn." In Z. Szarvas, ed., *Migráció és turizmus. Migrációs folyamatok hatása a helyi társadalmak változásaira*. Budapest: MTANLKI L'Harmattan, 19–57.

Kocsis, J.B. 2008. "Város menti agglomerációk átalakulása a globalizáció hatására." In G. Csanádi and A. Csizmady, eds., *Társadalom – Tér – Szerkezet*. Budapest: ELTE VRKK, 79–86.

Kocsis, J.B. 2009. *Városfejlesztés és városfejlődés Budapesten, 1930–1985*. Budapest: Gondolat.

Kocsis, J.B. 2012. "Lakáspolitika Budapesten 1960–1975 között." *Múltunk*, 57(1): 160–206.

Kocsis, J.B. 2013. "Urban Sprawl and Budapest: Polycentric Urban Development in Budapest Metropolitan Area." Paper presented at the European Network of Housing Research Conference, Tarragona, Spain, 19–22 June 2013.

Kocsis, J.B. 2015. "A nagyváros útjában: A helyi társadalmak átalakulása Budapest délnyugati agglomeriációjában." In A. Keszei and Z. Bögre, eds., *Hely, identitás, emlékezet*. Budapest: L'Harmattan, 163–84.

Kocsis, J.B., and A. Dúll. 2016. "A megújulás környezeti és társadalmi aspektusai." In J.B. Kocsis, ed., *Főutcák, üzletutcák: Megújulás és fejlesztés*. Budapest: L'Harmattan, 199–207.

Kocsis, J.B., and K. Tomay. 2018. *Törökbálint kutatási zárótanulmány*. Törökbálint: Törökbálint Város Önkormányzata.

Kondor, A. 2016. "Helyi konfliktusok Budapest szuburbán zónájában." *Földrajzi Közlemények*, 140(3): 216–28.

Konrád, G., and I. Szelényi. 1969. *Az új lakótelepek szociológiai problémái*. Budapest: Akadémiai Kiadó.

Koós, B. 2004. "Adalékok a gazdasági szuburbanizáció kérdésköréhez." *Tér és Társadalom*, 18(1): 59–71.

Kovách, I., and L. Kristóf. 2005. *Mobilising and Commercialising Rural Goods and Services: An Intermediate Actors Oriented Comparative Analysis*. Budapest: MTAPTI.

Kovács, Z., Z.J. Farkas, T. Egedy, A.C. Kondor, B. Szabó, J. Lennert, D. Baka, and B. Kohán. 2019. "Urban Sprawl and Land Conversion in Post-Socialist Cities: The Case of Metropolitan Budapest." *Cities*, 92: 71–81. https://doi.org/10.1016/j.cities.2019.03.018

Kovács, Z., and I. Tosics. 2014. "Urban Sprawl on the Danube: The Impacts of Suburbanization in Budapest." In K. Stanilov and L. Sýkora, eds., *Confronting Suburbanization: Urban Decentralization in Postsocialist Central and Eastern Europe*. Chichester, UK: John Wiley & Sons, 33–64.

Ladányi, J., and I. Szelényi. 1999. "Sozialrämliche Polarisierung und Suburbanisierung in Ungarn." *Zeitschrift für Wissenschaftsgeographie*, 43(1): 1–15. https://doi.org/10.1515/zfw.1999.0001

Laki, I. 2018. "Agglomeration Issues in Respect of Budapest." *Belvedere Meridionale*, 30(4): 160–80. https://doi.org/10.14232/belv.2018.4.10

Markley, S. 2018. "Suburban Gentrification? Examining the Geographies of New Urbanism in Atlanta's Inner Suburbs." *Urban Geography*, 39(4): 606–30. https://doi.org/10.1080/02723638.2017.1381534

Marshall, A. 2001. *How Cities Work: Suburbs, Sprawl, and the Roads Not Taken*. Austin: University of Texas Press.

Mumford, L. 1961. *The City in History*. San Diego: Harcourt.

Nemes, G., V. Csizmadiáné Czuppon, K. Kujáni, É. Orbán, A. Szegedyné Fricz, and V. Lajos. 2019. "The Local Food System in the 'Genius Loci' – The Role of Food, Local Products and Short Food Chains in Rural Tourism." *Studies in Agricultural Economies*, 121(2): 111–18. https://doi.org/10.7896/j.1910

Népesség.com. 2020. "Népesség." http://nepesseg.com

Phillips, M. 1993. "Rural Gentrification and the Processes of Class Colonisation." *Journal of Rural Studies*, 9(2): 123–40. https://doi.org/10.1016/0743-0167(93)90026-G

Porter, M.E. 1995. "The Competitive Advantage of the Inner City." *Harvard Business Review*, 73(3): 55–71. https://doi.org/10.1016/0024-6301(95)94327-u

Ricz, J., G. Salamin, A. Sütő, C. Hoffmann, and L. Gere. 2009. *Koordinálatlan városnövekedés az együtt tervezhető térségekben.* Budapest: VÁTI.

Sailer-Fliege, U. 1999. "Characteristics of Post-Socialist Urban Transformation in East Central Europe." *GeoJournal,* 49: 7–16. https://doi.org/10.1023/A:1006905405818

Salamin, G. 2004. "A gazdasági térszerkezet alakulásának legújabb folyamatai." *Falu Város Régió,* 11(9): 14–24.

Salamin, G., G. Kigyóssy, M. Borbély, B. Tafferner, B. Szabó, F. Tipold, and M. Péti. 2014. "A fejlesztéspolitika és területfejlesztés új koncepciójáról." *Falu Város Régió,* 20(1): 7–24.

Salamin, G., Á. Radvánszki, and A. Nagy. 2008. "A magyar településhálózat helyzete értékelés." *Falu Város Régió,* 3: 6–26.

Salamin, G., J. Ricz, A. Sütő, C. Hoffmann, and L. Gere. 2009. *Koordinálatlan városnövekedés.* Budapest: VÁTI.

Sassen, S. 2009. "Cities in Today's Global Age." *SAIS Review of International Affairs,* 29(1): 3–34. https://doi.org/10.1353/sais.0.0034

Speck, J. 2013. *Walkable City: How Downtown Can Save America, One Step at a Time.* New York: North Point Press.

Stanilov, K., and L. Sýkora, eds. 2014a. *Confronting Suburbanization: Urban Decentralization in Postsocialist Central and Eastern Europe.* Chichester, UK: John Wiley & Sons.

Stanilov, K., and L. Sýkora. 2014b. "Managing Suburbanization in Postsocialist Europe." In K. Stanilov and L. Sýkora, eds., *Confronting Suburbanization: Urban Decentralization in Postsocialist Central and Eastern Europe.* Chichester, UK: John Wiley & Sons, 296–320.

Stanilov, K., and L. Sýkora. 2014c. "Postsocialist Suburbanization Patterns and Dynamics." In K. Stanilov and L. Sýkora, eds., *Confronting Suburbanization: Urban Decentralization in Postsocialist Central and Eastern Europe.* Chichester, UK: John Wiley & Sons, 256–95.

Szabó, T., B. Szabó, and Z. Kovács. 2014. "Polycentric Urban Development in Post-Socialist Context: The Case of the Budapest Metropolitan Region." *Hungarian Geographical Bulletin,* 63(3): 287–301. https://doi.org/10.15201/hungeobull.63.3.4

Szelényi, I. 1996. "Cities under Socialism – and After." In G. Andrusz, M. Harloe, and I. Szelényi, eds., *Cities after Socialism: Urban and Regional Change and Conflicts in Post-Socialist Societies.* Oxford: Blackwell, 286–317.

Szirmai, V., Z. Váradi, S. Kovács, N. Baranyai, and J. Schuchmann. 2011. "Urban Sprawl and Its Spatial, Social Consequences in the Budapest Metropolitan Region." In V. Szirmai, ed., *Urban Sprawl in Europe: Similarities or Differences.* Budapest: Aula Kiadó, 141–86.

Tamáska, M. 2011. *A vidéki tér emlékezete. Az építészeti formaképződéstől a kulturális örökségalkotásig.* Budapest: Martin Opitz.

Tomay, K. 2019. "Az összekötő társadalmi tőke (újra)termelődésének terei. A városi és vidéki dzsentrifikáció szerepe a vállalkozói és innovációs ökoszisztémákban." *Replika*, 111: 43–62. http://doi.org/10.32564/111.4

Tomay, K., and G. Nemes. 2020. *Meghasadt valóságok – dilemmák a falusi turizmus szerepéről a területi fejlődésben.* Budapest: s.n.

Tosics, I. 2013. "From Socialism to Capitalism: The Social Outcomes of the Restructuring of Cities." In N. Carmon and S.S. Fainstein, eds., *Policy, Planning and People: Promoting Justice in Urban Development.* Philadelphia: University of Pennsylvania Press, 75–100.

Tsenkova, S., and Z. Nedović-Budić, eds. 2006. *The Urban Mosaic of Post-Socialist Europe: Space, Institutions and Policy.* Heidelberg: Physica-Verlag.

Varga, V., D. Teveli-Horváth, and G. Salamin. 2020. "A fiatal, képzett lakosságot vonzó potenciál a Budapest körüli csapágyvárosokban." *Területi Statisztika*, 60(2): 179–210. http://doi.org/10.15196/TS600204

10 Creativity in Contemporary Housing Estate Neighbourhoods: The Case of Kontula, Helsinki

JOHANNA LILIUS

The shopping centre is the heart of Kontula. It is a concentration of social life and encounters, and a necessary place of business and transit. For many, it is a cozy and lively familiar living room, but many are also afraid of it. It evokes perhaps the most contradictory feelings of all the parts of Kontula, both outside and in Kontula itself.[1]

– Routes to Kontula (2016)

Introduction

The shopping centre in the high-rise suburb of Kontula in eastern Helsinki is the point of reference for a multitude of desires: customers, passers-by, loiterers, workers, residents, outsiders, and lately real estate agents and planners. It is also a place entailing new spaces of creativity stemming both from inside the neighbourhood and from outside in the form of arts initiatives realized by non-residents. Just like the high-rise neighbourhood in which it is located, the shopping centre has been characterized in many ways, even stigmatized (Nironen, 2014). Yet, as is often the case with disparaged neighbourhoods, the point of view promoted by the media and the general public differs from that of the residents (Kokkonen, 2002; Ilmonen, 2014: 51; Rosendahl, 2014). The quotation at the beginning of this chapter is an extract from a collection of voices about the shopping mall in Kontula, produced during the Routes to Kontula project carried out by the Finnish National Theatre in the neighbourhood. It sets the scene for this chapter, which explores the meaning of cultural, social, and spatial creativity in high-rise suburbs, asking what can be created in the neighbourhood and for whom.

The chapter is set on the periphery of Helsinki in one of Finland's many high-rise suburbs. Developed during the first wave of urbanization in the 1960s and 1970s, when a large amount of affordable housing was needed, these decent but homogeneous estates were soon pigeonholed as dormitory towns in the Finnish media. By the 1980s, many families that were able to afford single family homes moved away to access more desirable housing. When the deep economic recession hit Finland in the early 1990s, the socio-economic profile of many of the high-rise suburbs, including Kontula, started to decline. Since then, these suburbs have experienced an increasing concentration of deprivation, including socio-economic decline, an increasing proportion of older residents, and a growth in immigrant populations (Stjernberg, 2017). Feelings of unrest and insecurity have increased in high-rise suburbs, with many residents wishing to move away (Vaattovaara et al., 2018). Although almost one third of all Finns live in a high-rise suburb, they are the least favoured living environments for the population (Strandell, 2017). The general imaginary of such high-rise housing estates in Finland is rather negative, and in the public debate and media, these neighbourhoods have become the symbol for urban deprivation – unemployment, abuse, and broken families (Roivainen, 1999; Hyötyläinen, 2019).

Today, however, there is a new interest in these high-rise estates, driven by what has been called the second wave of urbanization in Finland. Simultaneously, since the 1990s, the urban planning framework has undergone an ideological change. The modernistic ideal with a separation of urban functions has been replaced by a new paradigm in planning preferred by planners in Helsinki. The new paradigm emphasizes urbanity and the mixed urban qualities that can be found in urban cores. As elsewhere in Western cities, urban lifestyles, urban cultures, and the quality of the urban environment are considered necessary in order to generate economic wealth (Florida, 2006). Thus, one of the core tasks for urban planners is to make cities for the middle classes (Wyly and Hammel, 2001; Uitermark et al., 2007; Van den Berg, 2013).

In Helsinki, the middle classes have been outpriced from the city core, which is creating pressure for the development of new housing in the high-rise suburbs. Indeed, one of the core issues in the densification process in Helsinki is to increase the amount of middle-class housing in the suburbs. Using social mix as a strategy to raise the socio-economic profiles of neighbourhoods is a well-known strategy within neoliberal urban planning (see, for example, Lees, 2008; Hackworth and Smith, 2001; Wyly and Hammel, 2001). Consequently, there are already several studies showing that the socio-economic status of some suburbs in Helsinki has risen due to infill building, since the people moving in are generally better educated and have higher incomes than existing residents (Kytö and Kytö, 2014; Miettinen, 2018).

Despite the trends just discussed and despite physical closeness to the urban core through effective public transportation, the suburbs remain distant for the middle classes due to their negative image and socio-economic profiles. However, the master plan in Helsinki promotes densification in the suburbs and outlines the extension of urban lifestyles to these areas, which are described as "small towns" in the plan. One of the core physical strategies adopted by the city planning department for the high-rise suburbs is the demolition of old shopping centres, such as the one described in Kontula, and their replacement with new facilities. Research shows that this demolition and the development of new facilities has increased housing prices, upgraded service provision, and attracted higher income groups (Miettinen, 2018). Thus, attracting residents who can make suburbs more "urban" has become an important task for the planners (Lilius, 2019). An excerpt from the *Helsinki City Plan: Vision 2050* states:

> Urban centres in the current suburbs will see substantial development, enabling an urban lifestyle outside the traditional central areas of Helsinki. The new, more urban city comprises a network of original, attractive and distinctive small towns. (City of Helsinki, Planning Department, 2013: 16)

However, Landry and Bianchini (1998 [1995]: 12) have argued that the goal of urban planning to create a good city has failed because the efforts made to improve the urban environment have focused mainly on the physical elements. Planning a good city also requires a cultural dimension.

The suburbs, and Kontula in particular, as this chapter will show, have also been targeted by a number of top-down incremental urban regeneration programs seeking to increase liveability. Urban regeneration programs refer to local policies and strategies with an integrated approach to deal with urban decline, decay, or transformation (Lang, 2005). However, the results of the regeneration programs in Kontula have not always been convincing, as they have lacked a social and cultural dimension (Kokkonen et al., 2009). Haylett (2003: 69) claims that, while regeneration programs are typically culturally pitched, the standard discourse on social inclusion does not include positively naming and valuing the people and locations where these regeneration programs take place. According to Thörn and Holgersson (2016), regeneration programs often build a narrative in which an area is simultaneously stigmatized and upgraded as "cool" and "upcoming." Part of being cool and upcoming, then, is the type of creativity expressed by the so-called creative classes. By contrast, suburbs are typically understood as mundane places in which creativity is non-existent (Keil, 2017). As this chapter will show, in Helsinki a set of interventions disconnected from the urban planning framework have lately aimed at bringing the arts into the housing estates. These two different strands, planning goals on the one hand and creative interventions on the other, come into play together in Kontula.

One reason behind the growing number of art events held in Helsinki's suburbs is the recognition that cultural grants had been unequally distributed around the city, with a particular concentration on events in the inner city (Räisänen, 2014). Therefore, the city developed a new funding instrument, the Helsinki Model. The aim of this program is to encourage cultural institutions and arts professionals to "work outside their own walls" and to simultaneously make culture more accessible in areas that are lacking these services. According to Gibson and Klocker (2005: 94–5), creativity is typically targeted in policymaking because it is "positioned at the core of innovation, invention and enterprise culture." By contrast, the goal of the Helsinki Model seems to be more in line with Landry's (2000: 8–9) arguments; indeed, he is frequently invited by the City of Helsinki to give talks. Landry claims that culture can be expected to enhance a sense of community, increase self-esteem, improve life skills, and better the physical and mental well-being of people, as well as strengthen people's ability to become active citizens. In addition, Pratt and Hutton (2013: 91) identify one aspect of culture as social regeneration. More than civilizing society through high culture, this approach emphasizes betterment through participation in cultural activities. The Helsinki Model is developed as "a model for participatory local cultural work, which encourages art institutions and professional groups of artists to expand their operations beyond their own facilities and work in cooperation with the residents and communities of various neighbourhoods." While the city wishes to distribute the supply of culture around the city more equally, the model also has broader goals, such as improving the image of neighbourhoods (City of Helsinki, 2019). This approach echoes the argument of Kloosterman and Trip (2011), who claim that culture can contribute to the strengthening of local identity. Gibson and Klocker (2005: 100) further claim that the cultural turn in regional policy celebrates "elements of social life that bring vitality, vibrancy, surprise and emotional pleasure to a great many people." An important goal of the Helsinki Model is also to increase cultural participation among residents, thus contributing to their well-being. Yet, according to Bain (2018), the spatial politics of cultural production can endanger the social fabric in low-income neighbourhoods. This kind of creative destruction appears through gentrification processes when cultural workers direct their interest towards affordable real estate, while remaking the image of an area.

The core of this chapter is to understand how the image of Kontula is reframed through creative initiatives established by the city but implemented by the residents. Thus it asks the questions: What kind of creativity exists in Kontula? How is creativity perceived in the neighbourhood? Is this creativity being captured by city administrations? According to Gibson (2010), exploring creativity in places where typical inner-city urbanity does not necessarily exist means that creativity needs to be understood in more versatile ways. The chapter is thus built on a variety of data, including secondary sources such as statistics, planning documents, media articles, and research reports, as well as observation and interviews from a specific intervention in Kontula, namely the Routes to Kontula project (2013–15), funded by the Finnish National Theatre and the Helsinki Model. The author took part in a follow-up study of the project, funded by the Ministry of Education and Culture, which aimed to bring perspectives from the outside but also to ask critical questions about using culture as a way to enhance liveability in the suburbs. The follow-up was built around an ethnographic approach, which meant that observation was carried out from April 2015 to February 2016 at rehearsals, theatre performances, meetings, and feedback sessions, both during and immediately after the Routes to Kontula project. Additionally, a series of two semi-structured interviews, thirteen structured interviews, and sixteen short interviews were conducted. All written feedback from participants was also analysed. In this chapter, the Routes to Kontula project is used to examine the aims and goals of an art project in a high-rise suburb, as well as to consider how residents perceived their involvement in such a project and how it changed their perception of their neighbourhood.

The chapter is structured as follows. It first introduces Kontula and contextualizes interventions around culture and creativity in the suburb. It then elucidates the meaning of the Routes to Kontula program to the involved artists, but more particularly to the participants, both in terms of their well-being and as a way of strengthening their place identity. The chapter then explores the latest agents who are adding a new layer of creativity to Kontula. In conclusion, the interplay between planning goals and art interventions in Kontula is discussed.

Kontula at the Crossroad of Change?

While some of Helsinki's high-rise estates have already undergone socio-economic upgrading, it is not the case in Kontula – at least not yet. Current statistics show that Kontula's residents are less educated, more frequently unemployed, and earn less than the city average. Around 40 per cent of the housing in Kontula is social rental housing, a considerably higher percentage than in other parts of Helsinki. The proportion of residents with an immigrant background is one of the highest in the city (City of Helsinki, 2015). Yet, while visiting Kontula, the high number of retirees is striking. Many of the older inhabitants have been living in Kontula since it was constructed, which shows that it is not solely a place of immigrant newcomers. Apart from those who have been living in the neighbourhood for decades, there has also been a recent re-migration to Kontula. Since the beginning of the century, residents aged between eighteen and forty-four years have moved to Kontula (Lankinen et al., 2015). According to several interviewees, these incomers are often returnees moving back to the neighbourhood where they were born and raised.

The shopping centre, as stated at the beginning of the chapter, is the heart of the neighbourhood. It is the transportation hub and the place where people come for all of the services in the neighbourhood. Due to its central location and its diversity of users, it is also one of the most discussed places in Kontula. For a local centre, it has a great variety of services; all the large grocery store chains are present there, while other amenities and commercial services include flea markets, cheap general stores, clothing shops, and a newly renovated outdoor playground.

Nevertheless, it is the number of bars that has made Kontula's shopping mall (in)famous. Indeed, in 2014 the mayor of Helsinki suggested that these bars should be closed down to make Kontula more peaceful and safer. He claimed that the shopping centre "has become a place that everyone is afraid of, especially children," while it should be a place "where people come together" (Kuokkanen, 2014). When journalists from the major newspaper in Finland, *Helsingin Sanomat*, visited the bars to hear how the clients felt about such a proposal, they described a riotous atmosphere and underlined that people were already getting drunk before 11 a.m. One customer interviewed in the article was quoted as follows: "This is our living room, a social space, where we come for coffee and beer. We are all having fun here; the mayor should come here in person to proclaim his prohibitions" (Autio, 2014). The bars are highly visible at the shopping centre since (often drunk) crowds typically gather outside the bars to smoke. However, early Finnish research on the suburbs has emphasized the social function of such bars, as they are important gathering places for the locals (Sulkunen, 1985). Besides drinking, the bars also offer some activities such as board games and, particularly important in the Finnish context, karaoke. Thus, karaoke singing can also be heard echoing between the buildings of the shopping centre.

There are no statistics available concerning alcohol and drug users in Kontula; however, while moving around the shopping centre, it is difficult to avoid people who are under the influence of alcohol and/or drugs. The frequency of addicts around the shopping centre is not only related to the bars but also to the fact that a day centre for alcohol and drug users is located there. In an interview, the president of the Association of Somalis in Kontula explained that the situation with drugs is improving, and many third sector agents are working to keep youth off the streets. Yet, it is through drug use that Kontula also appears in the popular rap artist Steen1's music. In his 2004 song "Sinisiä rappuja ja punaisia hintalappuja" (Blue staircases and red price tags), he rhymes (in the Finnish version):

> If you have a drug problem, you are living in Kontula
> If you are Somali, you are living in Meri-Rastila
> Simultaneously, as I record this shit
> The government is putting all the marginalized in the same hoods
> Cheap social housing in the east, you can't say it's a coincidence
> Outpatients run around and create nuisance
> The gangs take heroin, and nobody wants to intervene
> Can you keep clean, when your neighbour keeps a drug flat? (translation)

Steen1, who spent his childhood and youth in Kontula, claimed in an interview for the Finnish Broadcasting Company (Yleisradio Oy [Yle]) that he would prefer to pay a higher rent in the city centre than to raise his child in Kontula because too many of his former friends in Kontula became addicts (Yle, 2006).

However, the image of Kontula and its shopping centre is evolving. Following the 2015 refugee crisis, a growing number of immigrant restaurants and shisha cafés opened at the shopping centre. This development was met with enthusiasm by the media. The broadcaster Yle, for example, headlined in 2016 that "the shopping centre in Kontula is living its multicultural boom" (Nelskylä, 2016). The bottom-up activities of the ethnic entrepreneurs were presented as one of the enriching effects of the refugee crisis. Yet, it is unclear whether the multicultural character of the shopping centre will survive the current densification pressure. In the attempts to regenerate Kontula, the city organized an architectural competition covering the area around the shopping centre. Apart from architectural criteria related to urban design, the design was also to be evaluated on its prospects to facilitate "community making," enhancing of "amenity and public perceptions of safety," and creating a "setting for arts and culture within the site" (SAFA, 2020: 14). The winner is an entry that proposes demolishing the shopping centre and building a new one. The jury found the winning entry particularly successful:

> Of all the entries, "Vaellus" [the winner] is the most consistent and balanced embodiment of the kind of civilized, restful, sympathetic, but energy giving place-making and ambience that would make for a successful renewal direction for Kontula. (SAFA, 2020: 62)

It is obvious that the current mix of people spending time at the shopping centre does not satisfy the planning authorities. However, where the current users of the shopping centre are supposed to go after demolition remains unclear.

Alongside these partly contradictory developments such as the statistically verified low socio-economic profile and the quest of urban planners to give the neighbourhood a more upscale profile, a more lifestyle-oriented portrayal of Kontula has recently emerged in the Finnish media. As if to underline that the cultural elite also live in Kontula and that the image of the neighbourhood is diverse, the celebrated Iraqi-born film director and writer Hassan Blasim said in an interview for *Helsingin Sanomat*: "I feel at home in Kontula. You can see contemporary Finland here" (Petäjä, 2019). Further, the middle-class magazine *Apu* recently portrayed Kontula through the story of a couple working in the cultural industries who moved from a gentrified city centre neighbourhood to Kontula. They talk about their satisfying everyday life in Kontula and emphasize that they recognize gentrification is advancing towards eastern Helsinki:

> It always starts in the centres and moves towards the suburbs, and sort of swallows them. Now it is heading here. Everyone wants to move to Helsinki and everyone is looking for cheap flats. They can be found here. Prices win over prejudice, and then the housing estate starts to change. (Hiltunen, 2020)

Culture in Kontula?

Apart from the creativity within, such as that represented by former resident rap artist Steen1, culture has also been brought to Kontula through different interventions, including those partly financed by the European Union. When the Urban II program (2001–6) was evaluated in 2007, residents were asked what they considered was important to improve in the neighbourhood. The answers included safety issues – the residents asked the authorities to pay more attention to order, alcohol and drug use, and the environment around the shopping centre (Korhonen, 2008: 68). Although the need for culture was not raised by the residents, the program funded a new music festival, proposed by a local resident activist. The vision was to create a cultural event in Kontula that would attract both local residents as well as crowds from beyond. The (free) Kontu-festival was highly appreciated by the locals (65). It is still an annual event in the neighbourhood and has become very popular, drawing more than 10,000 visitors to Kontula each year. Bands from eastern Helsinki, rap in particular, are well represented. In 2013, the festival was visited by the left-wing cultural minister. Since the start of the Kontu-festival in 2003, the prevalence of different top-down events and projects focusing on culture and creativity has been on the rise. While the aim of the Kontu-festival was to give outsiders a new view of Kontula, and connect it to good music, most of the new initiatives have focused on remaking the image of the neighbourhood, including enhancing the area's mental landscape by increasing the self-esteem of the residents towards their home surroundings.

Kontupiste, a small cultural space, was developed at the shopping centre during the Urban II program. As part of the program, residents were invited to show their view of Kontula through photographs, which were then published on the internet and exhibited at Kontupiste. In another attempt to display a more vivid imaginary of Kontula, residents were encouraged to make documentaries about the suburb. These were shown at the annual Helsinki documentary film festival, DocPoint. According to Broman (2008), these interventions were important, but they have not created anything lasting since the activities stop when the project funding ends.

Nevertheless, new project funding creates the possibility for new interventions to take place. The plethora of initiatives from respected cultural institutions shows there is a real interest in working with locals in Kontula. This interest could be termed an "institutionalization of creativity" in the high-rise suburbs. Many of the initiatives are built around the idea of a professional art worker teaching or creating art with different resident target groups. This institutionalization of creativity serves several interests. First, it creates (new) jobs for the artists. Second, there is a rather good supply of both private and public funding for projects with a social agenda. Third, in order to be funded by the Ministry of Culture, cultural institutions are required to make a societal impact, including cultural, social, and economic impacts. These criteria are monitored, for example, in terms of the project's level of civil society engagement in cultural activities, and its support of culture and ability to ensure democracy as well as access to cultural services (Strategy for Cultural Policy, 2009). In other words, it is in the interest of both artists and institutions to reach out to the suburbs. For example, the modern art museum in Helsinki, Kiasma, has organized street art courses for young residents in Kontula. The Helsinki Art Museum organized an intervention in Kontula called "Our East." The exhibition showed important places for Kontula residents and stories related to these places. The goal was to exhibit accounts of how residents in eastern Helsinki see their neighbourhoods. The Museum of Photography extended its festival of political photography to eastern Helsinki and displayed some works at Kontupiste (the venue at the Kontula shopping centre). Despite these many activities, few studies have addressed the meaning of these interventions for the residents. Therefore, the next section explores what an art intervention can mean for long-term Kontula residents and what kind of creativity it brings forth, drawing on a case study of the National Theatre's Routes to Kontula initiative.

Routes to Kontula

The National Theatre of Finland is one of the high-profile cultural institutions that have been funded on the basis of the Helsinki Model. Their project in Kontula lasted three years (2013–15) and employed a total of thirty-nine artists at different stages, including lighting and sound designers, actors, musicians, and writers. A community theatre director was employed half-time for the whole period of the project. The goal of Routes to Kontula was to bring theatre to the suburbs and to eliminate prejudices about suburbs. Through theatre, the program also intended to bring joy, a sense of community, and well-being to the Kontula neighbourhood. The project can be interpreted as a community arts project, where the actions take place in the local community but the program is effected through the supervision of an official institution within the arts (Verkasalo, 2012: 17). While community art is often linked to social aims such as enhancing social mobility (Rundkvist, 2015), that was not a specified goal of the Routes to Kontula project.

Nevertheless, undertaking community theatre work, and particularly in the neighbourhood of Kontula, was a new task for the National Theatre. Because of earlier interventions in the neighbourhood, there were existing networks that helped the community theatre director to get an overall picture and to contact associations and organizations operating in Kontula. During the program, the community theatre director was in contact with at least thirty-four of these institutions in the area. These included different residential and ethnic organizations and associations and private clubs, such as music schools for school children, demonstrating that there is a lot of creativity in the neighbourhood despite its dubious reputation. For the National Theatre, these contacts were crucial in order to effectively get an overall picture of the area and learn about the key players and the needs there.

The configuration of all the workers employed by the National Theatre was typical for suburban interventions. The cultural workers had not been in any close contact with Kontula before the program. Nevertheless, Kontula was no tabula rasa for them; they knew its reputation and were also aware that the reputation was the reason why they were working there. Through different kinds of events, workshops, and small-scale programs, the National Theatre slowly started to search for a group of people who would take part in the play that would be written about Kontula and shown at the National Theatre.

The program organized a number of workshops, storytelling sessions, and several participatory events combining different groups of people living in Kontula, for example, retired women and pupils from the primary school's assimilation class for newly arrived immigrants. In this way, the program sought to bring together people who would not otherwise come into contact with each other. Besides these events, performance pieces were written and performed at different venues in Kontula, such as the communal space Wanha Posti (Old Post), which is maintained by a local resident's organization. Because of the bad reputation of the shopping centre, one performance particularly addressed the centre. First, stories about the centre were collected from residents through mailboxes located at different points around the centre. Based on the stories, a performance piece was developed in which the participants organized a tour of the shopping centre for residents, recounting the collected stories. The idea behind the performance was to evoke thoughts and feelings about the centre, and the outcome was a plethora of viewpoints and memories, many of them highlighting the meaning of the centre for the residents in their everyday life.

The largest scale interventions, however, were two plays, based on residents' stories, which were performed in the area for a variety of audiences, including school pupils, as well as on the National Theatre stage in the city centre. Politicians and civil servants at the City of Helsinki were invited to see the plays, although very few attended. The first play was about Kontula and intended to break down prejudice about the suburb. It was performed by Kontula residents of different ages. The second was about "movers" and dealt with immigration, with a Finnish actor and immigrants from several different countries on stage. At the end of both plays, the audience was invited to discuss and to pose questions to the actors, which they actively did.

Because the National Theatre program was a pilot, feedback was constantly collected from the artists and participants involved. Moreover, specific feedback sessions were also held after each project in order to discuss what had been done together. The feedback sessions were held separately for artists and residents. The artists who had been involved in the program emphasized that they felt they did "something important" in Kontula and that their work – though not always easy because, for example, of language barriers – felt meaningful. It is understandable that the professionals structured their relationship from a professional point of view, adding social hierarchies to their work environment. Several of the artists specifically expressed that they felt what they produced was high-quality (artistic) work, which was important for them and also inspired them. They also learned from the difficulties they encountered while working with the people in Kontula. Many artists observed that, especially while co-creating with immigrants, they had to come to terms with people having a different perception of time and learn how and when to be in contact. Taking part in the hour-long plays required a lot of commitment from the participants, and so the director sometimes had to make a difficult decision when a participant repeatedly missed rehearsals. Yet, according to her understanding, her main task was to produce high-quality theatre. The feedback from the participants in relation to the creative workers was very positive. They were happy about their professionalism, their way of working, and their engagement. It was also very obvious from the feedback that the community theatre director was very much appreciated by the attendees, being thought of as a mother or mentor. And the social hierarchies mentioned earlier were not always visible to the participants. For example, one participant bewilderedly asked the community theatre director if she was paid to work in Kontula.

Finding people in Kontula willing to take part in the different events organized by the National Theatre was not always easy. Although it was fairly easy to get Finnish-born participants for the longer play about Kontula, it proved difficult to find immigrants willing to participate. Taking part in one of the longer plays required a commitment to be involved in rehearsals for many nights a week over several months. However, the difficulty in finding participants may relate to participation in similar high-rise suburbs on a broader level. Although there is participation through different bottom-up organizations, it does not necessarily relate to institutionalized processes (Luhtakallio and Mustranta, 2017). In order to find immigrant participants, the call was also extended to immigrants outside of Kontula, while immigrant organizations turned out to be of key importance in helping to recruit participants.

According to the attendees' feedback, the presence of the National Theatre in Kontula was very important image-wise. To many, it was meaningful that an eminent cultural player such as the National Theatre wanted to operate in Kontula. The collected feedback, the feedback sessions, and the interviews with participants proved that taking part in the organized activities enhanced their sense of community. It was obvious that many felt their neighbourhood was unjustly stigmatized, and they were glad to take part in events that aimed to give a fresh, more positive picture of the lived reality in Kontula. Top-down projects have often been criticized for not considering contextual differences and local nuances. The National Theatre program did create a sense of community, and this point was very much emphasized by the participants when they expressed their general feelings about the project. It could be claimed that a sense of community was produced in the *process* of making art together. The residents in this process creatively produced and reproduced notions of Kontula. Many participants also claimed that the events organized by the theatre provided something meaningful to do for the residents in the area, for those, according to two feedback comments, who "do not care to sit in the numerous bars in the area" or who want "something worthwhile to do besides sitting in the bar." These responses also showed the diversity of people who took part in the activities: both men and women of different ages, with a variety of backgrounds, interests, and work experiences / labour market positions. The immigrants, in addition, made connections across other ethnic groups. All in all, for most of the attendees, the programs organized by the National Theatre offered the possibility to get to know new people in the neighbourhood and perhaps, more specifically, people whom they would not otherwise have gotten to know. The immigrants in particular expressed the importance of the "theatre community" that had been built during the rehearsals and shows. Those participants who acted in the longer plays had often faced a challenge to go on stage. That they were able to do it had increased their self-confidence. In this way, it could also be suggested that Routes to Kontula brought out the creativity in some of the participants. Through the project, they were able to discover something new about themselves. The Helsinki Model approach is not to see creativity as human capital (Gibson and Klocker,

2005: 94) but rather to encourage new potential creativity. For example, one of the younger participants decided to apply to the Theatre Academy of Helsinki and was supported by the community theatre director.

As part of the National Theatre's activities in Kontula, free guided visits and affordable tickets to the theatre's city centre venue were offered. This initiative was highly appreciated by the residents, many of whom had never visited the National Theatre, despite having lived in Helsinki for their whole lives. This fact highlights the mental distance between the city centre and Kontula, despite its physical proximity – it only took participants a twenty-minute metro ride followed by a five-minute walk to reach the National Theatre. Luhtakallio and Mustranta (2017) have also shown that residents in Kontula do not typically visit museums or arts exhibitions in the centre because they simply lack the money or do not have the time or interest.

Many participants were sad when the project ended, and wished that the National Theatre could have stayed in the neighbourhood longer. As a continuation of the first full-length play, the participants founded an amateur theatre group. Although this group produced a few plays, and the community theatre director went out of her way to try to encourage the group to apply for external funding, it proved too difficult to sustain, with the actors feeling they needed one clear leader. Nevertheless, after the National Theatre left, the retired ladies who were connected with the immigrant children continued to keep in contact for some time. Many of the retirees said they have no natural way of getting to know the immigrant children and therefore were very happy about the program. During the feedback discussion, many expressed the importance of getting to know people from the many different cultures that inhabit Kontula. One lady said: "Now I have the courage to tell racists that children are wonderful no matter what colour they are."

As one of the goals was to change the image of Kontula through the project, it is also interesting to note that the broadcaster Yle particularly highlighted the connection between institutionalized high culture and residents in Kontula, including posting several stories about the Routes to Kontula project on their website. Their headlines proclaimed: "Prejudices about Kontula Are Fought on the Stage of the National Theatre" and "Residents from Kontula on the Stage of the National Theatre." The newspaper *Helsingin Sanomat* also wrote about the project (Mäkinen and Niiranen, 2014). As emphasized by Bain (2018: 328), media coverage also gives cultural workers an opportunity to receive acknowledgement for their work and increases its profile. However, did the project change the participants' notions of Kontula? In their feedback, many of the residents expressed their satisfaction with their residence in

Kontula, in particular pointing out that it is a much better place to live in than its reputation suggests. Many of the responses also suggested that the attendees had gotten to know their neighbourhood even better. But most emphasis was put on getting to know other Kontula residents.

Breaking the Suburban Spell

Since the departure of the National Theatre from Kontula, the emphasis of ongoing creativity in the area has changed somewhat. It appears that more spontaneous cultural events have also started to take place in Kontula. Kontula still has affordable spaces that can be used experimentally. Several of these events are funded by highly competitive and prestigious grants and are connected to residents from the city centre. Since 2016, the shopping centre has been the venue for the Kontula Electronic Festival, described as "an international high-quality festival which is anchored in the everyday reality of the Kontula shopping centre and gives space to local actors" (Tuominen, 2017). A research project financed by the Kone Foundation, an important funder of arts and research in Finland, is currently doing ethnographic fieldwork on the festival, surveying "the worst pitfalls such as the fear that hipsters from Punavuori and Kallio [two city centre neighbourhoods] will conquer the festival, that suburban residents will be dislocated, and that the festival will be a place for marvelling exotic otherness" (Tuominen, 2017). The festival has also gained media attention, with the broadcaster Yle presenting the Kontula Electronic Festival in the following manner:

> Kontula has had a dubious reputation for decades. Kontula has changed from a restless housing estate for rural settlers to a multinational place, where the kind of restlessness that is part of life exists, but even more humanity and open tolerance. Different cultures do not collide with each other here, but they enrich life. (Yle, 2020)

This coverage echoes the narratives produced during the Routes to Kontula project, where the media approached Kontula by turning its previously perceived weakness, such as the high number of immigrants, into a strength by emphasizing an open mindset and culture in the neighbourhood. Meanwhile, in 2017 the Museum of Impossible Forms was established in Kontula. It is funded by two key art foundation bodies and retains a space in the shopping centre. It represents a "dynamically open space that practises decolonial, queer and intersectional feminist values as a heterogeneous platform to engage with experimental, marginal, and migrant forms of expression and

as a laboratory for experiences, critical thought, and radical imagination." It is run by "art and cultural workers" and facilitates the
coming together of a collective of artists, curators, pedagogists, philosophers, and facilitators. The Museum of Impossible Forms program includes, for example, movie nights and events on ethnic music
(Museum of Impossible Forms, 2022). What is particular about the
new players in Kontula is that they draw distinct crowds from the
centre: hipsters who look for special foods and alternative music and
arts lovers. Moreover, they come specifically to the shopping centre
that will soon be demolished.

Nevertheless, creativity and voices from within, especially voices of
growing numbers of young people with immigrant backgrounds, are
also increasing and growing louder in Kontula. This change is something that the City of Helsinki has also emphasized on its My Helsinki
page, its tourist guide targeted at both locals and visitors. The site presents Kontula by quoting YouTuber Hassan Maikal:

> Kontula is home: you never feel that you wouldn't be part of the crowd.
> The coolest thing is that you have a crowd of people from different cul-
> tures and countries. Everyone together. (Lindroos, 2020)

Hassan Maikal, in his account and also in his rap music, is far more
positive about Kontula than his predecessor Sten1 was fifteen years
earlier. Akin to the artist Adam Tensta from the stigmatized neighbourhood Tensta in Stockholm, Hassan Maikal in his YouTube videos uses
the pseudonym "Kontulan-Hassan," which means " Hassan from Kontula." Using this pseudonym, he wants to emphasize his "hoods." Further, his YouTube videos dismantle prejudices about Kontula and about
his Somali ethnicity. Hassan Mikal's creativity broadens the view of
Kontula and its residents to the outsider, while simultaneously enhancing a positive atmosphere within the neighbourhood. This stance has
also been recognized by institutional players, as Mikal is constantly
employed by the City of Helsinki for different events taking place in
Kontula.

Conclusion: In Kontula, Problems and Possibilities Are Huge

This chapter set out to understand how the notion of Kontula has been
reframed through creative initiatives established by the city but implemented by the residents, and in this concluding section, it will discuss
the interplay between the creativity stemming from the neighbourhood
and the current urban planning goals.

Keil (2017) has put forward the idea that new forms and ways of life now emerge in the suburbs rather than in the inner cities. Suburbs to Keil have become the stages where urban "dramas" are played out, as they are the new scenes of multiple diversities and contradictions. The practice of taking advantage of this kind of "urbanity" in a neighbourhood and transforming it into a commodity for branding purposes is well known in urban theory (see, for example, Thörn and Holgersson, 2016). As this chapter has shown, the image of Kontula is constantly being contested by both outsiders and insiders. The contradictory image of the neighbourhood is well captured in the following line from an article in the newspaper *Helsingin Sanomat*, quoting a long-term resident and participant of the Routes to Kontula project, who said: "In Kontula, the problems and the possibilities are huge" (Mäkinen and Niiranen, 2014). This kind of nuanced imaginary of Kontula was also put forward by other residents involved in the different actions that comprised the Routes to Kontula project. This creativity from within presented an image of Kontula where unemployment and public alcohol and drug use coexist with the good life, closeness to nature, excellent services, and positive neighbour relations. Thus the project succeeded in improving the reputation of the neighbourhood, while at the same time enhancing the self-respect of its residents. Haylett (2003: 77) has emphasized that attaining social justice for the working classes "requires understandings about the complexity of working-class identity to be familiar and widespread, in particular understandings about ways of life that may be simultaneously positive and negative for those who embody them."

Complexity and contradictions, positive and negative notions of spaces also came into play when the Routes to Kontula project collected the residents' stories about the local shopping centre. As the chapter has shown, the shopping centre has become a place of new creativity involving elements from other localities, from the inner city, and also from new residents, mainly with immigrant backgrounds. It is an example of the creation of liveability from within. Yet, as the chapter has demonstrated, it has also become a place addressed as a "setting for arts and culture" by the urban planning authorities. The justification to demolish the current shopping centre presents a straightforward reversed stigmatization of the ongoing life at that shopping mall. The new centre is to be a "civilized, restful and sympathetic place," in other words something it is not today. Yet, it should also be an "energy giving place-making and ambience that would make for a successful renewal direction for Kontula" (SAFA, 2020: 62). There is a risk that, with demolition, the individual and collective identities and the creative processes of the shopping centre described in this chapter end up, as Gibson and Klocker (2005: 100) have put it, in a "singular interpretation of creativity," which is then "incorporated into a rather uncreative framework." The "cultural dimension" in planning, which Landry and Bianchini (1998 [1995]) describe as being so important for a good city, remains a physical feature in the urban planning context, although the physical is only part of its feature. By interpreting "setting for arts and culture" in a broader and more holistic way, as the chapter has shown, creative activities can enhance the self-respect, liveability, and social chances of the residents while simultaneously improving the suburb's reputation.

Nevertheless, as the chapter also showed, there is an ongoing "institutionalization of creativity" in the high-rise suburbs in which cultural institutions direct funds to facilitate community art projects there. As the creativity stemming from these projects is rooted in the community, there could be potential for these institutions to intervene in the urban planning processes and bring forth the uniqueness of Kontula and its residents. As studies in Kontula have shown, the residents are active but not in institutionalized processes. Therein lies the idea that the cultural institutions, by intervening in the planning process, could make a long-term difference to the spatial structures that the plan reworks. This intervention could present the possibility for the creativity from within to withstand the regeneration process that currently threatens to repress it.

NOTE

1 Translations from the Finnish are by the author.

REFERENCES

Autio, M. 2014. "Kontulan pubeista lähetetään haisevat terveiset Pajuselle: Älä tule kieltämään meidän baareja." *Helsinki Sanomat*, 1 October 2014. https://www.hs.fi/kaupunki/art-2000002765859.html

Bain, A. 2018. "Cultural Production in the Suburban Context." In B. Hanlon and T.J. Vicino, eds., *The Routledge Companion to the Suburbs*. London: Routledge, 323–32.

Broman, E.-L. 2008. "Eurooppalaista kaupunkipolitiikkaa soveltamassa – kokemuksia EU:n Urban II –yhteisöaloiteohjelmasta Helsingissä." In Nupponen, T., E-L. Broman, E. Korhonen, and M. Laine, eds., *Myönteisiä muutoksia ja kasvavia haasteita*. Helsinki: Helsingin kaupungin tietokeskuksen tutkimuksia 6/2008, 95–126.

City of Helsinki. 2015. *Helsinki Alueittain 2014*. Helsingin kaupungin tietokeskus. http://www.hel.fi/hel2/tietokeskus/julkaisut/pdf/15_02 _23_Hki_alueittain2014_verkko.pdf

City of Helsinki. 2019. "The Helsinki Model." https://www.hel.fi/kulttuurin -ja-vapaa-ajan-toimiala/en/about_us/culture-division/helsinki-model/

City of Helsinki, Planning Department. 2013. *Helsinki City Plan: Vision 2050*. Reports by the Helsinki City Planning Department General Planning Unit 2013:23. https://www.hel.fi/hel2/ksv/julkaisut/yos_2013-23_en.pdf

Florida, R. 2006. *Den kreativa klassens framväxt*. Translated by A. Sörmark. Göteborg: Daidalos AB.

Gibson, C. 2010. "Guest Editorial – Creative Geographies: Tales from the 'Margins.'" *Australian Geographer*, 41(1): 1–10. https://doi.org/10.1080/00049180903535527

Gibson, C., and N. Klocker. 2005. "The 'Cultural Turn' in Australian Regional Economic Development Discourse: Neoliberalising Creativity." *Geographical Research*, 43(1): 93–102. https://doi.org/10.1111/j.1745-5871.2005.00300.x

Hackworth, J., and N. Smith. 2001. "The Changing State of Gentrification." *Tijdschrift voor Economische en Sociale Geografie*, 92(4): 464–77. https:// doi.org/10.1111/1467-9663.00172

Haylett, C. 2003. "Culture, Class and Urban Policy: Reconsidering Equality." *Antipode*, 35(1): 55–73. https://doi.org/10.1111/1467-8330.00302

Hiltunen, K. 2020. "Kontulan lähiö – Näin Helsinki muuttuu: Poika toi koulusta venäläisiä kirosanoja – Äiti käänsi suomeksi: 'Näitä ei sitten käytetä.'" *Apu*, 21 January 2020. https://www.apu.fi/artikkelit/kontula -kansainvalistyi-ennen-muuta-helsinkia-lahioelamaa

Hyötyläinen, M. 2019. "Divided by Policy: Urban Inequality in Finland." PhD diss., University of Helsinki. https://helda.helsinki.fi/handle/10138/299277

Ilmonen, M. 2014. "Asuinalueiden brändäys – Voiko alueen mainetta suunnitella?" In M. Norvasuo, ed., *Elävän esikaupungin eväitä*. Aalto-yliopiston julkaisusarja. Espoo: Aalto yliopisto, 51–40.

Keil, R. 2017. *Suburban Planet: Making the World Urban from the Outside In*. Boston: Polity Press.

Kloosterman, R.C., and J.J. Trip. 2011. "Planning for Quality? Assessing the Role of Quality of Place in Current Dutch Planning Practice." *Journal of Urban Design*, 16(4): 455–70. https://doi.org/10.1080/13574809.2011.585863

Kokkonen, J., ed. 2002. *Kontula. Elämää lähiössä*. Helsinki: Suomen kirjallisuuden seura.

Kokkonen, J., M. Seppänen, and I. Haapola. 2009. *Lähiöiden kehittäminen Suomessa: Kokemuksia asukkaiden osallistumisesta*. Helsinki: University of Helsinki. https://www.ara.fi/download/noname/%7bcd073800-14a5 -434b-8bb5-eec9c2f54cd6%7d/22777

Korhonen, E. 2008. "Helsingin lähiöissä viihdytään – ongelmiakin on." In Nupponen, T., E-L. Broman, E. Korhonen, and M. Laine, eds.,

Myönteisiä muutoksia ja kasvavia haasteita. Helsinki: Helsingin kaupungin tietokeskuksen tutkimuksia 6/2008, 49–74.

Kuokkanen, K. 2014. "Kaupunginjohtaja Pajunen kieltäisi Itä-Helsingin häiriöbaarit." *Helsingin Sanomat,* 1 October 2014. https://www.hs.fi /kaupunki/art-2000002765737.html

Kytö, H., and S. Kytö, S. 2014. *Asuntotuotannon ja muuttovirtojen väliset kytkennät pää- kaupunkiseudulla vuosina 2001–2012.* Helsinki: Helsingin yliopisto. https://www2.helsinki.fi/sites/default/files/atoms/files /loppuraportti_vt_tdk_raportti.pdf

Landry, C. 2000. *The Creative City: A Toolkit for Urban Innovators.* London: Earthscan.

Landry, C., and F. Bianchini. 1998 [1995]. *The Creative City.* London: Demos.

Lang, T. 2005. "Insights in the British Debate about Urban Decline and Urban Regeneration." IRS Working Papers 32. Leibniz Institute for Regional Development and Structural Planning (IRS). https://ideas.repec.org /p/zbw/irswps/32.html

Lankinen, J., K. Pursiainen, T. Rajaniemi, and T. Nuutilainen. 2015. Kontula. Kurssityö Aalto yliopiston maankäyttötieteiden laitoksella, 8 April 2015.

Lees, L. 2008. "Gentrification and Social Mixing: Towards an Inclusive Urban Renaissance." Urban Studies, 45(12): 2449–70. https://doi.org /10.1177/0042098008097099

Lilius, J. 2019. "'Mentally, We're Rather Country People' – Planssplaining the Quest for Urbanity in Helsinki, Finland." *International Planning Studies,* 26(1): 78–80. https://doi.org/10.1080/13563475.2019.1701425

Lindroos, K. 2020. "Kontula." myhelsinki.fi. https://www.myhelsinki.fi /fi/näe-ja-koe/naapurustot/itäiset-naapurustot/kontula

Luhtakallio, E., and M. Mustranta. 2017. *Demokratia suomalaisessa lähiössä.* Helsinki: Into.

Mäkinen, T., and J. Niiranen. 2014. "Kontula muistaa Batmanin vierailun." *Helsingin Sanomat,* 8 January 2014. https://www.hs.fi/mesta/art -2000002700852.html

Miettinen, A. 2018. "Myllypuron omistusasuntojen hintakehitys esimerkkinä Helsingin aluekehittämistoimien vaikutuksista." *Kvartti/Helsinki Quart*erly, 6 April 2018. https://www.kvartti.fi/fi/artikkelit/myllypuron-omistusasuntojen -hintakehitys-esimerkkina-helsingin-aluekehittamistoimien

Museum of Impossible Forms. 2022. "About." https://museumofimpossibleforms .org/about

Nelskylä, L. 2016. "Syyrialaista ruokaa, irakilainen parturi ja imut vesipiipusta – Kontulan ostoskeskus elää monikulttuurista nousukautta." Yle, 4 August 2016. https://yle.fi/uutiset/3-9068570

Nironen, S. 2014. "Tarinoiden takapiha ja maineensa leimaama – Suomen tunne-tuin lähiö täyttää 50 vuotta." Yle, 18 March 2014. https://yle.fi/a/3-7143331

Petäjä, J. 2019. "Hassan Blasim kysyi lapsena, kuka on luonut Allahin ja sai tuolista päähän – nyt Blasim on ateisti, asuu Kontulassa ja kirjoittanut uskontoa rienaavan kirjan nimeltä *Allah99*." *Helsingin Sanomat*, 24 March 2019. https://www.hs.fi/kulttuuri/art-2000006045747.html

Pratt, A., and T. Hutton. 2013. "Reconceptualizing the Relationship between the Creative Economy and the City: Learning from the Financial Crisis." *Cities*, 33: 86–95. https://doi.org/10.1016/j.cities.2012.05.008

Räisänen, P. 2014. *Taide keskittyy keskustaan. Helsingin kulttuuri- ja kirjastolautakunnan avustusten alueellinen jakautuminen vuonna 2013*. Helsinki: Helsingin kulttuurikeskuksen julkaisu.

Roivainen, I. 1999. "Sokeripala metsän keskellä: Lähiö sanomalehden konstruktiona." Helsinki: Helsingin kaupungin tietokeskus.

Rosendahl, S. 2014. "'Täällä asun ja tällä pysyn!' – lähiömielikuvia." In M. Häyrynen, A. Wallin, S. Siro, and S. Forsell, eds., *Lähikuva lähiöstä! – itä-Porin kulttuurikartoitus ja toimenpidesuunnitelma*. Turun yliopiston Kulttuurituotannon ja maisemantutkimuksen Lähiön henki -tutkimushanke 2014. Turku: University of Turku, 42–52.

Routes to Kontula. 2016. "Matkakohteena Kontula –esitys." Reittejä Kontulaan (blog). http://reittejakontulaan.blogspot.com/p/matkakohteena-kontula -esitys.html

Rundkvist, P. 2015. "Kan kulturen hejda segregationen?" Transcript of lecture held in Stockholm, 2 October 2015, Finnish Association of Architects (SAFA).

SAFA (Finnish Association of Architects). 2020. *Kontula City Centre, Invitational Architectural Competition: Competition Jury's Evaluation Protocol*. https://www .safa.fi/wp-content/uploads/2020/03/Kontulan-keskusta-Arvostelupöytäkirja .pdf

Stjernberg, M. 2017. "Helsingin seudun 1960- ja 1970-lukujen lähiöiden sosioekonominen ja demografinen kehitys vuoden 1990 jälkeen." Research series 2017:1. Helsinki: City of Helsinki, Executive Office, Urban Research and Statistics.

Strandell, A. 2017. *Asukasbarometri 2016 – Kysely kaupunkimaisista asuinympäristöistä*. Suomen ympäristökeskuksen raportteja 19/2017. Helsinki: Suomen ympäristökeskus. https://helda.helsinki.fi/handle/10138/193009

Strategy for Cultural Policy 2025. 2009. Opetusministeriön julkaisuja 2009:12. Helsinki: Opetusministeriö, Kulttuuri-, liikunta- ja nuorisopolitiikan osasto.

Sulkunen, P., P. Alasuutari, R. Nätkin, and M. Kinnunen. 1985. *Lähiöravintola*. Helsinki: Otava.

Thörn, C., and H. Holgersson. 2016. "Revisiting the Urban Frontier through the Case of New Kvillebäcken, Gothenburg." *City*, 20(5): 663–84. https:// doi.org/10.1080/13604813.2016.1224479

Tuominen, P. 2017. "Kontula Electronic – lähiö muutosten keskellä." *AntroBlogi*, 18 April 2017. https://antroblogi.fi/2017/04/kontula-electronic-lahio-muutosten -keskella/

Uitermark, J., J.W. Duyvendak, and R. Kleinhans. 2007. "Gentrification as a Governmental Strategy: Social Control and Social Cohesion in Hoogvliet, Rotterdam." *Environment and Planning A: Economy and Space*, 39(1): 125–41. https://doi.org/10.1068/a39142

Vaattovaara, M., A. Joutsiniemi, M. Kortteinen, M. Stjernberg, and T. Kemppainen. 2018. "Experience of a Preventive Experiment: Spatial Social Mixing in Post-World War II Housing Estates in Helsinki, Finland." In D.B. Hess, T. Tammaru, and M. van Ham, eds., *Housing Estates in Europe: Poverty, Ethnic Segregation and Policy Challenges*. Springer Open, 215–40.

Van den Berg, M. 2013. "City Children and Gentrified Neighbourhoods: The New Generation as Urban Regeneration Strategy." *International Journal of Urban and Regional Research*, 37(2): 523–36. https://doi.org/10.1111/j.1468 -2427.2012.01172.x

Verkasalo, A. 2012. "Amatöörien elinympäristötaide lähiöissä." *Alue ja Ympäristö*, 41(1): 14–26.

Wyly, E., and J. Hammel. 2001. "Gentrification, Housing Policy, and the New Context of Urban Redevelopment." In K.F. Gotham, ed., *Critical Perspectives on Urban Redevelopment*. Vol. 6 of *Research in Urban Sociology*. Bingley: Emerald Publishing, 211–76.

Yle. 2006. Interview with Steen1 in the Program Vaunu. Yle, 27 August 2006. https://www.youtube.com/watch?v=C6B6q7pjJz4

Yle. 2020. "Helsingin Kontula ja Kontula Electronic." Yle, 26 May 2020. https://yle.fi/aihe/artikkeli/2020/01/03/helsingin-kontula-ja-kontula-electronic

11 The Fung Bros Rep the Ethnoburb

MARGARET CRAWFORD

Lemme tell you about a place out east
Just 15 minutes from the LA streets
Hollywood doesn't even know we exist
Like it's a mystical land, filled with immigrants …
Six-two-six young wild and free

– Fung Bros, *The 626*

Introduction

In the late 1980s, a new type of suburb appeared in California, the "ethnoburb," named by geographer Wei Li to classify a suburb with a large population of immigrants (Li, 1998, 2012). The first ethnoburbs were created by Asian immigrants in suburban towns in the San Gabriel Valley (SGV), east of downtown Los Angeles. They were the result of multiple factors, beginning with the 1965 Immigration Act, which opened up new categories of entry to the United States. Other factors included the lack of political and economic opportunities in Taiwan and geopolitical events such as the 1984 announcement of Hong Kong's return to China. In 1980, changes in refugee policies admitted numerous Vietnamese. After 2000, Chinese immigration regulations relaxed, encouraging the migration of growing numbers of mainland Chinese in search of business and educational opportunities. Ethnoburbs differed considerably from earlier urban ethnic enclaves such as Los Angeles's Chinatown, a tightly bounded area in the oldest part of the city. Displaced by construction of the new Union Station in 1939, a new Chinatown relocated nearby, but Chinese residents remained restricted by racist covenants. Chinatown was largely populated by Cantonese-speaking immigrants from Guangdong province, earlier arrivals at a time when further immigration was severely limited. Although widely evaded, the Chinese Exclusion Act, in effect from 1882 until 1943, prohibited all Chinese immigration. After this period, extremely limited quotas were in force until 1952. After 1965, as new groups of immigrants moved directly to suburbs, Chinatown began to shrink. Today, Vietnamese restaurants, hip art galleries, and non-Asian eateries have largely replaced traditional Chinese businesses. The ethnoburbs in this chapter were not the result of this type of segregation but were voluntarily chosen by middle-class, well-educated immigrants who could afford to purchase houses in suburbs. The unremarkable town of Monterey Park was the first ethnoburb, its transition from white and Latino to majority Chinese encouraged by Fred Hsieh, a local real estate agent, who advertised the town in Hong Kong and Taiwan as the "Chinese Beverly Hills." This title exaggerated the suburb's affluence, which was more middle than upper class, but it worked, and by 1990, Monterey Park became the first majority Asian municipality in the United States. As more immigrants arrived, they spilled over into neighbouring suburban towns such as San Gabriel, Alhambra, Rosemead, Arcadia, Temple City, and San Marino. By 2010, with twelve majority Asian-American towns, the SGV was the largest aggregation of Asians in the United States (Asian-Americans Advancing Justice, 2018).

Sociologists and geographers were fascinated by these new "suburban Chinatowns," publishing numerous scholarly studies that charted the valley's demographic transformation into "majority minority" suburbs. Their earliest studies focused on the racism that new residents encountered. Initially, long-time white residents in Monterey Park resisted the Asian presence with political initiatives to make English the official language, ban Chinese signage, and impose slow-growth restrictions clearly intended to limit Asian investment in business and real estate. But as numbers, citizenship, and voting grew, and Chinese-American politicians won city, state, and national political offices, the balance of power shifted. Once "white flight" to other suburbs diminished the backlash and lessened outright racism, the remaining residents stabilized a new multicultural balance. Roughly 60 per cent of residents were Asian, mostly Chinese with smaller numbers of Vietnamese and Filipinos, 30 per cent Latino, with very few whites and even fewer African-Americans. This new demographic mix, according to sociologist John Horton, produced a much stronger racial awareness than the colour-blind ethos that prevailed in white suburbs (Horton, 1995). American studies scholar Wendy Cheng optimistically saw this racial balance as a progressive non-white alliance between Asians and Latinos (Cheng, 2010).

Most scholars focused on political and social dimensions of this demographic inversion. While noting the transformation of the retail environment and the increasing number of Asian restaurants and shops, they devoted their attention to changes in local institutions such as municipal governments and schools, and the expansion of public events and celebrations to include the Lunar New Year along with Cinco de Mayo and the Fourth of July. No one mentioned creativity as part of the San Gabriel Valley. Yet, by 2010, the SGV was home to an emerging cultural phenomenon: Asian-American YouTube channels and personalities. Already investigating changes in the SGV's built environment, and familiar with the academic literature, I discovered this additional, innovative dimension from a Fung Bros YouTube video my daughter sent me. This video led me to pursue several lines of research. As my daughter was a college student in Los Angeles, many of her friends were from the San Gabriel Valley. Her introductions and other connections allowed me to interview twenty young people of all ethnicities. I also made a point of visiting popular shopping malls, local restaurants, tea shops, and night markets. Finally, I went online to analyse dozens of YouTube videos, interviews, and social media posts. These investigations provided the context for understanding the ways in which the Fung Bros and other YouTube personalities proposed a new form of suburban creativity, inspired by the reality of the San Gabriel Valley ethnoburb and its youthful Asian inhabitants. Remaking popular music videos and producing short dramas, they articulated and redefined Asian-American identities in opposition to existing stereotypes, taking advantage of mass-produced popular culture's ability to provide alternative communities and potential identities.

This chapter analyses videos made by the Fung Bros and other young Asian-American YouTube personalities, situating their meanings and implications in the interaction between the virtual setting of YouTube and the physical setting of the San Gabriel Valley ethnoburbs. Moving back and forth from representation to reality, I show how the Fungs repurposed existing music videos with Asian-American content, producing meanings in direct opposition to their previous content and performances. They borrowed this new content directly from the Asian-American youth subcultures fostered by the SGV suburbs, highlighting Taiwanese bubble tea and Asian food as their main symbolic representations. By making these practices and settings visible, the videos reinforced and expanded this new identity, portraying Asian-Americans as cool and fun-loving in a distinctively hyphenated way unlike young

Blacks, Whites, or Latinos. The prominence of the SGV suburbs (called "the 626" after the telephone area code) in the videos celebrated the important role that the suburban settings played in creating new Asian-American identities. The Fungs' message resonated across the 626 and beyond, recasting Asian-Americans and their suburbs as unique and desirable. To be sure, their approach had blind spots and limitations, but they made the SGV recognizable to its inhabitants, visible to the larger public, and placed it firmly on Los Angeles's map of exciting places.

YouTube as an Asian-American Cultural Space

Beginning in 2005, when YouTube appeared, young Asian-Americans quickly took advantage of the low-entry threshold to establish their own channels, sometimes while they were still in high school. By 2008, personalities such as Ryan Higa (nigahiga) and Kevin Wu (KevJumba) had millions of subscribers and in 2012 were at the top of YouTube's rankings, surpassing celebrities like Justin Bieber or Lady Gaga (Lopez, 2016: 143). Philip Wang, Wesley Chan, and Ted Fu started Wong Fu Productions in 2007 and by 2014 had 3.29 million subscribers and 500 million views. In 2011, David and Andrew Fung started their channel, attracting 2 million followers and 450 million views. Although they specialized in different formats, YouTube functioned as a showcase for what Wang called "young Asian creatives" whose output ranged from sketches, comedy routines, and music videos to short narrative films and series, utilizing singing, rapping, acting, and comedy (Kwon, 2018). They collaborated with each other in front of and behind the camera. Both Higa and Wu had roles in Wong Fu Production's first video serial. Singers Priscilla Liang and Jason Chen and comedian Richie Le play prominent roles in Fung Bros videos. Wong Fu Productions makes a point of showing their all-Asian writers, director, and crew at the end of every video. Functioning without traditional media gatekeepers, YouTube allowed young Asian-American performers to reach large audiences, mostly other young Asian-Americans.

Media critic Lori Kido Lopez points out the importance of YouTube as a venue for Asian-American identities and perspectives at a time when almost no young Asian-Americans were visible in films, television, or popular music (Lopez, 2016). Largely excluded from mainstream entertainment opportunities, their YouTube channels gave them complete creative control and the ability to become hyper-visible, even within an ethnically limited audience. Like the Fung Bros, other young Asian-Americans chose YouTube because "we realized this is the only way [to enter the media world]" (interview). Almost all of them follow the same business model, making a living from the percentage YouTube pays its top influencers, fees from live campus visits and performances, sales of their branded merchandise, and self-producing and distributing their films. Media scholars have observed YouTube's ability to act as a discursive arena, rather than simply a one-way entertainment channel (Novak and El-Burki, 2016). Many videos have a home-made quality, lessening the gap between performer and audience. KevJumba, for example, filmed many of his early videos in his bedroom or the backyard of his parents' house, inviting viewers into his personal space. User comments are the most important expression of YouTube's interactive potential. By posting responses to videos, viewers form dialogic relationships between performer and their audiences, and among themselves. The Fung Bros' or KevJumba's most popular videos can attract anywhere from two to ten thousand comments. While many are innocuous, others initiate long discussions or prompt highly divergent or even contradictory reactions, demonstrating the intensity of the viewers' engagement. The performers often take part in the conversation, increasing the interaction and fostering intimacy.

Discussions of ethnic stereotypes have an important function within this discourse (Novak and El-Burki, 2016). Unlike other minorities who suffer from negative stereotypes, Asian-Americans are typically defined by the "model minority" image, depicting them as high-achieving students who work hard to satisfy their demanding "tiger parents" and naturally good at math, science, and technology. In spite of academic and economic success, they are perceived as quiet and obedient, not rebelling or making waves. While Higa and Wu's videos often dealt with general subjects, and included white friends, they regularly identified Asian stereotypes but rarely overturned them. In KevJumba's video *I Have to Deal with Stereotypes*, he ridicules clichéd images of Asian teens but ends up by reinforcing them. He shows off his calculator to demonstrate he is not a nerd, his B+ grade to indicate he isn't a top student ("dishonouring" his family, invoking another stereotype), and ends up eating egg rolls and fried rice (KevJumba, 2007). Similarly, in *Yellow Fever*, Philip Wang explains the phenomenon of white men who only date Asian women not as a fetish but as a problem for Asian men who can't find partners. The narrative focuses on their perceived deficiencies, such as their lack of both body hair and confidence, reinforcing negative images of Asian masculinity as "weak" and "effeminate" (Wong Fu Productions, 2006).

By contrast, the Fung Bros took great pains to replace these existing stereotypes with newly positive representations of Asian-Americans.

Instead of the generic suburban environments of the earlier videos, some obviously in the SGV, the Fungs clearly identified their SGV settings, even including their names and signs. Andrew and David Fung, Chinese-American brothers, grew up in Kent, a largely white working-class Seattle suburb. The children of immigrants from Hong Kong and China, they had often visited their cousins in Monterey Park. There, they had been blown away by the vast array of available Asian food and the emerging Asian-American youth subculture. Coming from an area with few Asians, the brothers quickly noticed two things: one, Asian residents didn't appreciate their distinctive environment, and two, non-Asian people in the Los Angeles area were barely aware of the SGV. So the brothers, aspiring comedians and rappers, set out to represent both (Wei, 2012). Over the next five years, calling themselves the Fung Bros, they produced dozens of YouTube videos and made numerous personal appearances, most highlighting the suburban Asian-American experience.

Suburban Spaces with Chinese Characteristics

What the Fungs found so exceptional in the SGV was the way that Asian immigrants had embraced the form of the suburb while completely transforming its content. Although many newcomers came from dense cities like Taipei, Hong Kong, or Shanghai, they wanted to live in the suburbs. They readily adapted to Southern California's auto-oriented culture of shops and services lined up along commercial strips or in mini-malls. Like everywhere else in suburban Los Angeles, the SGV was organized along long boulevards of roadside businesses, surrounded by single family houses or small apartment buildings. Beginning in the 1980s, Chinese investors and entrepreneurs opened new restaurants, video and bookstores, acupuncture clinics, travel agencies, and Asian supermarkets. Founded by a Taiwanese immigrant, the Ranch 99 Markets became iconic – as large and well organized as their American counterparts but offering tanks of live fish, Chinese produce, and authentic prepared food (Hamilton, 1997). By 2015, in some towns, such as San Gabriel, Asian supermarkets had completely replaced local chains. Overseas and immigrant investments fuelled local commercial growth, financing everything from new developments such as San Gabriel Square, a vast stucco mall anchored by a Ranch 99 and multiple floors of restaurants, to updating former American chains with Asian imagery. Mini-malls grew to gigantic sizes with an ever-changing array of culinary options. Layered on top of the existing suburban landscape, the

SGV represented a new kind of hybridized suburban landscape with Chinese characteristics.

As new restaurants filled the SGV, it became the capital of Chinese food in North America (Wei, 2015). Although most Angelenos were not aware of the massive demographic, political, and cultural changes occurring on the city's eastern fringe, many had read about its restaurants in Jonathan Gold's columns in the *LA Weekly* and the *Los Angeles Times*. Gold, an omnivorous advocate of Los Angeles's ethnic food, caught on early to what was happening in the SGV. Beginning in 2002, he explored its boulevards and mini-malls, celebrating soup dumplings, noodle shops, and little-known regional specialties. His enthusiastic reviews and the increasing number of valley restaurants included in his yearly "101 Best Restaurants" list began to draw diners from other parts of the city, the first clue to outsiders that something important was happening there (O. Wang, 2018). The large numbers and diversity of SGV residents, from everywhere in the Chinese-speaking world, with disposable incomes and discerning tastes, supported a vast range of restaurant offerings, from modest Hong Kong–style cafés, to Asian chains like Din Tai Fung and Little Sheep Mongolian Hot Pot, to high-quality innovative spots like Chengdu Taste and Sichuan Impression; add to this list Taiwanese tea shops, restaurants specializing in pho and other Vietnamese foods, and every variety of Japanese cuisine. On Valley Boulevard alone, there are an estimated 200 restaurants, with 600 to 800 in the San Gabriel Valley (Wei, 2015). Critical evaluations by obsessed foodies on English and Chinese food sites kept competition keen, continually upgrading the quality and choice of offerings. Even that generic suburban staple, the mall, adjusted. Westfield Santa Anita remodelled its Arcadia mall, following the example of local Asian malls, by adding areas featuring high-quality Asian restaurants, most notably Din Tai Fung, the Taiwanese dumpling chain that is so popular customers wait up to two hours for a table (Li, 2017). *Food and Wine* magazine advised visitors without the time and patience to explore Valley Boulevard that they could still taste "authentic" SGV at the mall (Landsel and Wang, 2017).

Living the Bobalife in the 626

The Fungs, accomplished rappers, began their career on YouTube by posting a mixtape with their short-lived rap group, Model Minority. Although praised by critics for their ironic and politicized take on Asian-American stereotypes, they only got limited numbers of views (Wang, 2011). So they turned to another genre, parodies of popular videos, changing their content to focus on Asian-American life in the 626. They rapped over the instrumental tracks of hit songs with clever new lyrics. A mix of comedy and satire, parody videos had been popularized by "Weird Al" Yankovic. Beginning in 1976, Yankovic had produced more than 150 parodies. His specialty was humorously deflating the original song with absurd new characters and lyrics. One of his most popular efforts was *Amish Paradise* (Yankovic, 2009), in which he transformed Coolio's #1 hit *Gangsta's Paradise*,[1] a soulful rap about despair and redemption in the gang lifestyle, into an Amish farmer's lament about his chores. The Fungs went further than Yankovic, using their parodies to create a more complex relationship with the original material. In many respects, this approach resembles the practice of "détournement." Introduced by avant-garde groups such as the Lettrists and the Situationist International, this artistic technique involves "rerouting" or "hijacking" existing material for new purposes. It turns existing representations against themselves by creating tension between the old material and the new oppositional meanings. Although some might argue that there was less originality than in the Fungs' authored raps, the remakes established a dialogic relationship with popular culture and with their Asian-American audience, displaying very different authorial intentions that reflected a postmodern ethos of appropriation. By choosing highly recognizable popular songs, they made it clear that their intent was not to copy or ridicule the originals but to replace their content with new, implicitly critical meanings specific to Asian-American and SGV experiences. While far from the culturally revolutionary messages employed by the Situationists, the remakes nonetheless functioned effectively to defamiliarize the racial content of popular music videos.

The Fungs' big YouTube video hit, *Bobalife* (Fung Bros, 2013, 20 February), gave a name to the local youth culture, centred on drinking boba or bubble tea in one of the countless tea shops that covered the valley. The Fungs rap about boba while Kevin Lien, Priska, and Aileen Xu sing new lyrics to the bouncy tune of the hit song *Good Time* by Owl City and Carly Rae Jepson.[2] They equated drinking boba, "living the bobalife," with being a young Asian-American in the 626. Its all-Asian cast and specific SGV locations made an explicit contrast with the original video, which featured a largely white cast with a few African-Americans in generic urban and rural settings. *Bobalife* offered a new image of young Asians, as Kevin abandons his studies (and the stereotype of the quiet, diligent Asian) to join a group of fun-loving young Chinese-Americans having a good time socializing at a boba café. Another scene depicted young Asian women in tight halter tops walking down the sunny streets of the SGV drinking boba while Aileen sings that boba is slang for big breasts in Taiwan. The Fungs' video is far more dynamic and engaging than the mellower original. Its lively choreography and the Fungs' emphatic rapping gave it an energy and intensity missing in *Good Time*.

The Fungs interpreted life in the SGV in a very specific way, as a simultaneous fusion of multiple identities. The Fungs' rapping style clearly owed a debt to Black culture. The young people in the video are also visibly American in their clothes, speech, and behaviour (the video begins with Andrew apologizing for not speaking Chinese). But the Asian cast and many of the lyrics insisted on boba's connections with Chinese culture. The video begins with a vignette of David, dressed as an elderly Chinese immigrant, bringing his granddaughter to a boba shop to order tea (in Mandarin), followed by lines such as "It's Taiwanese"; "In Mandarin zhenzhu naicha"; or "It's Chinese culture to stay with a cup of tea, we updated it with pearls and a sealing machine." They took pains to distinguish bobalife from the hard-drinking culture of their white peers, suggesting that a tea-centred social life is superior to the alcohol-fuelled partying of white teens: "No alcohol, it's a sober night"; and "You won't get no STDs [sexually transmitted diseases] or unwanted pregnancies. When you drink that boba tea." In fact, many young Asians avoid alcohol because of the "Asian flush," a mild allergic condition that causes the face, arms, and shoulders to turn red. In interviews, the Fungs noted the lack of bars in the SGV, claiming that if people are really into food, drinking gets in the way, and "there's a lot of food nerds in the 626" (Wei, 2012). The video's identifiable setting, the Factory, a well-known SGV tea shop, the catchy tune, humorous lyrics, and energetic cast of wholesome ABCs (American-born Chinese) promoted a new and exciting image of young Asian-Americans and the ethnoburbs where they lived. Some viewers objected to these depictions, complaining: "Not only are they putting Asians back into a box, they objectify women" (quoted in Vuong, 2013); or ridiculed the video: "them: dancing and singing, me: cringing" (Cleaps, 2020). But many others immediately identified with them. Sandman (2014) commented: "Oh my god, I'm Asian, but I have to say, THIS IS THE MOST ASIAN VIDEO I'VE SEEN."

The Fungs did not see themselves as social critics but rather the opposite, celebrating an existing subculture and advocating for its visibility. David Fung pointed out: "I didn't create that lifestyle; I just gave them a push to own it" (quoted in Wei, 2012). Boba or bubble tea was already an important signifier of identity for ABCs as well as a hallmark of the SGV. The name "boba" refers to the warm black tapioca pearls added to different varieties of tea and other beverages. It was invented in Taiwan and first arrived in the United States in the SGV. By the 2000s, nearly every block had a boba shop, either independent or a branch of a Taiwanese chain. These cafés offer many choices, with constantly changing flavours of tea and toppings. Customers can personalize the temperature, sugar content, and amount of ice. Many also serve Chinese snacks or desserts. In the SGV, boba shops are important social spaces for young ABCs, who spend hours hanging out there. Students go after school to study, on dates, and to meet up on weekends. Many shops offer couches and comfortable seating, and even, as shown in the video, provide board games and decks of cards to entertain their customers (Dinh and Li, 2017). Reminiscing about growing up as a Taiwanese-American kid in the SGV, Clarissa Wei called boba shops her "sacred gathering grounds ... You could tell what kind of person you were by the type of boba shop you frequented." She and her friends cheered when *Bobalife* appeared, "perfectly capturing a cross-section of our culture that the mainstream had ignored" (quoted in Wei, 2017). She later said: "It was as if, for the first time, people were able to define what the subculture was. Because before ... no one knew how to describe what was happening" (quoted in Zhang, 2019). Phil Wang, of Wong Fu Productions, said of boba: "I was like, wow, this is something uniquely Asian-American" (quoted in Zhang, 2019). Boba is so popular in the SGV that the first Starbucks in San Gabriel didn't open until 2018 (Wong and Yee, 2018).

The Fungs also produced *The 626*, a music video featuring singer Jason Chen (Fung Bros, 2012). A showcase for the 626 food culture, it opens with a collage of local streets, Asian restaurants, and dishes: "From Shanghai to Saigon, Tokyo to Taiwan, Beijing to Hong Kong, something to bite on." Eating Hainan chicken and dim sum and drinking tea, they celebrate the authenticity of local foodways: "We're far enough from LA, nobody's acting' phony." They list the characteristics of "real" Asian restaurants: low prices, family style, multiple generations eating together, only accepting cash, and even getting a B rating from the health department. "B means better." Rapping over Snoop Dogg and Wiz Khalifa's *Young, Wild and Free*,[3] they explicitly countered the original video's images of debauched Black and white high school students, customized cars, girls in bikinis, and massive quantities of marijuana with the food-centric activities of Chinese-American teens in the 626. Instead of "So what we get drunk, so what we smoke weed,"[4] the chorus goes: "So what we hang out, so what we drink tea, we jus' eating good, in the SGV" – "where kids drink more milk tea than liquor, and they roast more duck than Swishers" (flavoured cigarillos used to smoke marijuana). The video ends with a series of shout-outs over the background of Jay-Z and Kanye West's *Ni**as in Paris*,[5] retitled by the Fung Bros in *The 626* as "Chiggas in Paris." The Fungs' list includes SGV towns, restaurants open until 3 a.m.: "So what we eat late? That's how it's 'sposed to be," other Chinese-American YouTube performers, and local high school hip-hop and breakdancing crews, highlighting this popular pastime among young Asian-Americans.

Like the Fungs, young locals embraced Asian food as an important element in their lives. Many observers attributed the continued evolution of Chinese food in the valley to their enthusiastic patronage (Wei, 2015). For example, after a group of young Asian-Americans created the first 626 Night Market, modelled after Taiwanese outdoor food venues, 15,000 people showed up on opening day, mostly locals under twenty-five (Xia, 2012). Since immigrants had settled there in search of safe, quiet neighbourhoods to raise families, the SGV had a preponderance of young people. Teens and young adults, aged fifteen to thirty-five years old, represented nearly 29 per cent of the population in the SGV, an unusually high number (Yee, 2018). An important attraction was the high-quality public school system in the area. By 2000, some school districts, such as Arcadia or San Marino, were as much as 70 per cent Asian. In most schools, Asian students dominated the college preparation and AP (advanced placement) courses, student government, and the honour roll. After graduating, so many of them enrolled at different campuses of the University of California (UC), that they too became majority Asian. Although they only make up 15 per cent of the state's population, Asian-Americans are the single largest ethnic group among UC's 173,000 undergraduates. In 2008, drawn from all over the state, they accounted for 40 per cent at UCLA and 43 per cent at UC Berkeley – the two most selective campuses in the UC system – as well as 50 per cent at UC San Diego and 61 per cent at UC Irvine (NBC News, 2009).

As they attracted more publicity and a wider audience, the Fung Bros gave the 626 and its young Asian inhabitants, for the first time, an aura of "coolness." According to sociology professor Oliver Wang, "there's now a generation of young people in the SGV who want to have a stake in a local identity. I think this generation of youth wants to be able to 'rep the hood,' as they say" (quoted in Xia, 2012). Viewers responses posted on YouTube underline this claim: Milkteaful (2015) posts: "I love it when I tell people from UCSB [University of California, Santa Barbara] that I'm originally from 626 LA and they recognize the area omg. Night Markets + Boba Paradise"; while amyzzzism (2013) comments: "You made me so proud of being an asian we've got great food!! I'm Taiwanese I Boba everyday lol." This identity was so appealing that it even resonated with non-Asians; Halima (2015) wrote: "The 626 looks awesome. Loving the chorus to this too. I wish I was Asian now :/"; and Kali-Gold (2015) posted: "I might be the last white guy in the 626 … lol … j/k. I love it here. Favorite places to eat are at San Gabriel Square and Alhambra."

These videos carved out a racially distinct niche in a popular culture where young Asian-Americans were previously invisible, located somewhere between the bland white-bread style of Carly Rae Jepson and Owl City and the transgressive obscenity-laced rap culture. The Fungs' music videos and listicles like *Things Asian Parents Do* and *18 Types of Asian Girls* assumed an inclusive pan-Asian identity contradicted by the videos themselves. Although there are important differences in ethnicity, languages, culture, politics, and immigration history among Asians, the Fungs' videos were almost exclusively populated by first-generation Chinese-Americans with an occasional Vietnamese. (Many of the Vietnamese in the SGV are Chinese-Vietnamese.) Some young Chinese-Americans found this objectionable, such as UCLA freshman Calvin Lam: "Their checklist way of promoting Asians reduces the rich diversity of Asians to the few, numbered characteristics they describe. Why call the videos 'Things Asian Guys Like' if it only applies to the Asian guys they observe in their own life?" (quoted in Vuong, 2013). Even their depiction of Chinese-Americans only applies to those whose parents are immigrants. According to David, "we're immigrant children. Our parents come from Asia, and we're all trying to find our place in Western society" (quoted in Vuong, 2013). One group not included was recent immigrants from mainland China, often called FOBs (fresh off the boat), who had begun to populate the valley as immigrants and international students. The video *Don't Hate FOBs* (Fung Bros, 2011) reminded viewers that everyone's parents are FOBs but did little to alter the image of FOBs as either ping-pong playing nerds or spoiled rich kids, both distinctively foreign. Yet, for many first-generation Chinese-Americans, these videos captured the hyphenated specificity of their identities. Aileen Yu, a young woman who grew up in the SGV, commented: "You don't fit anywhere, so you create something new … A lot of us don't necessarily connect to our homeland. We're not from China. We speak English" (quoted in Xia, 2012).

Moving beyond youth culture, the Fungs' appreciation for the richness and complexity of the SGV's suburban landscape reached a peak in *Garvey, Valley, Main, Huntington*, with singer Priscilla Liang (Fung Bros, 2013, 21 March). Rapping over Macklemore's *Thrift Shop*,[6] the video maps the SGV's diverse social and spatial geography. Starting at a Ranch 99, the Fungs and Priscilla travel the long east-west boulevards that cut through the valley, lined with the Southern California suburban staples of stucco storefronts, strip malls, mini-malls, and lots of parking lots. They interpret this landscape as a framework that contains intricate economic, social, and cultural meanings. On Garvey, they note the changing demographics: "You can't find this mix anywhere on the planet, Canto's the language, a little Viet and Spanish. But almost everyone speaks Mandarin. Cuz' all the mainland Chinese are moving in"; class differences: "They say it's kinda' ghetto, I call it working class"; and the array of small mom and pop restaurants: "From chiu chow noodles to the pho in El Monte. You know the waitress'll remind you of ya auntie!" Standing on Valley Boulevard (according to them, the most exciting place in the 626), they enumerate the apparently infinite number of Asian restaurants. Alhambra's Main Street is more "multi-grain," prompting the ironic comment "Cuz' this is America, sometimes I need a burger and sometimes I want to speak English to my server." The video ends on upscale Huntington Drive in San Marino, the wealthiest community in Southern California. Holding a baguette and a bottle of wine, Priscilla predicts that these symbols of whiteness will soon disappear as the town's commercial landscape becomes as Asian as its residents. Widely viewed, these videos added the 626 to the short list of Los Angeles suburbs in the city's popular imagination, along with the San Fernando Valley, bastion of white teenagers, and Compton, birthplace of gangster rap.

As the Fungs attracted attention from all over the region, their local audience expanded to include officials and business leaders. Although they appreciated the business the new image brought to the SGV, they also saw the Fungs' creativity as an important antidote to stereotypical representations of Asians. "I think they're creative, they're on to something," asserted Monterey Park Mayor Mitchel Ing. "I'm a fan." Ing started showing the Fungs' YouTube videos at city department meetings. "What the Fung brothers have done has surpassed what the Chamber of Commerce and what we as the city have done in promoting ... the businesses in the community" (quoted in Xia, 2012). "The Fung Bros took a creative approach to promoting the region's business, culture and diversity while increasing a sense of pride for the SGV, or should I say, 'the 626' ... It exposes the populace to the way Asian-Americans think and also shows the creative side of us," Ing said. "We're not just the math nerds and the science nerds. It exposes mainstream (society) to the fact that Asian-Americans are creative" (quoted in Vuong, 2013). In 2013, the Fungs won the San Gabriel Valley Economic Partnership's

Innovative Marketing Award, "honouring outstanding people and businesses in the San Gabriel Valley who are making a significant contribution" (Smith, 2013).

Conclusion: Beyond the Asian Bubble

In 2014, the Fungs released a short film *The Asian Bubble*, acknowledging problems with the Asian-American identity they had done so much to define (Fung Bros, 2017). By this time, the Asian population dominated the SGV numerically and culturally. New versions of "white flight" emerged there and in other ethnoburbs as white families left since their children could not compete academically with Asians (Lung-Amam, 2017). Instead of finding common ground with Latinos, as Wendy Cheng hoped, it appeared that many Chinese had simply replaced whites at the top of the racial hierarchy, relegating other minorities, including Filipinos, Southeast Asians, and Latinos, to lower positions. Their demographic superiority and pervasive cultural, commercial, and culinary presence allowed them to minimize, as whites had done in the past, their interaction with other races and ethnicities. The Fungs named this situation the "Asian bubble." Their short film begins with Andrew being rejected by his crush for staying in "the bubble" that is the SGV (Fung Bros, 2017). Growing up there, it was possible to live in a nearly total Asian world, belonging to an Asian church, having all-Asian friends in high school and college, joining Asian fraternities or sororities and dance crews. Challenged to move out of his comfort zone, Andrew and his friends leave the SGV to visit Venice Beach, Hollywood, and downtown Los Angeles, trying to interact with people of different races and cultures. Although they are only partly successful, the film ends with him concluding that life outside the bubble is "more interesting." The Asian bubble struck a nerve among young ABCs. Several of my interviewees made a point of explicitly saying that they were not in the Asian bubble, adding remarks such as "I went to Community College, so I know a lot of Latinos" or "I chose to go to a White college" (that is, not a UC campus).

Making this point even clearer, Wong Fu Productions' short films and series present the next iteration of Asian-American identity, in many ways a corrective to the some of the Fungs' blind spots. Unlike the Fungs' sometimes sexist "bro" personas, Wong Fu's funny but thoughtful, character-based narratives expanded the SGV Asian-American world to include characters of other races and a more nuanced view of Asian-American identities. *In-Between* (Wong Fu Productions, 2017, 7 December), for example, made by Chinese-Americans whose parents are not immigrants but multigenerational Americans, addresses the assimilation issues of a group "not Asian enough for Asians but too Asian for Americans." Even humorous parodies like *Asian Bachelorette* (Wong Fu Productions, 2017, 2 August) carry messages by emphasizing the wide variety of backgrounds and ethnicities within the racial category of Asian. The 2018 series *Yappie* (young Asian professionals), subtitled *(Re)Model Minority*, tackled a range of societal issues faced by valley ABCs as they grew up and moved out of the bubble (P. Wang, 2018). Episodes follow the everyday lives of Andrew (Philip Wang) and his friend group (Asians and one Caucasian). Andrew is an ABC everyman, a naive software engineer who grew up in the SGV and is educated by the informed and "woke" women in his life. The series tracks his evolution as he is confronted with more complex issues facing Asians, such as the histories of discrimination against different groups of Asians, the difficulties of interracial dating (his girlfriend is "blasian," Japanese/African-American), and comparative racism (Asian-American versus African-American; Pan, 2019). Moving beyond the Asian bubble, Wong Fu Productions' features construct self-aware and intersectional Asian-American identities (Lee, 2018).

After fifteen years, there are signs that Asian-American suburban creativity might be locked into its own bubble. Young suburban Asian-Americans have become slightly more visible in movies and TV shows such as *Fresh off the Boat* or *Nora from Queens*, with its popular star Awkwafina. But none of the early YouTube personalities have crossed over into mainstream media. "YouTube is a place where Asian Americans have done well and have established audiences, but for some reason that hasn't given Hollywood enough evidence or proof that they can go ahead and start a whole series with Wong Fu or Ryan Higa," said Nancy Wang Yuen (quoted in Lee, 2018), a Biola University professor who examined Asian-American representation on television (Yuen et al., 2017). There is a large audience for Asian-American content as widely popular global Facebook groups such as "Subtle Asian Traits" demonstrate. Asian-American platforms like Wong Fu Productions still attract followers, but as YouTube expands, having millions of subscribers is no longer rare. The Fungs themselves stepped out of the bubble in 2015, relocating first to New York and then back to Seattle, moving into food criticism and sports commentary. Their move can be interpreted as acknowledging that they had exhausted the creative opportunities available in the 626. During their five years in the SGV, they had pushed the image of young suburban Asian-Americans to its limits. Unlike earlier rebellious youth cultures, such as the British punks that Dick Hebdige analysed in his classic study of subcultures (Hebdige, 1979), the Fungs did not want to create an oppositional identity but only a more exciting and visible Asian-American image. Rather than employing the complex assemblages of commodities (clothing, hairstyles, vehicles, and music) that Hebdige identified as subcultural statements, they accomplished their goal with limited means: appropriated music videos, boba tea, and Asian food. In contrast to Black and Latino subcultures, where ethnicity and class intersected to create subaltern subject positions, the Fungs were aware that Chinese-Americans rank at the top of the educational and professional ladder, even if their salaries and positions are not always commensurate with those of their white colleagues. Nonetheless, the Fungs' achievement cannot simply be dismissed as merely celebrating a superficial consumer lifestyle. They successfully employed their creativity to distinguish young Asian-Americans from their peers and to reposition the middle-class suburb as a racialized terrain fostering more than just white cultural identities. They revealed that low-density dispersed suburban environments of commercial strips and shopping malls could be dynamic, culturally complex environments and exciting places to live. Finally, as the United States moves towards a future where minorities will be the majority, they offered an example of how to effectively fashion and communicate their unique identities to themselves and others.

NOTES

1 For Coolio's *Gangsta's Paradise*, see Coolio (1995).
2 For *Good Time*, see Owl City and Jepson (2012).
3 For *Young, Wild and Free*, see Snoop Dogg and Khalifa (2011).
4 Ironically, Snoop Dogg lives in Diamond Bar, a town in the eastern San Gabriel Valley.
5 For *Ni**as in Paris (Explicit)*, see Jay-Z and West (2012).
6 For *Thrift Shop*, see Macklemore and Lewis (2012).

REFERENCES

Amyzzzism. 2013. "You made me so proud of being an asian we've got great food!! I'm Taiwanese I Boba everyday lol." [Comment on the Fung Bros video *The 626*]. YouTube. https://www.youtube.com/watch?v=3n3HQ9uge 0g&lc=UgwBiI909GUpmcxuqRt4AaABAg

Asian Americans Advancing Justice. 2018. *A Community of Contrasts, Asian Americans, Native Hawaiians and Pacific Islanders in the San Gabriel Valley.*

https://archive.advancingjustice-la.org/sites/default/files/A_Community_of_Contrasts_SGV_2018.pdf

Cheng, W. 2010. "'Diversity' on Main Street? Branding Race and Place in the New 'Majority-Minority' Suburbs." *Identities: Global Studies in Power and Culture*, 17(5): 458–96. https://doi.org/10.1080/1070289X.2010.526880

Cleaps. 2020. "them: dancing and singing; me: cringing." [Comment on the Fung Bros video *Bobalife*]. YouTube. https://www.youtube.com/watch?v=zccNQPH7Xe0&lc=Ugxfr0w2iqeOw5XgmLt4AaABAg

Coolio. 1995. *Gangsta's Paradise*. Directed by Antoine Fuqua. Propaganda Films. https://imvdb.com/video/coolio/gangstas-paradise

Dinh, S., and A. Li. 2017. "Living in a Boba Bubble." *The Matador*, 26 April 2017. http://thematadorsghs.us/index.php/2017/05/26/living-in-a-boba-bubble/

Fung Bros. 2011, June 16. *Don't Hate FOBs*. Fung brothers. Official music video. YouTube. https://www.youtube.com/watch?v=XSmn4crVI2A

Fung Bros. 2012, 19 February. *The 626*. Fung brothers, featuring Jason Chen. Official music video. YouTube. https://www.youtube.com/watch?v=3n3HQ9uge0g

Fung Bros. 2013, 20 February. *Bobalife*. Fung brothers, featuring Kevin Lien, Priska, Aileen Xu. Music video. YouTube. https://www.youtube.com/watch?v=zccNQPH7Xe0

Fung Bros. 2013, 21 March. *Garvey, Valley, Main, Huntington*. Fung brothers, featuring Priscilla Liang. YouTube. Music video. https://www.youtube.com/watch?v=kzzUaHim1Vg

Fung Bros. 2017. *The Asian Bubble*. Directed by Jon Maxwell. Film. YouTube. https://www.youtube.com/watch?v=jn1gRSOcN7A

Halima. 2015. "The 626 looks awesome. Loving the chorus to this too. I wish I was Asian now :/" [Comment on the Fung Bros video *The 626*]. YouTube. https://www.youtube.com/watch?v=3n3HQ9uge0g&lc=UghSTTj9Gf-ms3gCoAEC

Hamilton, D. 1997. "99 and Counting." *Los Angeles Times*, 27 April 1997. https://www.latimes.com/archives/la-xpm-1997-04-27-fi-52939-story.html

Hebdige, D. 1979. *Subculture: The Meaning of Style*. London: Routledge.

Horton, J. 1995. *The Politics of Diversity: Immigration, Resistance and Change in Monterey Park, California*. Philadelphia: Temple University Press.

Jay-Z and Kanye West. 2012. *Ni**as In Paris (Explicit)*. Music video. YouTube. https://www.youtube.com/watch?v=gG_dA32oH44

Kali-Gold. 2015. "I might be the last white guy in the 626 ... lol ... j/k. I love it here. Favorite places to eat are at San Gabriel Square and Alhambra." [Comment on the Fung Bros video *The 626*]. YouTube. https://www.youtube.com/watch?v=3n3HQ9uge0g&lc=UghmWGtte5KeX3gCoAEC

KevJumba. 2007. *I Have to Deal with Stereotypes*. YouTube. https://www.youtube.com/watch?v=nbZ9zJ22WfQ

Kwon, J. 2018. "Navigating the Lives of Young Asian Professionals: A Sit-Down Interview with Wong Fu Productions' Phil Wang." *Infusian News*, 23 September 2018. http://vaff.org/wong-fu-interview/

Landsel, D., and A. Wang. 2017. "The San Gabriel Valley for Dummies" *Food and Wine*, 22 September 2017. https://www.foodandwine.com/travel/san-gabriel-valley-dummies

Lee, T.G. 2018. "After a Decade on YouTube, Wong Fu Productions Still Has a Story to Tell." *NBC News*, 19 June 2018. https://www.nbcnews.com/news/asian-america/after-decade-youtube-wong-fu-productions-still-has-story-tell-n881606

Li, S. 2017. "As Other Malls Die Off, This One in Arcadia Focuses on Asian Shoppers." *Los Angeles Times*, 2 April 2017. https://www.latimes.com/business/la-fi-santa-anita-westfield-mall-20170331-story.html

Li, W. 1998. "Anatomy of a New Ethnic Settlement: The Chinese Ethnoburb in Los Angeles." *Urban Studies*, 35(3): 479–501. https://doi.org/10.1080/0042098984871

Li, W. 2012. *Ethnoburb: The New Ethnic Community in Urban America*. Honolulu: University of Hawaii Press.

Lopez, L. 2016. *Asian American Media Activism: Fighting for Cultural Citizenship*. New York: NYU Press.

Lung-Amam, W. 2017. *Trespassers: Asian-Americans and the Battle for Suburbia*. Berkeley: University of California Press.

Macklemore and Ryan Lewis. 2012. *Thrift Shop*. Featuring Wanz. Music video. YouTube. https://www.youtube.com/watch?v=QK8mJJJvaes

Milkteaful. 2015. "I love it when I tell people from UCSB that I'm originally from 626 LA and they recognize the area omg. Night Markets + Boba Paradise." [Comment on the Fung Bros video *The 626*]. YouTube. https://www.youtube.com/watch?v=3n3HQ9uge0g&lc=Ughf7VMn7_ucdngCoAEC

NBC News. 2009. "Asian-Americans blast UC Admissions Policy." NBC News, 24 April 2009. https://www.nbcnews.com/id/wbna30393117

Novak, A., and I.J. El-Burki. 2016. *Defining Identity and the Changing Scope of Culture in the Digital Age*. Hershey, PA: IGI Global, 150–60.

Owl City and Carly Rae Jepsen. 2012. *Good Time*. Official music video. YouTube. https://www.youtube.com/watch?v=H7HmzwI67ec

Pan, Y. 2019. "Phillip Wang Discussed 'Yappie' after CWRU Showing of Limited Series." *The Observer* (Case Western Reserve University), 1 February 2019. https://observer.case.edu/phillip-wang-discusses-yappie-after-cwru-showing-of-limited-series/#photo

Sandman. 2016. "Oh my god, I'm Asian, but I have to say, THIS IS THE MOST ASIAN VIDEO I'VE SEEN." [Comment on the Fung Bros video *Bobalife*]. https://www.youtube.com/watch?v=zccNQPH7Xe0&lc=Ugj7YsCtSHMhWHgCoAEC

Smith. K. 2013. "San Gabriel Valley Gala Honors Business, Community Leaders." *San Gabriel Valley Tribune*, 7 September 2013. https://www.sgvtribune.com/2013/09/07/san-gabriel-valley-gala-honors-business-community-leaders/

Snoop Dogg and Wiz Khalifa. 2011. *Young, Wild and Free*. Featuring Bruno Mars. Official video. YouTube. https://www.youtube.com/watch?v=Wa5B22KAkEk

Vuong, Zen. 2013. "The Fung Brothers Explain Asian Life in the 626." *Pasadena Star-News*, 21 September 2013. https://www.pasadenastarnews.com/2013/09/21/the-fung-brothers-explain-asian-life-in-the-626/

Wang, O. 2011. "Model Minority: Three Chinese Americans Shuffle between Racially Colored Humor and Politics." *Los Angeles Times*, 22 March 2011. https://www.latimes.com/archives/blogs/pop-hiss/story/2011-03-22/model-minority-three-chinese-americans-shuttle-between-racially-colored-humor-and-politics

Wang, O. 2018. "The Geography of Gold." *BOOM California*, 24 July 2018. https://boomcalifornia.org/2018/07/24/the-geography-of-gold/

Wang, P. 2018. "Wong Fu: No We Are Not Sell-Outs – Why We Created 'Yappie.'" *NextShark*, 13 June 2018. https://nextshark.com/wong-fu-no-not-sell-outs-self-hating-asians-created-yappie

Wei, C. 2012. "Q&A with the Fung Brothers: Food Nerds, Asian Vegetable Superiority and the Lack of Drunk People in the 626." *LA Weekly*, 2 April 2012. https://www.laweekly.com/q-a-with-the-fung-brothers-food-nerds-asian-vegetable-superiority-and-the-lack-of-drunk-people-in-the-626/

Wei, C. 2015. "How L.A. Became a Powerhouse for Chinese Food." *First We Feast*, 20 January 2015. https://firstwefeast.com/eat/2015/01/how-l-a-became-a-powerhouse-for-chinese-food

Wei, C. 2017. "How Boba Became an Integral Part of Asian-American Culture in Los Angeles." *LA Weekly*, 16 January 2017. https://www.laweekly.com/how-boba-became-an-integral-part-of-asian-american-culture-in-los-angeles/

Wong, K., and C. Yee. 2018. "San Gabriel's Very First Starbucks Is Now Open." *San Gabriel Valley Tribune*, 28 June 2018. https://www.sgvtribune.com/2018/06/28/san-gabriels-very-first-starbucks-is-now-open-could-it-mean-more-national-chains-are-coming/

Wong Fu Productions. 2006. *Yellow Fever*. YouTube. https://www.youtube.com/watch?v=P2ojpefxk6o

Wong Fu Productions. 2017, 2 August. *Asian Bachelorette*. YouTube. https://www.youtube.com/watch?v=ag1IisyP1ak

Wong Fu Productions. 2017, 7 December. *In Between*. YouTube. https://www.youtube.com/watch?v=wflQGe3Bzi0&t=9s

Xia, R. 2012. "Asian American Youth Culture Is Coming of Age in 'the 626.'" *Los Angeles Times*, 27 August 2012. https://www.latimes.com/local/lanow/la-xpm-2012-aug-27-la-me-valley-asians-20120827-story.html

Yankovic, "Weird" A. 2009. *Amish Paradise*. Official Parody of *Gangsta's Paradise*. YouTube. https://www.youtube.com/watch?v=lOfZLb33uCg

Yee, C. 2018. "More Asian-Americans Live in San Gabriel Valley than in 42 States, Report Says." *Pasadena Star-Ledger*, 22 February 2018. https://www.pasadenastarnews.com/2018/02/22/report-san-gabriel-valley-asian-american-population-larger-than-those-in-42-states/

Yuen, N.W., C.B. Chin, M.E. Deo, F.M. DuCros, J.J.-H. Lee, and N. Milman. 2017. *Tokens on the Small Screen: Asian Americans and Pacific Islanders in Prime Time*. https://www.nancywyuen.com/aapis-on-tv.html

Zhang, J.G. 2019. "The Rise (and Stall) of the Boba Generation." *Eater*, 5 November 2019. https://www.eater.com/2019/11/5/20942192/bubble-tea-boba-asian-american-diaspora

12 Grounding Suburban LGBTQ+ Vernacular Creativities in the Toronto City-Region

ALISON L. BAIN

Introduction

In Canada and many other countries, within a few generations, public discourse has changed around sexual and gender minorities from exclusion to increased socio-political protection and equality under the law. For LGBTQ+ (lesbian, gay, bisexual, trans, and queer) Canadians, these inclusions consist of the decriminalization of homosexual acts (1969), protections against discrimination based on sexual orientation (1996) and gender expression (2017), as well as same-sex benefits and pensions (1999), marriage (2005), adoption rights (1995), and household tabulations in the national census (2001). Concomitantly, the centres of Canada's largest cities have acted as sanctuaries for many LGBTQ+ people, hosting Pride celebrations and marking territorial and social service gains in the formation of residential neighbourhoods, commercial districts, and community organizations. While such accomplishments are significant, social change is seldom a linear narrative of progress. In Canada, the experiential benefits of such legal and material gains are socially and geographically unevenly distributed; they predominantly benefit white, male, cis, middle-class, and able-bodied gays and have a spatial bias towards the centres of large city-regions.

This chapter reorients the scholarly investigative gaze away from central cities. It focuses instead on the contemporary production of queer social spaces in the informal "underground" and "middle ground" of suburbs, paying particular attention to micro-creative community practices of LGBTQ+ organizational collaboration. The significance of queer suburban spaces and creative practices has been largely overlooked within disparate metronormative creative city and geographies of sexualities literatures that have each respectively privileged the creativity and queerness of inner-city districts and neighbourhoods. In Richard

Florida's (2005) theorizations of a creative capital model of urban development, for example, he develops a controversial "Gay Index," an instrument of measure that correlates tolerance and acceptance of the LGBTQ+ community through the spatial quantification of resident gays and lesbians (a selection of those deemed to represent "alternate" and "other" sexual preference while negating the spectrum of queer identities) with economic prosperity. As Lang and Danielsen (2005: 204) note, "it is not that members of the creative class are gay; rather gays are symbolic – like canaries in a coal mine – that the atmosphere is healthy for alternative lifestyles." The consumptive dimensions of homosexuality and its tethering to culturally progressive politics, then, are rendered emblematic of contemporary urban – and not suburban – prosperity and diversity. But even then, it is a misapplied proxy of diversity, signalling the suitability of downtown neighbourhoods for a privileged selection of wealthy, educated males to enjoy. Given that both creativity and queerness in cities are understood by scholars to be inherently spatial and relational (Florida, 2005; Podmore and Bain, 2020), this chapter prioritizes an examination of LGBTQ+ relational interconnectivities – the complex connections, interactions, and entanglements of individuals and organizations within specific societal, cultural, and political contexts – on the peripheries of the Toronto city-region.

Toronto, Montreal, and Vancouver are Canada's three largest census metropolitan areas (CMAs). These city-regions are the case studies of a larger research program that this chapter contributes to, investigating LGBTQ+ suburban lives in ten peripheral municipalities through media representations, GIS census mapping, policy analysis, interviews, and focus groups (see, for example, Bain and Podmore, 2020a, 2020b, 2020c, 2021). The empirics of this chapter come from 65 semi-structured informational interviews undertaken by the author in peripheral municipalities in the Toronto CMA in 2018 – Mississauga (21), Brampton (11), Markham (21), and Ajax (12) – in their corresponding regional municipalities (Peel, York, and Durham) with LGBTQ+ community service providers and civic leaders.[1] With a population of approximately 6.4 million people in 2016, the Greater Toronto Area (GTA) is organized into four regional municipalities (Halton, Peel, York, and Durham) and twenty-five municipalities under the authority of the provincial government. Over the last four decades, Toronto has intensively suburbanized – non-central population and economic growth in concert with urban spatial expansion (Ekers et al., 2012) – driven by rising incomes, high levels of immigration, innovations in transportation and communication technologies, and the decentralization of employment (Bain, 2019). Economic and population diversity along with "the immigrant experience" now characterize much of suburban Toronto and Canada (Keil and Addie, 2015: 895). Such social diversity is attributed to reductions in housing discrimination, the socio-spatial assimilation of visible minority groups, the centrifugal settlement patterns of immigrants, and the construction of demographically targeted and financially accessible housing developments (Hall and Lee, 2010; Keil, 2015). To date, the dimensions of social diversity that Canadian suburban studies researchers have prioritized are gender, ethnicity, racialization, religion, and socio-economic status to the neglect of sexuality, yet same-sex households can be found throughout the Toronto city-region and not just in the City of Toronto's downtown.

While suburbs were conceived of and built as bastions of heteronormative reproductive futurity, they are not just sexually homogenous places of heteronormative homeownership, atomized nuclear family life, and patriarchal gender relations (Tongson, 2011). Following Smith (2004) and Hubbard (2008), heteronormativity is a set of organizational structures (social institutions and policies) and everyday practices that present heterosexuality and the gender binary as hegemonic norms. Certainly, heterosexual sex – for the purposes of reproduction within marriages and acquiring mortgages – drove the North American post-war demand for suburban housing, but suburbs have also always been places in which non-normative sexualities are constituted and where sexual and gender minorities live (Cooper, 2009). Yet suburbs tend not to register in the scholarly and public imaginations as queer places.

This chapter unpacks some of the queerness of suburbs by examining the production of a selection of LGBTQ+ suburban social spaces in the Toronto city-region. It demonstrates that, even though LGBTQ+ social spaces may appear ephemeral and invisible within extensive suburban geographies, they are not afterthoughts but rather the products of strategic collaborations and insider activism. Following Browne and Bakshi (2013), insider activists are understood to be LGBTQ+ people who use their employment to further an LGBTQ+ equalities agenda. This chapter argues that the queer underground and intermediary middle ground of suburbs is where LGBTQ+ adult suburbanites and their allies – as individual activists and within organizations – undertake significant collaborative creative work to produce safe(r) and supportive social spaces for sexual and gender minorities, especially youth, to exchange information and tacit knowledge, and to build interpersonal relationships. An appreciation of such LGBTQ+ suburban collaborative practices is essential to any understanding of the cultural dynamics of the peripheries of city-regions and the vernacular creativities they shelter.

The chapter begins by theorizing vernacular creativity and its relationship to physical peripherality and social marginality. It then discusses scholarly conceptualizations of the structural socio-spatialities of creative cities, highlighting how the relational dynamics of the underground, middle ground, and upper ground within local innovative milieus inform an understanding of suburban LGBTQ+ vernacular creativities. The ensuing empirical analysis of a sample of LGBTQ+ suburban social spaces from across the Toronto city-region is contrasted with Toronto's downtown gay village in order to accentuate particular suburban challenges, namely the management of spatial distance, the negotiation of visibility, and the sharing of power, which together illustrate the collaborative dynamics in the suburban underground, middle ground, and upper ground. Through this empirical discussion, the chapter raises fundamental questions about the longevity and sustainability of suburban LGBTQ+ vernacular creative initiatives into the future.

Theorizing Peripheral Vernacular Creativities

Creativity drives knowledge-based societies and economies, producing innovative, "novel and valuable contributions[s] to a particular domain" (Hautala and Ibert, 2018: 1688). Within the social sciences and humanities, creativity is understood as "a collective process rather than an output" and therefore is "inherently social and interactive" (1689). Creativity is also appreciated as the by-product of marginality. For Hautala and Ibert, it is "the presence of challenging problems and irritating worldviews that provide opportunities for creative responses" (1691). Although their observation relates to the work of scientists and artists whose involvement in "radically new ventures" often generate "experience[s] of exclusion and denial from the community of peers and contemporaries" (1693), it nevertheless has provocative implications for historically marginalized social groups. It is useful to consider the lived experiences of marginality for LGBTQ+ suburbanites, how they manifest in particular "challenging problems," and what complex, everyday, vernacular "creative responses" and intricate (inter)actions have been undertaken to address these problems by individuals and organizations. On a daily basis, LGBTQ+ suburbanites grapple with intersecting discriminations, exclusions, and innocuous erasures, consciously challenging the heteronormativity of suburbia by improvising inclusive spaces to safely be themselves and to socialize.

When creativity is appreciated as "practices that produce change," it is not limited to particular occupations or industries (Ettlinger, 2010: 43) but applies to many different social groups and is enabled through access to and mobilization of resources and networks within particular socio-spatial contexts. It is cities, however, with their positive externalities that are publicly heralded as loci of creativity (Arvidsson, 2007). Suburbs are the supposed "Other" to this characterization, possessing many of the deficits of peripheries – lack of resources, services, and spaces for informal social interaction. However, over the last decade, a growing body of scholarship on suburban cultural economies has helped reverse prevailing understandings of peripheries as hindrances to creativity by emphasizing how features of affordability, diminished power and control, and less mainstream influence can be leveraged (Gibson et al., 2010). Recent research by Grabher (2018: 1786), for example, determines that "the absence of local ... buzz emanating from high levels of local interactivity" on peripheries can be strategically mitigated by individuals and community groups through mobility and collaboration. Such a finding, while derived from an appreciation of creativity as the novelty and valuation of products from within socially constructed evaluative regimes, holds analytical promise for understanding the vernacular creative practices of LGBTQ+ suburbanites. Nevertheless, it also requires acknowledgment of the constraints wrought by immobility – many LGBTQ+ suburbanites, particularly youth, do not own cars and rely on infrequent, poorly synchronized, and expensive public transit services to navigate significant distances.

There is a growing body of research on alternative forms of hidden and ignored everyday creativities – what is often referred to as vernacular creativity – and their role in the socio-cultural life of places that is instructive here. The "everyday" refers to a multitude of micro-spaces, daily practices, and mundane activities of ordinary people who may not occupy positions of power (Ettlinger, 2010). To theorize creativity from a vernacular perspective necessitates the employment of different vocabularies and concepts; "it also means being attuned to the uncertainties, multiplicities and constant forms of becoming which define all urban spaces" (Mosselson, 2017: 1292).

Vernacular creativity depends upon "oral forms of knowledge production and stems from personal passions, reliant on generosity and sharing, as well as the quest to perform idiosyncratic elements of personal identity" (Warren and Gibson, 2011: 2707). While identification as LGBTQ+ is not idiosyncratic, the production of queer social spaces is associated with "informal social networks" and queer-identified bodies (Rantisi and Leslie, 2009). Much like the working-class custom-car designers whom Warren and Gibson (2011) document, LGBTQ+ suburbanites are creative in how they are adaptive, resourceful, collaborative, and social; these articulations of creativity are born, in part, out of necessity because of the lack of formal civic policy and financial and program support. It is a creativity derived from personal experience, refined through activity, and accessed through informal social interactions. Such LGBTQ+ vernacular creativities have an initial underground quality that sometimes percolates upwards, sometimes gaining both visibility and organizational support.

Grounding Suburban LGBTQ+ Vernacular Creativities

The "essence" of any creative city for Cohendet et al. (2010: 92) is found in the "middle ground" – the intermediate layer between the informal underground (individuals) and the formal upper ground (institutions) within local innovative milieus. Constantly navigating between informal and formal worlds, they frame the middle ground as "the main repositories of innovative micro-ideas" (97). By tapping "into the fertile soil of the creative city" and the lived experiences of individuals, these authors assert that the middle ground codifies knowledge from the underground and makes it externally legible and useful to macro-level institutions and market forces (96). In these ways, the middle ground becomes "indispensable loci where spontaneity is progressively structured" (97).

When applied to the development of artistic spaces in Hong Kong, Charrieras et al. (2018: 135) describe a middle ground that is heterogeneous, relational, and constantly shifting – "composed of different actors with different agendas and goals; it is a place of contestation and negotiation" between the upper ground and underground. They draw upon Arvidsson's (2007: 14) articulation of an underground as a "subcultural scene" where "autonomous processes of cultural production" unfold beyond standardizing logics of capitalist urban neoliberalism. At yet a lower level, it is the "closed and exclusive environments" of the "deep underground" (17), with its niche audience appeal, anti-business sentiments, free labour, and cutting-edge authenticity, that can trickle-feed inspiration to actors higher up the cultural production hierarchy. Crucially, Arvidsson attributes "network entrepreneurs" in their position at the top of the hierarchy of the underground with the endowments of "contacts, subcultural capital, respect and the general biopolitical capacity that enables them to recruit and mobilize desired forms of life" (20). Cohendet et al.'s (2010) and Charrieras et al.'s (2018) tripartite theorization of the structural socio-spatialities of creative cities in conjunction with Arvidsson's (2007) notion of the "deep underground" are particularly useful for understanding suburban LGBTQ+ vernacular creativities and their challenges.

The production of queer suburban social spaces in the Toronto city-region faces three key challenges. First, the sheer spatial extent of suburban geographies separates and isolates LGBTQ+ residents from each other and appears to dilute, rather than concentrate, their presence. Yet proximity and frequent interaction are deemed necessary conditions to embed an underground cultural scene in a particular milieu and foster its expansion (Cohendet et al., 2010). This chapter then, considers how spatial distance is managed by suburban LGBTQ+ residents in the present and with what personal tolls for individual insider activists, raising fundamental questions about the longevity and sustainability of vernacular creative initiatives.

Second, LGBTQ+ visibility politics is contentious and contradictory, particularly for intersections of immigrant, racialized, youth, and older adult populations. Visibility is a key concept in liberal gay identity politics; a tactic and a goal, the means and the end of gay activism, visibility works within and against dominant cultural formations and often involves (un)mediated public (re)presentation to achieve emancipation and recognition (Ritchie, 2010). For Brouwer (1998: 118), visibility politics is the "theory and practice which assumes that 'being seen' and 'being heard' are beneficial and often crucial for individuals or a group to gain greater social, political, cultural or economic legitimacy, power authority, or access to resources." Visibility politics problematically assumes that to be visible – whether in the media, the built environment, or civic discussions – is inherently liberating, but that is not the case for some racialized queer and transgender people because it invites heightened surveillance and enforced hypervisibility (Brouwer, 1998). Thus, this chapter reflects upon how visibility is negotiated in suburbs by LGBTQ+ individuals and organizations.

Third, the upper ground – namely the institutional governance regime of a municipality – is predominantly engaged in assimilationist and tokenistic forms of social inclusion[2] that "names" different social groups to show "diversity" but rarely includes the voices of the LGBTQ+ community in policy development and implementation or questions how interlocking sources of power create different lived experiences. Thus, in shifting suburban landscapes of emerging and disappearing queer grassroots initiatives, politicians and their staff in peripheral municipalities have grown complacently reliant upon the informal "creative" work of LGBTQ+ individuals and communities. While municipal political leaders may individually demonstrate support for Pride flag-raisings and celebrations once a year in June or July, beyond construction of some gender-inclusive washrooms in civic facilities and the strategic placement of rainbow crosswalks, there are few examples of formal municipal planning, policy, funding, and human resource investments that serve sexual and gender minority suburban publics. This chapter attends to how, if at all, power is shared in the suburban upper ground. In the ensuing discussion, these three main challenges are interwoven into the discussion of three examples in order to illustrate the collaborative dynamics of vernacular creativities in the suburban underground, middle ground, and upper ground that have contributed to different LGBTQ+ suburban social spaces.

Collaborating to Produce LGBTQ+ Suburban Social Spaces

From almost imperceptible beginnings in the 1950s, through the 1970s, Toronto's LGBTQ+ population gradually received increasing social tolerance and acceptance, with the downtown gay village emerging "despite rather than because of gay movements' efforts" as a "national and international oas[i]s of so-called queer culture" (Nash, 2006: 2). Over the last half century, the Toronto gay village became a consolidated and visible LGBTQ+ commercial and residential presence along Church Street. At its north end, the 519 Church Street Community Centre was established in the mid-1970s. As Canada's foremost community centre with a LGBTQ+ mandate, the 519 continues to serve segments of the LGBTQ+ community with a wide range of services and drop-in programs for queer and transgender newcomers and refugees, older adults, parents, youth, and children (Micallef, 2017). With its colourful murals, street lamp banners, and signage, the 519 remains a strong physical anchor in the urban landscape.

Variously a place of celebration, political protest, social networking, and consumption in its community centre, coffee shops, restaurants, bars, night clubs, and bathhouses, Toronto's gay village that surrounds the 519 is a gentrifying social and cultural centre for those sexual and gender minorities who can afford it and whose bodies are read as desirable and respectable. For individuals who live at the intersections of youth, women, queers of colour, transgender, gender non-conforming, and housing-precariousness, the gay village is seldom experienced as an enclave of social inclusivity but rather one of exclusions, violence, and addictions (Rosenberg, 2019, 2017). Even with the influx of heterosexual homeowners and renters, and concomitant "de-gaying" of this gay enclave in an era of post-gay attitudes (Ghaziani, 2014a, 2014b; Nash, 2013a), the gay village remains the imagined and material heart of white, middle-class, gay cis-male political and social life in Toronto (Nash, 2013b). Such a homonormative gay village model of LGBTQ+ social space provision, however, is not emulated on Toronto's ethno-culturally diverse periphery. Homonormativity is a concept that encapsulates the problem of privilege within a very diverse LGBTQ+ community as it intersects with whiteness, capitalism, sexism, trans-misogyny, and cissexism.[3] It is a term used by Duggan (2003: 50) and other scholars to describe the mainstreaming of gay and lesbian politics over the last two decades and the prominence of an agenda that does not challenge dominant heteronormative assumptions and institutions but rather seeks inclusion within them, assuming that equality is found in a "privatized, depoliticized gay culture anchored in domesticity and consumption."

In Toronto's suburbs, there are no gay villages or community centres, like the 519, with a specific LGBTQ+ mandate. Instead, there is an eclectic scattering of services and irregular programming in shared spaces and at limited times in a week or month with often long waiting lists and minimal public visibility. Some LGBTQ+ suburbanites, who can afford the time and cost of travel, will commute to downtown Toronto for services and programming; others remain closer to home, tapping into the local LGBTQ+ underground and middle ground to access what is nearby. As the following three examples illustrate, when "physical space for queers is so limited" in suburbs and "not designated for queer and trans bodies," it has to be strategically and collaboratively "carved out" (non-profit administrator, 10 July 2018).

QTBIPOC Sauga

In 2016, two community activists (interview, 10 July 2018), who felt like "lonely queer[s] in Mississauga" and who still experience "extreme isolation" as LGBTQ+ organizers, initiated QTBIPOC Mississauga Meetup by emailing an invitation to friends, acquaintances, organizations, and listservs. They waited in a coffee shop to see who from the queer, trans, Black, Indigenous, and people of colour community would join them. Even without a formal outreach strategy, five years later the group still meets monthly, in different publicly undisclosed locations (most recently members' homes) in order to protect individual's privacy and safety.

As a grassroots organization without official non-profit registration status, it is difficult, if not impossible, to access municipal- and community-based funding. Initially, a small grant from Rainbow Health Ontario provided ten months of funding for food and public transit for the seven to fifteen QTBIPOC Sauga attendees, but that money has long-since run out and the organizers now cover the costs themselves. In their organizing, they have created a meeting structure "that allows for people to move in and out" of the group, prioritizing "building together without people spending more time on public transit than together" (Verma, 2020). The convening of this group is a deliberate "response to the trauma of losing community, plus the small everyday traumas of being in a place where reaching across the expanse of the city, and of the region, is so hard," pitted with "micro-aggressions" of "racism, homophobia, transphobia, and ageism" (Verma, 2017: 226, 228). The monthly gatherings are a reminder that "there is a long, complicated history of queer and trans organizing in the burbs" that "lives in the minds and bodies of individual people"; to write and talk about it is to construct "a living archive, and a new future for QTBIPOC communities, outside of the oppressive structures of non-profit organizations" (228). These comments come from first-hand experience of the sometimes-exploitative edge of non-profit organizational work; invariably, LGBTQ+ programs within organizations are contested political terrain where people struggle over the meaning of community, service, and inclusion. Despite initiating a suburban parking lot Queer It Up! event in 2008 and carrying organizational momentum from it into coordinating the region's first Pride Week in Peel march in 2011 and a Pride Week in Peel festival in 2013 in front of city hall in Mississauga's Celebration Square, in the years since, this organizer has struggled to obtain sustained municipal political support (beyond a tokenistic rainbow flag-raising to mark Peel Pride in July, requested by deputation in 2012 but not achieved until 2016) to create space for racialized LGBTQ+ residents. There has also been limited support or recognition of the political work of queer and trans people of colour from a predominantly white, middle-class, cis-gender, and politically networked Peel Pride committee that organized separate Pride festivities in the north of the region from 2002 until 2020. Lived experience has taught this activist that the regional municipality of Peel

> doesn't need a 519! How would you pick one central location? The concentration of resources in one or two organizations also doesn't serve the LGBTQ community well. There needs to be more options outside of the non-profit industrial complex. (interview, 10 July 2018)

Rather than wait for LGBTQ+ opportunities to materialize for QTBIPOC communities, she became an "insider activist" (Browne and Bakshi, 2013), leveraging a contractually limited position as a Community Arts Curator for Social Change with the Art Gallery of Mississauga

(AGM) to foster collaborations. As an upper ground public institution, the AGM has an equitable access mandate to support diversity and foster the inclusion of marginalized groups protected by the Ontario Human Rights Code.

Her first collaborative initiative with the AGM was to curate the 2018 exhibition *UNRULY*, which put two local racialized queer and transgender emerging artists with disabilities in conversation with one another about the (im)possibilities of thriving "in the suburbs, in 'community,' and in the season(s) when they are expected to be 'proud'" (Verma, 2018). The exhibition was held in the small XIT-RM gallery (intended for people who experience barriers in accessing traditional art space) in June 2018 as part of the summer series *Border Crossings: Travelling the In Between*. Her emergent curatorial practice is informed by a social justice agenda that seeks to disrupt power relations:

> I wanted to trouble binaries and the ways in which QTBIPOC people are asked to articulate our identities in a really narrow way, especially in the suburbs. In my own experience of being in spaces with South Asian people, the things I need to disrupt are that all South Asians are homophobic and transphobic ... and that there are no queer or trans people in the suburbs. (interview, 10 July 2018)

In addition to staging a public event in the Mississauga Civic Centre that affords critical insight into the complexities of suburban queerness, this activist has deliberately engaged a collaborative model of curation:

> We had a Google doc that we had going together, and we had Google hangout and Google chats and text messages. I had many conversations with them individually and together. It took me a long time to come up with the title and curatorial statement. I checked in with them about all of it, which I understand is a bit unusual for curators to do. I think in the capital A art world (despite contemporary critiques that it has perpetuated whiteness, cis-ness, and straightness) it is hard to disrupt those practices in the day-to-day ... [but] I have learned to believe and value the people you work with regardless of credentials or previous experience. (interview, 10 July 2018)

Her second collaborative initiative was to use this exhibition as an opportunity for the AGM and QTBIPOC Sauga to co-host QTBIPOC Crafting events throughout the summer wherein themes of pride and borders could be interactively explored through art and conversation. In so doing, the public profile and resource opportunities of QTBIPOC

Sauga have, if only temporarily, been enhanced. Leading colleagues (interview, 10 July 2018) at the non-profit AIDS service organization (ASO) Moyo Health and Community Services (formerly the Peel HIV/AIDS Network, PHAN) admiringly highlight the work of QTBIPOC Sauga as "a model of peer inclusion that utilizes a skill base that exists on a minimal funding basis." Hope and admiration are simultaneously expressed by colleagues that such activism can be a lasting model. Certainly, the collaborative, creative micro-interventions of the QTBIPOC Sauga co-founder demonstrate how suburban LGBTQ+ grassroots activism can bypass some obstructionist middle ground organizations and connect with the membership of others in order to create temporary QTBIPOC social space and, in the process, briefly reach a receptive portion of an upper ground public institutional audience. However, it remains questionable whether such a model of LGBTQ+ activism, which is rooted in the embodied and emotional labour of a few dedicated individuals, can be sustained over a long period of time.

When a resistive stance is taken on an issue, burnout and exhaustion are inevitable, particularly when they are compounded by a general lack of support. As Brown and Pickerill (2009: 28) note, "constantly feeling 'different' and apart from society adds a particular emotional pressure to activism and requires a high degree of emotional reflexivity in order to overcome or cope with this dissonance." The conflicting emotional demands of activism, then, often make it inherently unsustainable. The suburban LGBTQ+ vernacular creativity of QTBIPOC Sauga is rooted in highly personal and embodied labour that not only relates to specific temporalities of "growing up in the suburbs" but also to an understanding of suburbanization as a particular stage of urban development that may be later accompanied by more robust social service infrastructure and social inclusion policies that better support the lives of sexual and gender minorities.

My House Rainbow Meeting Spaces

Like its Peel counterpart PHAN, the AIDS Committee of York Region (CAYR) is a small ASO (there are now over 125 of these across Canada, many of which were established during the 1980s to provide information sharing and peer support during the AIDS pandemic) that collaborates widely in its region with other non-profit organizations and performs a key multipurpose service role not just for people living with HIV/AIDS (PWAs) but also for the LGBTQ+ community more generally. Initiated in 1993 as a volunteer organization, CAYR became a formally incorporated registered charitable organization in

1996 and has grown steadily as a community-based non-profit organization through the acquisition of grants. With seven full-time staff, its mandate is "to create safe, confidential, and inclusive spaces and services in the York Region where people can access dignified support and meaningfully engage in self-determined pathways to well-being" (https://cayrcc.org). Similar to PHAN, CAYR is also seeking to move beyond a disease model of care and to attend to the social and recreational dimensions of being LGBTQ+. As a staff member (interview, 22 June 2018) explains, in the regional municipality of York "it doesn't make sense to have a hub model like the 519 because we're an extensive region and we need to take the programs to where the people are." CAYR's solution has been to absorb and extend a local, grassroots "My House" community youth program and rebrand it as "Rainbow Rooms," offering a place for LGBTQ+ young people to safely and anonymously socialize.

Established in 2012 with some municipal funding and the intention of extending the success of the Gay-Straight Alliance[4] model in schools, My House shared office space with CAYR from its inception and worked from the same philosophies of harm reduction, anti-oppression, and inclusion. In 2017, My House was absorbed into CAYR, helping it to expand its programming scope by engaging members of the LGBTQ+ community in sexual health education but also to respond to requests "for social space especially in suburban and rural areas where there are no gay bars to hang out in to meet people" (staff, interview, 22 June 2018). Facilitated by a sexual health outreach worker, My House operates bi-weekly September to June as an after school, peer-led, drop-in program for youth in the safe, inclusive, and relatively anonymous lounge space of CAYR in Richmond Hill, on the second floor of a shopping plaza. An impermanent satellite space was also added in Markham and has moved between the transit-accessible spaces of a public library (described by staff [interview, 27 August 2018] as "not the most comfortable") and the York Region branch of the Canadian Mental Health Association. Each group attracts between ten and fifteen youth per session, many of whom are in their late teens, with more racialized gay young men in Markham and more white young people who variously identify as queer, transgender, and gender non-binary in Richmond Hill. These two "intentional" and "organic" LGBTQ+ spaces for youth to eat, hang out, socialize, and find solidarity have served upwards of 2,000 young people since their inception (staff, interview, 27 August 2018). The goal, however, is to establish Rainbow Rooms across this region through collaboration with community partners (for example, 360°

Kids; Community and Home Assistance to Seniors) and, in so doing, extend their access in the daytime to seniors who simultaneously value a sense of community and anonymity, and help, over the long-term, to eliminate the social isolation experienced by many LGBTQ+ suburbanites.

Intentional, yet flexible spaces with low barriers to access, Rainbow Rooms are seen by CAYR staff (interview, 22 June 2018) as a cost-effective, self-sustaining, "extreme partnership" model for delivering LGBTQ+ social space to youth (and, in the future, seniors) across an extensive geography. Success depends upon continuing to secure grants to hire community outreach program coordinators, to foster sustainable community organizational partnerships through which space can be shared, and to build political relationships with civic leaders. But there can be significant employee turnover and, with it, weakened social networks that then have to be built up again from community ties in the underground up to the upper ground. Some of that "relationship brokering" (social worker, interview, 22 August 2018) in the middle and upper ground of the York regional municipality has occurred through the creation in 2016 of a Rainbow Network Table, chaired by a superintendent insider activist at the York Regional Police Service. The superintendent has used her "grassroots" lived experience as a lesbian within the police organization and her seniority within the police hierarchy to initiate change in LGBTQ+-police relations. With quarterly meetings, this roundtable has afforded social service agencies with LGBTQ+ programming and mandates, like CAYR, which have been operating quite independently, an opportunity to discuss issues, build trust, and establish some foundations for future collaborations.

The history of LGBTQ+-police relations is one that is contradictory and tense; it is negatively overshadowed by acts of police brutality and targeted violence on LGBTQ+ bodies in public and private spaces, which criminalized homosexuality and gender diversity and associated LGBTQ+ people with deviance and perversion (Dwyer, 2014). In the contemporary period, police have sought to foster partnerships with LGBTQ+ individuals and groups in an effort to make policing practices more transparent, supportive, and human rights–focused. Such partnerships have necessitated some significant shifts in police culture, which have been brought about, in part, through social inclusion and sexual diversity training (often run by LGBTQ+ non-profit organizations in the middle ground as a necessary funding stream) in conjunction with targeted LGBTQ+ recruitment of new officers and LGBTQ+ youth outreach programs.

Colours Youth Group

In the regional municipality of Durham, the police also play a LGBTQ+ community-building role. Once a month, the Durham Regional Police Service (DRPS) facilitates the Colours Youth Group at the Carea Community Health Centre (CCHC) in Ajax. Designed to be a "safe and social drop-in" for youth ages thirteen to twenty, Colours was started by a CCHC health promoter in 2013. When she began working at this registered charitable organization, she realized that, other than the local Parents, Families, and Friends of Lesbians and Gays (PFLAG) chapter established in 2003 and the AIDS Committee of Durham Region, there were no drop-in groups or resources for LGBTQ+ youth. She commented:

> What was happening was happening in little silos and folks weren't working together. With the geography in Durham region and just the lack of public transportation making it hard for people to get around, people were quite isolated. Our LGBTQ community was quite invisible … We ended up having someone say: "How can we work together and pool our resources and offer more coordinated support?" Let's bring people out of the shadows and get them involved in programs. (social worker, interview, 22 August 2018)

LGBTQ+ community capacity building began in 2008 with a one-day youth conference entitled "SAY WHAT?! Exploring Sexual and Gender Diversity among Youth in Durham Region" at the University of Ontario Institute of Technology. It was followed by the hosting of the Durham Pride Prom (that has occurred annually since) in collaboration with a range of social service agencies and the Durham District School Board, PFLAG's annual summer Camp Rainbow Phoenix (2012), and the newly initiated Youth Pride Durham (2018) by the local Children's Aid Society. Youth have become a collaborative rallying point for LGBTQ+ inclusion activism by individuals and organizations in Durham.

The Colours Youth Group is run by two youth outreach workers one evening a week for an hour and a half in a community health centre activity room. It is designed as a safe place for young people to hang out, learn about different topics (for example, love and relationships, "coming out," addressing homophobia, and sexual health), build friendships, and be mentored. Attendance varies, but at its maximum there are usually about twenty youth, half of them racialized. Once a month, the Durham Regional Police Service facilitates an evening. Its involvement was initiated by an ally officer. In reactive frontline patrols and investigations, she had witnessed a disproportionate number of hate crimes against and self-harm service calls from LGBTQ+ youth and wanted to proactively help build resilience.

On small grants that have grown from $2,600 to $5,500, she spends a minimum of $200 an evening providing food, organizing activities (for example, meditation, professional make-up, bowling, concerts), running workshops on topics of interest (for example, perceptions of safety, social media bullying, sexual assault and abuse, tactical and canine unit operations), and having private conversations. The volunteer officers come in uniform and in their squad cars. A few officers identify as LGBTQ+, but many see it as "a good place to get exposure to, and education from LGBTQ+ youth" (police sergeant, interview, 28 August 2018). The intent is to "break down barriers of what the uniform represents," to offer a consistent and reliable presence, and to build youth rapport such that they feel comfortable asking questions, working through problems, reporting crimes, and reaching out after they leave the program (police sergeant, interview, 28 August 2018). The Colours Youth Group has received several Police Appreciation Awards for community service and has been written up as a key initiative of the DRPS Equity and Inclusion Unit.

The relationship between police services and LGBTQ+ communities is markedly different between the centre and peripheries of Toronto. In the City of Toronto, police surveillance and harassment of LGBTQ+ communities have been leitmotifs in its queer history – especially the aggressive entrapment of gay men in cruising grounds such as parks, theatres, public washrooms, and bathhouses (Bain and Nash, 2007). Since the 1980s and into the present, parks across Toronto, including its peripheries, have been subject to police surveillance operations, where, as agents of the state, police have charged men for public indecency and engaging in sexual acts (Bain et al., 2020). As the protests by Black Lives Matter Toronto during the 2016 Pride Parade (Davis, 2021) and the serial murders of South Asian and Middle Eastern gay men in the gay village (many of them residing in peripheral municipalities) attest, experiences of institutions, such as the police, are disparate. It is racialized people, transgender people, sex workers, and the poor who continue to be on the receiving end of police brutality and neglect. But the residue of these historical and contemporary LGBTQ+ experiences has not played out on the periphery in quite the same way as it has in the City of Toronto, where uniformed police officers were banned from marching in the Pride Parade from 2017 to 2019 and police-LGBTQ+ relations remain strained. Beyond wrapping police cruisers in rainbows, wearing rainbow pins and panels on flak jackets, and raising the rainbow flag in June, police services in the regional municipalities of Peel, York, and Durham facilitate LGBTQ+ programming and are widely seen as "working harder than Toronto police at being respectful and checking in with the community" (social worker, interview, 22 August 2018). In peripheral municipalities in the Toronto city-region, the police, as an upper ground institution, have over the last few years begun to play a key role sponsoring LGBTQ+ initiatives to foster greater safety and social inclusion but also to bridge LGBTQ+ individuals and organizations hard at work in the underground and middle ground. Such collaborative mechanisms and the LGBTQ+ social spaces they help create need to be appreciated as alternative forms of creativity that contribute to civic cultural and social life in suburbs.

Conclusion

Across a recent history, this chapter has documented the informal collaborative production of LGBTQ+ suburban social spaces in the Toronto city-region as important expressions of peripheral vernacular creativities that have been largely overlooked within dominant heteronormative scholarly and public understandings of suburbia. Queer suburban vernacular creativities are deeply embodied and personal expressions of creativity; they are reliant upon tacit knowledge that is distributed and accessed socially, networks that share limited resources and opportunities, and information communication technologies that enable a virtual space for communication and collaboration (Felton et al., 2010). Despite social equalities formalized in legal protections, these queer suburban vernacular creativities continue to be rendered vulnerable to intersecting micro-aggressions of homophobia, transphobia, racism, and ageism. Thus, any queer underground suburban "scene" that is created is necessarily eclectic and organizationally hybridized. It is temporally and geographically specific, heavily reliant upon the grassroots work of individuals who have grown up in the suburbs and the interpersonal relations that constitute their social networks.

On small, grant-driven budgets with few paid staff, some non-profit organizations have collaborated with volunteers and each other to provide social spaces for LGBTQ+ youth. This sense of duty and responsibility to queer youth is admirable, but there remain few options available in the suburbs when young people age out of these programs. Many of these collaborations – variously artistic, educational, religious, health, and policing – are quite recent, having emerged within the last decade. Such collaborations are inevitably mutable and can follow different trajectories, with some developing and gathering traction and more formalized institutional commitments while others fall by the wayside, garnering little support or investment. Inevitably, some collaborations are of limited duration due to unstable funding, staff and volunteer turnover, lack of permanent space, differing agendas, and competing inclusion priorities; while others produce tensions and frictions that result in LGBTQ+ community fractures and scars in the suburban underground and middle ground. It is the self-organizing underground and middle ground, however, that need ongoing financial and political support, as it is there that ideas accumulate and combine to challenge, enrich, and renew the institutional upper ground that has consolidated around hegemonic heteronormative imaginaries.

Supporting and sustaining queerness in the suburbs requires different strategies than those deployed in the central city. A homonormative model of a visible and concentrated LGBTQ+ service, resource, and social infrastructure in a gay village and accompanying community centre is an awkward fit with the spatial distance of and racialized differences cross-cutting Toronto's periphery, where there are few, if any, LGBTQ+-dedicated bars or social spaces. Certainly, some networks connect the suburbs to the central city in a conventional hub-and-spoke configuration to provide access – with significant investment of time and money in commuting – to resources and services that are absent on the periphery, but this thinking also "privileges the inner-city as *the* locus" of queerness (Felton et al., 2010: 67).

As this chapter has shown, however, suburbs have a countercultural queerness of their own, and its creative collaborations, hybridizations, and challenges need to be appreciated in their own right and deserve greater scholarly scrutiny. It is a queerness that is active and resourceful, repurposing public facilities and commercial premises to temporarily meet needs for socializing and informal networking. These LGBTQ+ temporary and modest transformations of material circumstances at the micro-scale through bottom-up underground initiatives intersect with middle ground organizational capacities and constraints, and upper ground opportunities and durabilities. They are articulations of suburban LGBTQ+ vernacular creativities that are inherently fragile, yet persistent, in part because they are less subject to market dynamics and mainstream co-optation. Appreciating such underground, middle ground, and upper ground (mis)alignments can be a useful strategy for understanding how suburban LGBTQ+ vernacular creativities with no market value are produced and can be better explicitly supported in peripheral municipalities.

NOTES

1 The interviews varied in length from forty-five minutes to two hours and were undertaken on the phone or in person in places of employment or volunteer service and occasionally in coffee shops or at public events. They were digitally recorded and selectively transcribed based on the coding of handwritten notes taken during the conversations.
2 Assimilationist inclusion involves claiming that LGBTQ+ communities are the same as everyone else (Smith, 2007). However, policy regulates and normalizes sexuality (Concannon, 2008), which means that, if LGBTQ+ communities are included in policy under the claim of sameness, this claim is in relation to the dominant heterosexual norm. Tokenistic inclusion, the weakest form of inclusion, entails naming groups in policy (Strid et al., 2013).
3 Cissexism is prejudice or discrimination against transgender people. Cisgender refers to people whose gender identity and expression align with the sex they were assigned at birth, while cis-normativity refers to organizational structures and everyday practices that naturalize this category.
4 Gay-Straight Alliances (GSAs) are in-school, extra-curricular student organizations hosted by an adult teacher-sponsor that provide a sense of safety, inclusion, and support for lesbian, gay, bisexual, transgender, queer, and two-spirit (LGBTQ2S) students, those who are questioning their sexuality, and their straight allies (Bain and Podmore, 2020a). The first GSAs were formed over thirty years ago, first in the United States and a decade later in Canada, to undermine the hegemonic heteronormativity of public educational institutions.

REFERENCES

Arvidsson, A. 2007. "Creative Class or Administrative Class? On Advertising and the 'Underground.'" *Ephemera: Theory and Politics inOrganization*, 7(1): 8–23.
Bain, A.L. 2019. "Suburban Cultural Production." In B. Hanlon and T.J. Vicino, eds., *The Routledge Companion to the Suburbs*. New York: Routledge, 323–31.
Bain, A.L., and C.J. Nash. 2007. "The Toronto Women's Bathhouse Raid: Querying Queer Identities in the Courtroom." *Antipode*, 39(1): 17–34. https://doi.org/10.1111/j.1467-8330.2007.00504.x
Bain, A.L., and J.A. Podmore. 2020a. "Challenging Heteronormativity in Suburban High Schools through 'Surplus Visibility': Gay-Straight Alliances in the Vancouver City-Region." *Gender, Place, and Culture*, 27(9): 1223–46. https://doi.org/10.1080/0966369X.2019.1618798

Bain, A.L., and J.A. Podmore. 2020b. "Relocating Queer: Comparing Suburban LGBTQ2S Activisms on Vancouver's Periphery." *Urban Studies*, 58(7): 1500–19. https://doi.org/10.1177/0042098020931282

Bain, A.L., and J.A. Podmore. 2020c. "Scavenging for LGBTQ2S Public Library Visibility on Vancouver's Periphery." *Tidjschrift voor Economische en Sociale Geografie*, 111(4): 601–15. https://doi.org/10.1111/tesg.12396

Bain, A.L., and J.A. Podmore. 2021. "More-Than-Safety: Co-Creating Resourcefulness and Conviviality in Out-Of-School Spaces for Suburban LGBTQ2S Youth. *Children's Geographies*, 19(2): 131–44. https://doi.org/10.1080/14733285.2020.1745755

Bain, A.L., J.A. Podmore, and R.D. Rosenberg. 2020. "'Straightening' Space and Time? Peripheral Moral Panics in Print Media Representations of Canadian LGBTQ2S Suburbanites, 1985–2005." *Social and Cultural Geography*, 21(6): 839–61. https://doi.org/10.1080/14649365.2018.1528629

Brouwer, D. 1998. "The Precarious Visibility Politics of Self-Stigmatization: The Case of HIV/AIDS Tattoos." *Text and Performance Quarterly*, 18(2): 114–36. https://doi.org/10.1080/10462939809366216

Brown, G., and J. Pickerill. 2009. "Space for Emotion in the Spaces of Activism." *Emotion, Space and Society*, 2(1): 24–35. https://doi.org/10.1016/j.emospa.2009.03.004

Browne, K., and L. Bakshi. 2013. "Insider Activists: The Fraught Possibilities of LGBT Activisms from Within." *Geoforum*, 49: 253–62. https://doi.org/10.1016/j.geoforum.2012.10.013

Charrieras, D., S. Darchen, and T. Sigler. 2018. "The Shifting Spaces of Creativity in Hong Kong." *Cities*, 74: 134–41. https://doi.org/10.1016/j.cities.2017.11.014

Cohendet, P., D. Grandadam, and S. Laurent. 2010. "The Anatomy of the Creative City." *Industry and Innovation*, 17(1): 91–111. https://doi.org/10.1080/13662710903573869

Concannon, L. 2008. "Citizenship, Sexual Identity and Social Exclusion: Exploring Issues in British and American Social Policy." *International Journal of Sociology and Social Policy*, 28(9/10): 326–39. https://doi.org/10.1108/01443330810900176

Cooper, A. 2009. "Point Chev Boys and the Landscapes of Suburban Memory: Autobiographies of Auckland Childhoods." *Gender, Place, and Culture*, 16(2): 121–38. https://doi.org/10.1080/09663690902795720

Davis, K.D. 2021. "Transnational Blackness at Toronto's Pride: Queer Disruption as Theory and Method." *Gender, Place, and Culture*, 28(11): 1541–60. https://doi.org/10.1080/0966369X.2020.1819208

Duggan, L. 2003. *The Twilight of Equality? Neoliberalism, Cultural Politics, and the Attack on Democracy*. Boston: Beacon Press.

Dwyer, A. 2014. "Pleasures, Perversities, and Partnerships: The Historical Emergence of LGBT-Police Relations." In D. Peterson and V. Panfil, eds., *Handbook of LGBT Communities, Crime, and Justice*. New York: Springer, 149–64.

Ekers, M., P. Hamel, and R. Keil. 2012. "Governing Suburbia: Modalities and Mechanisms of Suburban Governance." *Regional Studies*, 46(3): 405–22. https://doi.org/10.1080/00343404.2012.658036

Ettlinger, N. 2010. "Bringing the Everyday into the Culture/Creativity Discourse." *Human Geography*, 3(1): 49–59. https://doi.org/10.1177/194277861000300104

Felton, E., C. Collis, and P. Graham. 2010. "Making Connections: Creative Industries Networks in Outer-Suburban Locations." *Australian Geographer*, 41(1): 57–70. https://doi.org/10.1080/00049180903535576

Florida, R. 2005. *Cities and the Creative Class*. New York: Routledge.

Ghaziani, A. 2014a. "The Radical Potential of Post-Gay Politics in the City: A Reply to Molotch, Deener, Tavory, and Pattillo." *Environment and Planning A: Economy and Space*, 47(11) 2409–6. https://doi.org/10.1177/0308518X15613228

Ghaziani, A. 2014b. *There Goes the Gayborhood?* Princeton, NJ: Princeton University Press.

Gibson, C., S. Luckman, and J. Willoughby-Smith. 2010. "Creativity without Borders? Rethinking Remoteness and Proximity." *Australian Geographer*, 41(1): 25–38. https://doi.org/10.1080/00049180903535543

Grabher, G. 2018. "Marginality as Strategy: Leveraging Peripherality for Creativity." *Environment and Planning A: Economy and Space*, 50(8): 1785–94. https://doi.org/10.1177/0308518X18784021

Hall, M., and B. Lee. 2010. "How Diverse Are US Suburbs?" *Urban Studies*, 47(1): 3–28. https://doi.org/10.1177/0042098009346862

Hautala, J., and O. Ibert. 2018. "Creativity in Arts and Sciences: Collective Processes from a Spatial Perspective." *Environment and Planning A: Economy and Space*, 50(8): 1688–96. https://doi.org/10.1177/0308518X18786967

Hubbard, P. 2008. "Here, There, Everywhere: The Ubiquitous Geographies of Heteronormativity." *Geography Compass*, 2(3): 640–58. https://doi.org /10.1111/j.1749-8198.2008.00096.x

Keil, R. 2015. "Towers in the Park, Bungalows in the Garden: Peripheral Densities, Metropolitan Scales, and the Political Cultures of Post-Suburbia." *Built Environment*, 41(4): 579–96. https://doi.org/10.2148/benv.41.4.579

Keil, R., and J.-P.D. Addie. 2015. "'It's Not Going to Be Suburban, It's Going to Be All Urban': Assembling Post-Suburbia in the Toronto and Chicago Regions." *International Journal of Urban and Regional Research*, 39(5): 892–911. https://doi.org/10.1111/1468-2427.12303

Lang, R., and K. Danielsen. 2005. "Review Roundtable: Cities and the Creative Class." *Journal of the American Planning Association*, 71(2): 203–20. https:// doi.org/10.1080/01944360508976693

Micallef, S. 2017. "Town Squares and Spiritual Hearts." In S. Chambers, J. Farrow, M. FitzGerald, E. Jackson, J. Lorinc, T. McCaskell, R. Sheffield, T. Taylor, and R. Thawer, eds., *Any Other Way: How Toronto Got Queer*. Toronto: Coach House Press, 45–50.

Mosselson, A. 2017. "'Joburg Has Its Own Momentum': Towards a Vernacular Theorisation of Urban Change." *Urban Studies*, 54(5): 1280–96. https://doi.org/10.1177/0042098016634609

Nash, C.J. 2006. "Toronto's Gay Village (1969–1982): Plotting the Politics of Gay Identity." *Canadian Geographer*, 50(1): 1–16. https://doi.org/10.1111/j.0008-3658.2006.00123.x

Nash, C.J. 2013a. "The Age of the 'Post-Mo': Toronto's Gay Village and a New Generation." *Geoforum*, 49(2): 243–52. https://doi.org/10.1016/j.geoforum.2012.11.023

Nash, C.J. 2013b. "Queering Neighbourhoods: Politics and Practice in Toronto." *ACME: An International Journal for Critical Geographies*, 12(2): 193–219.

Podmore, J.A., and A.L. Bain. 2020. "'No Queers Out There?' Metronormativity and the Queer Suburban." *Geography Compass*, 14(9): e12505. https://doi.org/10.1111/gec3.12505

Rantisi, N., and D. Leslie. 2009. "Creativity by Design? The Role of Informal Spaces in Creative Production." In T. Edensor, N. Rantisi, and D. Leslie, eds., *Spaces of Vernacular Creativity*. London: Routledge, 33–45.

Ritchie, J. 2010. "How Do You Say 'Come Out of the Closet' in Arabic? Queer Activism and the Politics of Visibility in Israel-Palestine." *GLQ*, 16(4): 557–75. https://doi.org/10.1215/10642684-2010-004

Rosenberg, R. 2017. "The Whiteness of Gay Urban Belonging: Criminalizing LGBTQ Youth of Color in Queer Spaces of Care." *Urban Geography*, 38(1): 137–48. https://doi.org/10.1080/02723638.2016.1239498

Rosenberg, R. 2019. "Young, Queer, and on the Streets: Homeless LGBTQ2 Youth in Toronto's Gay Village. PhD diss., Department of Geography, York University.

Smith, M. 2004. "Questioning Heteronormativity: Lesbian and Gay Challenges to Education Practices in British Columbia, Canada." *Social Movement Studies*, 3(2): 131–45. https://doi.org/10.1080/1474283042000266092

Smith, M. 2007. "Queering Public Policy: A Canadian Perspective." In M. Orsini and M. Smith, eds., *Critical Policy Studies*. Vancouver: UBC Press, 91–110.

Strid, S., S. Walby, and J. Armstrong. 2013. "Intersectionality and Multiple Inequalities: Visibility in British Policy on Violence against Women." *Social Politics*, 20(4): 558–81. https://doi.org/10.1093/sp/jxt019

Tongson, K. 2011. *Relocations: Queer Suburban Imaginaries*. New York: New York University Press.

Verma, A.R. 2017. "Mississauga Meetup: Queer Organizing in the 905." In S. Chambers, J. Farrow, M. FitzGerald, E. Jackson, J. Lorinc, T. McCaskell, R. Sheffield, T. Taylor, and R. Thawer, eds., *Any Other Way: How Toronto Got Queer*. Toronto: Coach House Press, 226–8.

Verma, A.R. 2018. *UNRULY*. Mississauga: Art Gallery of Mississauga.

Verma, A.R. 2020. "Queer, Trans, Suburban: A Sex Salon Workshop." Sex Salon Speaker Series. University of Toronto, 29 January 2020. https://ischool .utoronto.ca/news/queer-trans-suburban-a-sex-salon-workshop/

Warren, A., and C. Gibson. 2011. "Blue Collar Creativity: Reframing Custom-Car Culture in the Imperilled Industrial City." *Environment and Planning A: Economy and Space*, 43(11): 2705–22. https://doi.org/10.1068/a44122

PART 4

Creating Suburbia

13 From Artistry to Agency? Transactional Architecture for the Creative Fashioning of the Antwerp Suburbs in the Early Twentieth Century

TOM BROES AND MICHIEL DEHAENE

Introduction

Looking at suburbanization through the lens of creativity is not only an invitation to look at the lives lived in the suburbs with greater care and appreciation but also to examine the process of suburbanization itself. Not only have suburbanites been labelled as "lesser" than urban; the suburbanization process itself has been depreciated as a degenerate form of urbanization. The most clear example is the general labelling and equating of suburbanization to sprawl and dross (Lerup, 2001; Berger, 2006; Ingersoll, 2006).

Spatially, suburbanization has been associated with notions such as fragmentation (Burgel, 1989; GUST, 2002), subordination within centre-periphery relations, loss of solidarity (Soja, 2000; Corijn, 2006), and loss of legibility (Lynch, 1966; Sieverts, 1997). The question of the aesthetic value of the suburbs and the need for aesthetic coding may also be placed within this list (Duany et al., 2000). Especially in architecture and design circles, the question of suburbanization has also been associated with the loss of control, not only pointing to the lack of development control but also lamenting the loss of the architect's control over the general formal appearance of urban development (Tunnard and Pushkarev, 1963).

In this chapter, we look at the interwar period and at the production of the Antwerp suburbs as a period of profound change in the roles taken up by architects. We try to look beyond the general narrative of loss (of control) to the shifting positions and the creative disposition of architects in carving out new roles for themselves. In doing so, we try to break out of a planning history that still tends to be strongly aligned with the rhetoric and apologetics of the players involved in the making of the suburbs. Rather than debunking the positions of planners as "ideological constructs" by playing them against the political economic reality of urbanization (Choay, 1965; Castells, 1975), we take the ideological production of planners, architects, and designers seriously but try to show the plurality of positions and the creative disposition that can be read in them. Our story tries to demonstrate that there is a whole range of positions between the rhetorical stances that historically played "the city scientific" (Ford, 1913) against "city planning according to artistic principles" (Sitte, 2002 [1909]) or, alternatively, started playing the positions of grand regional design against the petty architectural focus on the building.

By looking at the suburbs through a lens of creativity, we join the effort to trade the study of the city in its reified forms in favour of the interpretation of the process of urbanization and its ever particular, highly diverse, and spatially differentiated historical manifestations. At the same time, we want to move away from ways of studying the process of urbanization that gloss over the agency of the many players active in its dialectical making. Creativity, to put it simply, gives us the tools to study the practice of urbanization, the craft that goes into its making. If a political economy–informed perspective, for example, may reveal how processes of commodification are played out and surplus capital is circulated and absorbed within the dynamics of urbanization (Harvey, 1985), the creativity lens shows us how the production of this abstract order goes hand in hand with the creation of a lived reality upon which the very process of urbanization depends. The processes of urbanization that political economy describes, such as the commodification of land and housing, come in many forms (Henderson, 2004), which work together with the construction of the possibility to develop that land and to sell that possibility (both virtual and real) as a way to live a future life on that land. Rather than playing the lived reality of the suburbs against the abstractions of planning, we are interested in the lived reality of the city-making process itself, of which the work of planners and architects is an integral part. We look beyond the general narrative of suburban living and try to understand the hard work of making that promise available, in concrete locations, step by step, in places that at the time were still far removed from the possibility of leading an urban life.

We try to shed light on this piecemeal process by studying the ways in which architects in Antwerp during the interwar period tried to carve out for themselves entrepreneurial space, a veritable "transactional space" in which concrete jobs could be taken up, money could be earned, and yet a concrete reality could also be creatively produced. At the same time, we will also demonstrate how this suburban phase necessitated the creative refashioning of the architectural profession at that time – a change that opposed the dominant architectural practices and ideals with which the bourgeois city had been shaped and built throughout the nineteenth century.

In concreto, we will study the endeavours of a young generation of architects who referred to themselves as the "young watchmen" (*De Bouwgids*, 1911; Migom, 2004). We will argue that this generation took advantage of a strong phase of suburbanization in early twentieth-century Antwerp to reinvent their roles in times of radical socio-economic change. Like many other European cities, the city of Antwerp was caught up in the grand historical transition from "confined histori-cal city to massive (sub)urbanization" at the beginning of the twenti-eth century. This transition was fueled by massive state investments in Belgium's main (colonial) port city, causing a huge increase in popula-tion, which resulted in a law passed in 1906 aimed at demolishing the nineteenth-century fortification walls in order to relieve pressure on the historic city. One year later, in 1907, a provincial study committee (Study Committee for the Development of the Antwerp Agglomeration, SCAA) was appointed by Royal Decree with the aim of managing the urbanization processes in the Antwerp suburbs (Broes and Dehaene, 2017a). The SCAA had a strong technocratic approach, focusing mainly on the coordination between the allocation of public funds for the con-struction of infrastructures and the creation of private development and investment potential – a common economic recipe for capitalist suburbanization (Gottdiener, 1977; Harvey, 1985; Harris, 2013; Phelps, 2015). Yet it was within the slipstream of the SCAA's activities that a new generation of architects was mobilized in order to render, build, and sell the Antwerp suburbs as a plausible future habitat for a quickly urbanizing society at large (De Vos-Van Kleef, 1912; Laureys, 2004) – not by pursuing an "abstract utopia" formulated in the "distant order of urbanism" but by exploring new roles and positions from which they were able to build a "concrete utopia" conceived from the "near order of architecture" (Lefebvre, 2014). As such, they managed to imbue the suburbanization process with creative and spatial capital, turning the suburbs piecemeal and project after project into a pleasant *cadre de vie*.

We will start with a short portrayal of these "young watchmen" – who they were, how they related to each other, why they distanced themselves from their predecessors, and most importantly, how they related to the Antwerp suburbs. We then continue by stressing that, in order to succeed, the young watchmen were obliged to move beyond the dominant perspective of the architect as a liberal artist, operating from within the privacy of the hermetic design studio. Instead, we will argue that these architects began to adopt a more transactional perspective on architectural practice and production (Bridge, 2013). We will show how they adhered to a pragmatic approach, in the philosophical sense of that word, defining meaningful social-political positions for architects from and within everyday practice (Hickman et al., 2009; Bridge, 2013). Consequently, their practice can be described as spatial agency, informed by mutual knowledge, more than as liberal or expert artistry (Awan et al., 2011). By examining a wide variety of sources, such as those architects' own journals and publications, building applications, leaflets of construction companies, advertisements in journals and newspapers, publications in the *Belgisch Staatsblad* (Belgian Official Journal), published studies of public authorities, and also personal data on the individuals involved, we will distinguish four spaces of transaction. We successively look at how these architects recalibrated their creative roles in response to (1) the changing public sphere, (2) their involvement in the continued growth of real estate development into a full-fledged profession, (3) their dealing with the emergence of general contracting at the expense of individual craftsmanship, and (4) their involvement in public policy, joining the conversation on the regulation of private construction and the need for coherent building codes. We will argue that it was in the overlap of all these activities and transactions that the young watchmen explored concrete pathways that allowed them to creatively imbue Antwerp's urbanizing suburbs with a shared and civic sense of aesthetics and spatial capital.

From City to Suburbanization, from Old Guard to Young Watchmen, and from Stylistic Artistry to Spatial Agency

The architects who referred to themselves as the "young watchmen" belonged to a generation born roughly in the 1880s (Figure 13.1). They looked to the older architect Dirk de Vos-Van Kleef as their mentor. This group of about fifteen architects would gradually distance themselves from their nineteenth-century predecessors, whom they somewhat disdainfully referred to as the "old guard" (De Vos-Van Kleef, 1912; Migom, 2004). In general, these men had all received a very classical architectural training at Antwerp's Académie Royale des Beaux-Arts, where they had, according to the fundamental pedagogical tenets of a beaux-arts education, been trained as liberal artists engrained with the classical mores of an autonomous architectural culture centred on the "cult of genius," "appropriate form," and "good taste" (Casus, 1911a, 1911b; Chafee, 1977; Duany, 2004). Yet this new generation of architects began to challenge the conventional wisdom of their beaux-arts education.

Figure 13.1. Some architects of Antwerp's "young watchmen." Next to the architects shown, we should also mention Egide Van der Paal (1891–1953), John Van Beurden (1880–1967), Hendrik Wittockx (1893–1965), Edward Leonard (1890–1981), and Gerard De Ridder (1878–1958). Source: Collage by the authors, with photos from the Architectural Archives of the Province of Antwerp.

Displeased with their "outdated education" at the Académie Royale des Beaux-Arts in Antwerp, some members of the "young watchmen" founded the Circle for Building Arts (CBA) in 1899 in order to re-educate themselves in architecture through study trips and meetings (Nevejans, 2011). In this way, they tore themselves away from the more established Royal Society of Antwerp Architects (RSAA),[1] the oldest architectural association in Belgium in which the "old guard" set the tone. But it was only after the law of 1906 that the actions of the young watchmen took on a specific focus and sense of purpose. The prospect of suburbanization beyond the historic city literally offered a new space for these young watchmen to put their new ideas into practice. It is worth noting that some of the protagonists of the young watchmen were suburbanites themselves, or quickly moved out to the suburban environments that they helped shape. The law of 1906, officially opening up suburban territories for urbanization, directly triggered some of the young watchmen to publish a series of articles and books, and to launch CBA's magazine *De Bouwgids*. It was mainly this magazine that the young watchmen initially used as a common platform to promote new architectural approaches (De Vos-Van Kleef, 1907; *De Bouwgids*, 1911).

In an early 1911 *De Bouwgids* article, the author(s) under the pseudonym "Casus" complained that "public education [had] remain[ed] unchanged, entirely based on an outdated assembly of conventional formal theories, taught as a dogma, without the slightest relation to the real demands of the new age ... This kind of teaching has nothing in common with our conceptions of everyday life and our morals; it merely gives us half-formed master builders with a lifeless art education" (Casus, 1911a: 18–19).[2] The role of the architect as an artist trained in neoclassical styles and mainly concerned with aspects of monumentality, ordinance, and representation was fiercely criticized. These old stylistic views may have been in tune with the tastes of nineteenth-century urban elites, but by 1911, *De Bouwgids* architects claimed that they "left the public completely indifferent" (19). Apparently, the urbanizing suburbs no longer catered to the image of the city as an unambiguous "enwalled entity" that could be shaped according to the relatively stable taste of one dominant socio-economic group (Benevolo, 1975). The dissolving urbanity of the agglomeration became a home for "aristocrats, bourgeois, and working poor alike" (De Vos-Van Kleef, 1907: 39). This socio-spatial transition, these young architects understood, was in need of new urban and architectural imaginaries in order to get suburban development going.

The prospect of urbanization in the Antwerp suburbs literally offered a refuge for these architects to dispose of what they called the "academic tampering and toil in neo-Greek and neo-Roman styles" with which their predecessors had built (and still occupied) the city centre (*De Bouwgids*, 1911: 1). Instead, this new generation claimed the urbanizing suburbs to explore architectural practices that were not rooted in such ancient artistic narratives and traditions but rather in the aesthetics of everyday life: "We should aim for a more integral aesthetic approach and vary possible solutions according to the case, balance the abstract and the concrete, the classical and the accidental in the hope of one day achieving the perfect fusion of reality and imagination" (D.P.S., 1910: 17). In pursuing this situated, relative, and relational approach, the young watchmen slowly but surely exchanged their role of liberal artists for a role that could be described as "spatial agents" – agents in the production of (social) space who were "neither completely free as individuals, nor ... completely entrapped by [societal] structure [but rather] negotiators of existing conditions in order to partially reform them" (Awan et al., 2011: 31).

We will demonstrate that the architects of the "young watchmen" tried to anchor their ideals within concrete logics of urbanization. They were open towards negotiation and the principles of mutual and shared knowledge not directly aligned with norms or expectations specific to a particular profession (Giddens, 1984, 1987). This attitude expanded their opportunities to move beyond the continued extension of substandard urbanization protocols into the future, enabling them to work steadily towards an (admittedly) imperfect but (at least) practicable ideal – a concrete utopia. In what follows, we will isolate four kinds of agency that moved beyond the (normative) tenets of beaux-arts artistry, which architects started to creatively explore as soon as the 1906 law accelerated the urbanization of the Antwerp suburbs. Arguably, not all of these forms of agency must be understood as a clear break with the past, but they took on a professionalized shape after the turn of the century, especially in the context of suburban development. In that sense, the Antwerp suburbs can be seen as an example in which we see the functioning of, as Harriet Hawkins explains, "margins, as a site for diverse creative practices that are able to escape or challenge prevailing creativities" (Hawkins, 2017: 25).

Building Transactions within an Emerging Suburban Public Sphere: Towards a Renewed Civic Sense of Suburban Aesthetics

Debates on city planning in the early twentieth century were animated by competing visions between the conception of planning as a normal science and the revival of aesthetic planning principles (Rabinow, 1989; Picon, 1992; Talen, 2005). Artistic points of departure became less and less self-evident in urban design, partly because the consensus on these points of departure was increasingly falling apart. Several authors have argued that style as a category of the first modernity (of modernity as tradition, of bourgeois aesthetics) steadily disintegrated from the twentieth century onwards (Colenbrander, 1993).

Several authors, however, have argued that suburbanization processes are only likely to materialize if the technocratic making of those processes is accompanied by cultural imaginaries that render daily life in these emerging urban worlds tangible, socially acceptable, and desirable (Gottdiener, 1977; Schorske, 1980; Harvey, 1989; Smets, 1995). Apart from such an instrumental explanation, others have pointed to the "ethical" requirement of treating urban aesthetics as a genuine urban question in its own right, simply because the aesthetic quality of an environment codetermines what it means to live in a particular place and renders the experience of place more enjoyable and meaningful (Gottmann, 1962; Harvey, 1973; Sieverts, 1997). This notion may explain the widespread creative search for new images and guises of suburbanization beyond the historic cities at the beginning of the twentieth century. Howard's *Garden City*, Geddes's *Valley Section*, Abercrombie's *Regional Planning*, or Schumacher's *Schemata der natürlichen und wirklichen Entwicklung des Organismus Hamburg*, Schwartz's *Stadtlandschaft*, Hilberseimer's *Grosstadt Architektur* to Le Corbusier's *Ville Radieuse*, only to name a few, have all been understood as efforts to ward off the alienating effects of the industrial metropolis through aesthetic reform (Mantziaris, 2014).

In the context of Antwerp, it was mainly international masters such as Prost and Stubben who tried to portray the future urban expansion, but little or nothing came of their plans (Lombaerde, 2008). However, we will argue that the "young watchmen" would take care of the creative imaging of the suburbs, borrowing some of the aesthetic aspects from the emerging urban models in Europe without necessarily ascribing to the ideological positions of the movements through which they were promoted. Instead, they used those aesthetic aspects to construct architectural images and projections that enabled the materialization of concrete suburbanization dynamics in the Antwerp agglomeration – dynamics that were in many ways fostered by the SCAA's activities. In addition, they not only used these images themselves but also started to share and discuss them with a wider audience.

In the period 1906–39, the "young watchmen" launched two local architecture magazines centred on finding proper imaginaries for Antwerp's urbanizing suburbs. First, *De Bouwgids* was published, which ran between 1909 and 1933 as the monthly journal of the CBA. From 1922 onwards, the CBA became the RSAA's youth department, after which the young watchmen began to flow on to the RSAA themselves (Nevejans, 2011), trying to renew the organization, mostly through the introduction of their own publication channels. In 1930, they launched the *Monthly Bulletin of the RSAA*, which ran until 1939. The main goal of these magazines was not that of the typical professionally oriented architectural periodical; rather they tried to target a wider public. The editorials of the first editions of both magazines make these intentions very clear. *De Bouwgids* wanted to "reach private owners and individuals … in order to have influence on the public's mind and taste" (*De Bouwgids*, 1911: 1), while the *Monthly Bulletin of the RSAA* promised that reintroducing "the art of architecture in the midst of public life, as a main component and necessity of it, [was] the [journal's] ultimate goal" (KMBA, 1930: 1). From 1909 onwards, and in the slipstream of the 1906 law and the establishment of the SCAA, the young watchmen seized the opportunity to start a public debate about the future development of suburban Antwerp. It was a debate they would moderate until 1939, through publications, lectures, and exhibitions, until these "young watchmen" slowly became an "old guard" themselves.

Whereas *De Bouwgids* mainly promoted the image of the garden city, the more modernist-inspired *Monthly Bulletin of the RSAA* found more connection with the modernist movement and the architectural programs of the expanding metropolis. *De Bouwgids's* first articles were directly inspired by De Vos-Van Kleef's remarkable 1907 book *Le nouvel Anvers: cité-jardin*, which portrayed the future Antwerp surroundings as one large *garden-like* extension of the historic city. The image of the garden city referred to the architectural language of the British Garden Suburbs designed by Unwin and Parker, rather than to the garden city concept as it had been defined in the programmatic writings of Ebenezer Howard. On the one hand, this language can be interpreted as pragmatic, but at the same time, precise thought was given to how a vocabulary of concretely designed neighbourhoods, including amenities, could be used to gradually turn suburban land into an urban agglomeration in the making.

The *Monthly Bulletin of the RSAA* was launched on the occasion of the 1930 world exhibition in Antwerp and envisioned a more metropolitan architecture that was mainly applied along the major arterial avenues both inner-urban and suburban. These reflections are similar to the ideas propagated by Hilberseimer in his *Grosstadt Architektur* (1927), a book which tried to give form to the development of the city beyond its historical core through the architectural definition of the large programs that could establish a coherent architectural vocabulary of urban types.

But rather than choosing between those seemingly opposing models, many of the young watchmen tried to combine both perspectives, echoing two complementary logics of urbanization that were also promoted through the SCAA (see Figure 13.2). On the one hand, the SCAA negotiated with (mostly aristocratic) landowners to provide their land with roads and amenities at their own expense, which became part of the public domain only after completion. This process often resulted in picturesque street patterns suitable for the development of garden suburbs, which also matched the rather conservative tastes of the landowners and many (often Catholic) suburban municipalities. On the other hand, the SCAA had started a program to substantially widen supra-local arterial roads running through the city and its suburbs. The resale of expropriated land was to finance these operations. It was along these boulevards, both in the city and its suburbs, that the first experiments of "metropolitan architecture," upon which the *Monthly Bulletin*

Figure 13.2. Complementary suburban development logics. Upper: the widening of public state roads and the rise of metropolitan suburbia. Building by contractor François Amelinckx in the middle. Lower: public-private cooperation for local roads. Source: Antwerp City Archives.

of the RSAA commented, were built. These exercises in *Grosstadt Architektur* answered to the needs of new programs such as offices, banks, century hotels, and apartment buildings, which were mainly commissioned by more progressive industrial elites and entrepreneurs. Mostly, the newly established suburban boulevards were considered the perfect playground for these scaled-up metropolitan experiments, giving shape "instantly" to new suburban milieus where little else was yet in place.

Architects such as Portielje & De Braey, Cols & De Roeck, Smolderen & Van Beurden, Van der Paal, De Ridder, and Van Steenbergen all switched freely between these two currents, depending on the developmental dynamics they were part of (see Figure 13.3). They were prone to a certain pragmatism, which did not, however, lead them to build just anything. Their creative energy was not put towards individual artistic expression or signature pieces but rather towards the pursuit of

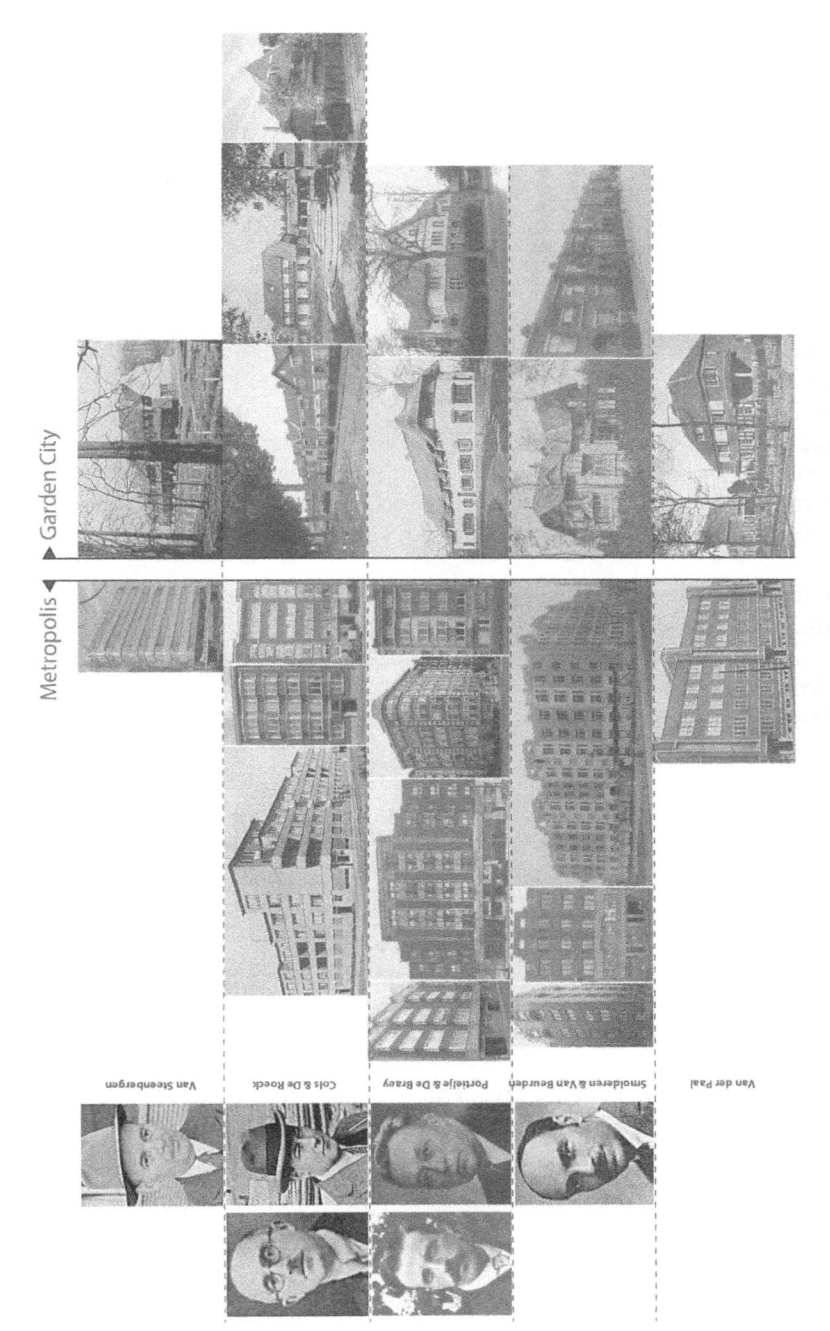

Figure 13.3. Architecture beyond individual ideology and the young watchmen's simultaneity of garden-suburb and metropolitan architecture. Source: Collage by the authors, with photos from *De Bouwgids*, Inventaris Onroerend Erfgoed, Antwerp City Archives, and Jubelboek der KMBA.

contrasting but internally coherent design that could define the physiognomy of these newly established suburban milieus.

In addition, these architectural imaginaries were widely applicable and responded to the aspirations of different social classes. It was, for instance, argued that some premature version of the garden city had always existed in the Antwerp suburbs and that "the garden city [responded] to the aspirations of the greatest number of people" (De Vos-Van Kleef, 1907: 16). In addition to the usual private villas for wealthy clients, the young watchmen quickly started working for a variety of clients such as housing cooperatives, real estate companies, and social housing companies. Both in the garden city and in the metropolitan experiments, the young watchmen were able to build for broad layers of society. Moreover, the architects' contributions were not limited to housing alone but also included a whole range of public projects such as schools, taverns, swimming pools, park pavilions, soccer stadiums, leisure domes, an airport, and the like. And so *De Bouwgids* architects massively contributed to building a mixed, yet shared suburban milieu, a setting for the *vivre-ensemble* of different social classes, characterized by a tactile, lively, and convivial red-brick architecture – a social achievement that had been considered difficult to achieve in the early 1920s, as critics claimed that "it [was] tempting to think that the layouts of the plans (cozy living room, bathroom, small garden) [were] purely theoretical and even somewhat utopian" (De Heem, 1921: 6) for the working poor to access.

By linking different architectural images to different developmental logics, the young Antwerp architects were moving away from the idea of architecture as a praxis of individual artistic inspiration and creativity, and refashioning architecture as a social practice. A 1935 Royal Society of Antwerp Architects publication put it in these words:

> The master builders, who some fifty years ago were mainly involved in the design of buildings of some importance, have kept pace with the times. They started contributing to the solution of social issues, of cheap housing, sanitary requirements and comfort in housing, economy in execution, and more specifically in terms of apartment buildings and the construction of garden districts. (KMBA, 1935: 109)

In the fashioning of suburban Antwerp, creativity not only coincided with the individual and pure impulses of visionary master builders within the privacy of their own studios but also aligned with the public quest for new and shared imaginaries that had to be intertwined with very concrete development dynamics rooted in the mess of the city-making process.

Building the Suburbs: Transactions between Architects and Contractors

From the end of the nineteenth century, the Belgian construction industry underwent a deep transformation that led to the gradual replacement of specialist craftsmen by general contractors (Bertels, 2018). Founded in 1865, the firm Blaton-Aubert, for instance, was a true pioneer, and through this transformation was able to refashion itself as one of Belgium's most notorious building companies (Culot et al., 2018). In the context of Antwerp, large building firms like Van Riel & Van den Bergh (1904) and Vooruitzicht N.V. (1905) devoted themselves to a building's entire construction process (Entreprises générales, 1930; Hypotheekbank Vooruitzicht N.V., 1930). In general, the established architects deplored this evolution, an opinion also voiced in *De Bouwgids* (De Vos-Van Kleef, 1912). They feared that the extinction of craftsmanship would lead to architectural impoverishment and negatively affect their autonomy and personal creativity.

However, the architects of the young watchmen did not support these views. In his sharp style, De Vos-Van Kleef wrote that it would only lead to "more work for the architect for less pay" (De Vos-Van Kleef, 1912: 54). He concluded that the old guard architects had to be "*übermenschen*" who placed the "*true principles of architecture* far above the worldly concerns of its servants," fulfilling "no less than the role of the Messiah, driving the cheapskates out of God's Temple!"[3] (De Vos-Van Kleef, 1912: 55). The young watchmen seems to have been more open towards working with general contractors – an openness that would especially affect their "suburban" creativity.

As it was not yet mandatory at the time to engage an architect in order to get a building permit, these general contractor firms, along with all kinds of land surveyors, were literally building urban development all by themselves. This practice was particularly true for the suburbs, where very few building regulations applied and new buildings did not have to conform to a pre-existing urban context. "As soon as a meadow, no matter how big or small, that is close to a new tram-line becomes available, all sorts of earthworms start eating it away and land surveyors start drawing the highest number of plots on the smallest surface. And then ... once the meadow has been parcelled out, one sees the infamous yellow houses rise up from the toy box filled with blocks" (Van Reeth, 1912: 81). Concerning the latest expansions of the suburban hamlet Sint-Mariaburg, De Vos-Van Kleef said: "There are also loads of houses there that are full of Blaton-Aubert, and they all have a gloomy and sad appearance" (De Vos-Van Kleef, 1907: 67). Whereas the old guard turned its back on these practices, the young watchmen were more opportunistic, but also seem to have believed that a rapprochement with such contracting firms was necessary in order to improve the quality of these projects. Their ambition reflected a more general rapprochement between architects and contractors in the Belgian context during the interwar period (Horta and Van de Velde, 1933) – a rapprochement that gained even more momentum with the rise of modernism in the early 1930s, which started, among other things, from a strong fascination for the aesthetics of (industrial) building processes and construction techniques.

An excellent example of this early rapprochement is the fact that even a contractor such as François Amelinckx (renowned in Belgium for his utterly banal post-war housing production deprived of any architectural consideration) cooperated with fairly interesting architects in suburban Antwerp (Broes, 2018). Most insightful, however, is the fascinating 1930 publication *Entreprises générales de construction Van Riel & Van den Bergh*, which bears witness to an almost structural collaboration between the firm of Van Riel & Van den Bergh and several young watchmen. The company was a concession holder of the French company Bétons armés Hennebique and a forerunner of applications in reinforced concrete in Antwerp. Architects such as Francken (with whom Van Riel had built Antwerp's first high-rise apartments, the Goliath Residences in the Helenalei), Portielje & De Braey, Van Beurden, Dens, and Ceurvorst all appear as (regular) clients of the firm. It is therefore no coincidence that the company became one of the first advertisers in the *Monthly Bulletin of the RSAA*. Their 1930 brochure displays a wide variety of buildings, ranging from metropolitan apartments, banks, cafés, theatres, and garden villas to social housing, factories, and even dry-docks. In other words, the firm was almost literally "building the urbanization process" at the turn of the twentieth century, both in the city and its suburbs.

Portielje & De Braey, whose projects are the most represented in the brochure, was especially very close to the building company. In the 1920s, Van Riel & Van den Bergh had founded the company Le Confort in order to build three apartment complexes, which were all designed by Portielje & De Braey (Agentschap Onroerend Erfgoed, 2020). It seems that their close collaboration with this contractor brought them to places and commissions for which they would otherwise never have been considered – including the building of suburban factories that are usually missing from these architects' historiographies. When studying Portielje & De Braey's projects from the brochure, it is striking that they became much more experimental as the context became more suburban (Figure 13.4). Whereas their inner-urban projects still conform to the tempered scale, intricate parcels, and style of the historic city, their suburban projects become much more free and experimental in

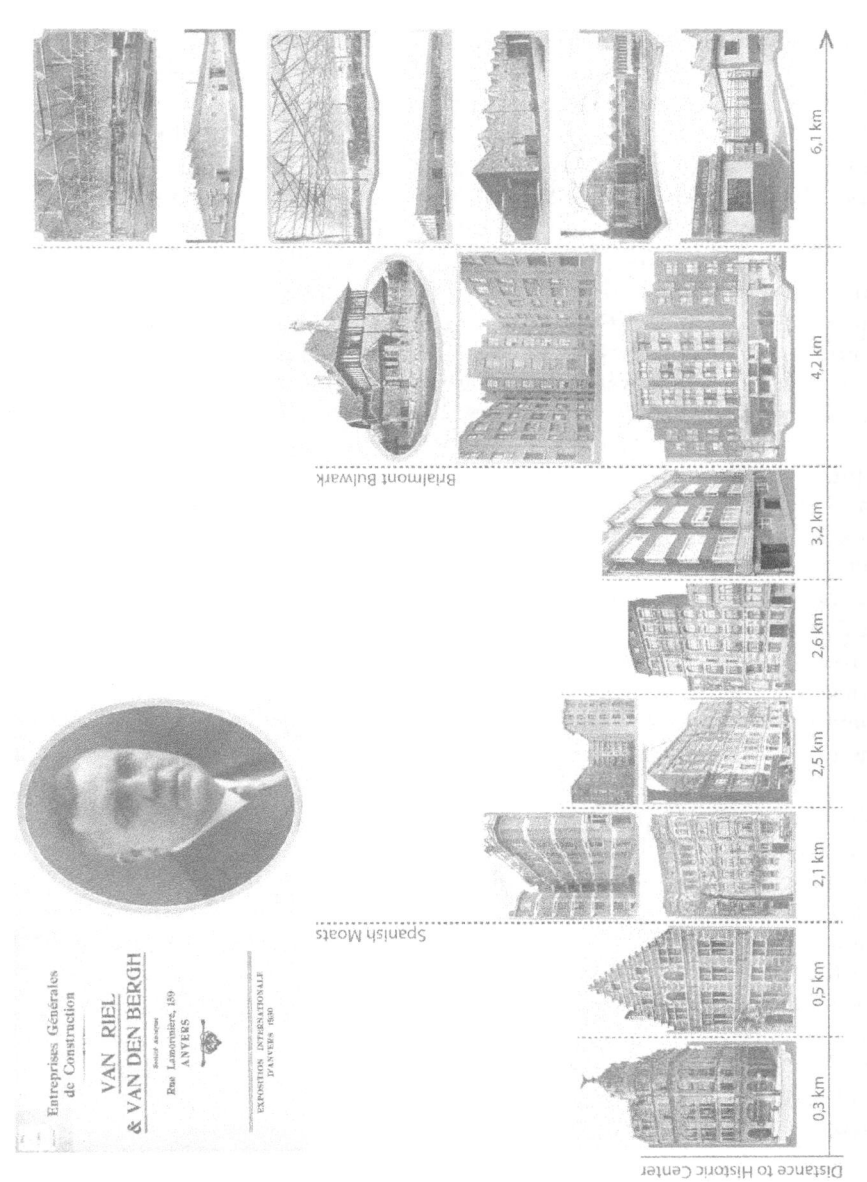

Figure 13.4. Projects by Portielje & De Braey with contractor Van Riel & Van den Bergh: the further removed from the city centre, the greater the freedom for architectural renewal and experiment. Source: Collage by the authors, with photos from Van Riel & Van den Bergh, 1930.

terms of scale, program, figure-ground, and style. Architectural creativity in these suburban projects was no longer rooted in the mastering of abstract style paradigms that somehow had to be executed but in mastering the act of building itself. Or, in the words of the famous modernist Ludwig Mies Van der Rohe: "Our task is precisely to liberate building activity from the aesthetic speculation of developers and to make it once again the only thing it should be, namely, BUILDING" (Mies Van Der Rohe, 1923). Arguably, it is this kind of building aesthetics that characterizes Portielje & De Braey's more suburban projects for Van Riel & Van den Bergh, such as the factories for the companies Bell Telephone and Minerva.

The suburbs functioned as a test bed for new construction methods stimulating a joined quest of contractors and architects for new typologies and aesthetics that were rooted in the act of "building." The sheer scale of these new general contracting firms made them look for spacious plots in the suburbs themselves that were suitable for large ateliers and sufficient stockroom. From 1929 onwards, for example, the company Van Riel & Van den Bergh settled on a site of more than 6,000 square metres in suburban Edegem. Large construction firms simply became part and parcel of urbanization dynamics in the suburbs, which may partially explain why the large majority of modern building experiments appeared in suburban Antwerp during the interwar period (Kennes et al., 1992). These experiments became the expression of a new material culture and aesthetic made of the creative processing of brick, steel, glass, and reinforced concrete, realized through the exchange between young architects and new creative entrepreneurship in the field of general contracting.

The Architect-Developer: In Search of Control over the Architecture of Suburbanization

The law of 1906 and the prospect of large scale suburbanization gave a boost to the real estate market in and around Antwerp at the beginning of the twentieth century and led to the creation of numerous and sometimes rather obscure development companies (Bunge, 1907; Delbeke, 1908; Schobbens, 1910; KMBA, 1935). As turning rural land into urban land was very lucrative, most of these companies targeted the suburbs as their preferred terrain of action. In general, architects feared – not unjustifiably – that promoting spatial and aesthetic qualities was not among the main motives of these groups. But, at closer look, the great variety of development companies also left room for architects to creatively engage.

Browsing through the founding deeds of many real estate companies reveals that quite a lot of them had architects among their shareholders, such as, for example, the firms Société anonyme pour l'Exploitation commerciale et industrielle de la banlieue d'Anvers (1905), Extensions et Entreprises Anversoises (1910), Groenborger Tuinstede (1911), Edeghem Extensions (1912), Ter Rivieren Extensions (1922), Compagnie des habitations Conforta (1925), Compagnie immobilière du Nord d'Anvers (1927), and Te Boulaer Extensions (1928).[4] More research is needed to clarify the exact role these architects played in those real estate companies. Yet the architectural and spatial qualities of many of these companies' development schemes, ranging from high-end villas to worker's housing, is striking (Figure 13.5). Arguably, their position within these larger companies allowed them to monitor spatial and aesthetic qualities beyond the individual building – to monitor, in sum, the architecture of suburbanization at large.

A number of young watchmen went one step further in gaining control over the architecture of suburbanization by founding their very own real estate companies – an idea that was occasionally put forward in *De Bouwgids*. In 1921, *De Bouwgids* editor De Vos-Van Kleef engaged in such an experiment in his suburban hometown Kapellen. Through a local notary, he had come into contact with a "consortium of landowners … who wanted nothing better than to use their beautiful forests to design a garden district" (De Vos-Van Kleef, 1924: 165). "And on the other hand," he continued, "I was able to secure the cooperation of six architects, who also wanted to take a financial interest, and so the N.V. Tuinwijk-Vredeburg came into being on 20 August 1921" (165). Among these architects were young watchmen members and *De Bouwgids* editors Paul Smekens, Flor Van Reeth, and Egide Van der Paal. As such, the architects managed "to start by designing a completely new street network on which all the buildings would be accurately marked with front gardens, squares, streets, and alleys" (165). In addition to housing, the general development schedule also included a number of public facilities, such as schools, sports facilities, and shops. At the time of publication, in 1924, the first ten country houses were built by architect Smekens. It was hoped that they would be sold before 1925, when Flor Van Reeth was asked to design the next ten houses. But it did not get that far; only these first ten houses were ever built (168). Despite the proximity of a station and a direct connection to Antwerp, there was no strong housing demand in remote Kapellen. The project turned out to be an act of poor judgement by architects whose beaux-arts education hadn't really prepared them for commercial entrepreneurship.

Figure 13.5. The agglomeration as an urban-suburban continuum. Architects infiltrated real estate companies and got a grip on the architecture of suburbanization. Projects by De Vos-Van Kleef, Van der Paal, and Smekens for building society N.V. Tuinwijk-Vredeburg (left); Jos Hertogs and Eugène Touret for building society S.A. Extensions et Entreprises Anversoises (middle); and Gerard De Ridder for building society N.V. Ter Rivieren (right). Source: Collage by the authors, with photos from *De Bouwgids*, Hendrik Conscience Bibliotheek, and Antwerp City Archive.

In the apartment sector as well, similar kinds of cooperation between architects and developers emerged. Along the Jan Van Rijswijcklaan, architects such as Portielje & De Braey, Stynen, and Ceurvorst all worked in close cooperation with real estate firms such as the Compagnie Nationale Immobilière, Home Van Rijswijck N.V., and others. What's more, some architects themselves started to act as project developers. Jan De Braey, for instance, developed a building on the Jan Van Rijswijcklaan, which he moved into himself and partly rented out (Broes and Dehaene, 2017b). Paul Smekens, in turn, developed a rental apartment complex with a commercial base in suburban Kruishofstraat (Laureys, 2004: 237). Most of the time, however, this dual role remained a one-off experiment. Nevertheless, a young man called Jean-Florian Collin (1904–85), founder of the development company Etrimo, would end up taking the principle of the architect-developer to a completely different level.

It is worth noting that Collin didn't study at Antwerp's Académie Royale des Beaux-Arts. As a graduate of the military engineering school in Berchem in the early 1920s, he nevertheless went down in history as an architect (Van Loo, 2003). But he followed an alternative trajectory. His reading of Sir Harold Bellman's (1928) *The Silent Revolution*, concerning the influence of building societies on the modern housing problem, drove him to undertake a journey to Great Britain – and later through the whole of Europe – to study the ways in which building societies had been able to literally build the urbanization process at large.[5] At a time when the architectural avant-garde travelled to Rome or Athens in search of the foundations of a veritable modern architecture, Collin was wandering through Manchester's grey streets to experience for himself what had come of the English building societies. He reported his conclusions in his intriguing 1938 book *L'épargne immobilière et sa fonction sociale*.

Collin's book reads as an open plea for mutual enterprise between developers, architects, and public authorities. Large savers' cooperatives would unite individuals and give them the capital strength to realize "a harmonious urbanization plan unique in the world" (Collin, 1938: 273). For this purpose and in order to curb speculation, municipal and regional design competitions need to be organized and building regulations installed requiring the cooperation of local political boards, architects, urbanists, and engineers (273–8). What Collin had in mind was a form of suburbanization requiring minimal state investment that was profitable for developers but would be built according to shared spatial and aesthetic standards and aspirations of urbanists, architects, and even future residents.

Collin's utopian plan was never systematically implemented on a large scale, but it is symbolic of the experimental climate and to some extent representative of actual practices during the interwar period, a climate in which entrepreneurial architects were encouraged to explore alternative modi operandi for suburban development in transaction with an emerging scene of real estate agents. New creative connections between architects and real estate developers encouraged a deeper understanding and recalibration of the link between economic and aesthetic capital in urban development.

Transactions with Public Authorities: Building Regulations as a Mould for the Suburban Landscape

At the beginning of the twentieth century, only established historical cities like Brussels and Antwerp had proper building codes in place systematically regulating urban development. In the suburban territory beyond these cities, local authorities had little means of controlling urban growth. As far as the regulation of urban form was concerned, let alone its aesthetic qualities, municipalities could only intervene on the basis of police law concerning health and safety. A 1914 law changed this situation, but it was infrequently applied, so "aspects of good layout and beauty in cityscapes still had such limited significance that never too much attention was devoted to it by public administrations" (Oostvogels, 1937: 27).

Throughout the 1920s, the focus seems to have been more on postwar economic recovery than on aesthetic beauty per se, but from the 1930s onwards, the question of urban aesthetics and building codes again became a general matter of concern, especially in the suburban context (Houtart, 1938; Verwilghen, 1938). In the *Monthly Bulletin of the RSAA*, architects increasingly expressed their fear that "the harmonious development of Greater Antwerp was threatened by the lack of proper building codes" (Suykerbuyk and De Ridder, 1937: 165). The alarming conclusion was drawn that, "of the 151 municipalities in the province, at most about 30 [had] some kind of building regulations" (Wittockx, 1937: 98). The young watchmen saw a solution through infiltrating the public administration by adding a number of architects to the SCAA, which was, after all, responsible for the development of the Antwerp suburbs. In a 1932 opinion piece, the architects offered to share their expertise with the rather technocratic committee in order to assist the SCAA in terms of "architecture, hygiene, aesthetics, building styles, etc., in order to allow the committee to take more broadly informed decisions" (KMBA, 1932: 208). As we will see, some of the young watchmen, such as De Braey, Wittockx, and Van der Paal, were prepared to take on policy roles to explore architectural practices beyond the act of building and act in an advisory capacity, especially in the suburban context, which lacked a clear spatial and legal framework.

However, the young watchmen never made it onto the SCAA staff. Their concerns were answered in a different way. In 1934, a provincial Committee for the Development of Building Regulations was founded. SCAA secretary and registrar of the province, Joseph Schobbens, chaired the committee, which aimed at remedying the "ensuing unhealthy and insane speculation on land [that has] led to such a situation that many of these complexes fall far short of what can and should reasonably be expected in terms of public health, morality, and beauty" (Oostvogels, 1937: 7). Three RSAA architects (De Braey, Ritzen, and Wittockx) were appointed to assist in creating the new provincial building regulations. As the provinces were mainly responsible for making overarching policy for small municipalities beyond the historical cities, it is fair to state that these architects mostly engaged in suburban policymaking. From a few articles in the *Monthly Bulletin of the RSAA*, it becomes clear that this task required a different kind of creativity from the architects. Architect Van der Paal wrote in 1936:

> Whereas in the past, a great deal of freedom was permissible and arguments of pure art fantasy prevailed, the question now arises quite differently ... Consideration should be given to, among other things, professionalism and well-considered compositions; financial possibilities; technical ingenuity; etc. ... Urban imaginaries, street and building profiles, in one word, the neighbourhoods and cities, will be based on order, arrangements, and stringent application of ordinances in such a way that will confirm the modern beauty in urbanism. (Van der Paal, 1937: 289)

Again, highly personal and artistic views of creativity made way for more relational approaches, facing up to many contingencies of the real world. As such, an architectural practice emerged that explored different imaginaries of suburban development on the basis of which a meaningful ordinance could be built – and vice versa: what kind of suburban landscapes were produced by which regulations?

The architects were thrilled and convinced "that the creation of the provincial building commission [would] contribute to the improvement of the architectural and aesthetic character of the constructions of town and country alike" Wittockx (1937: 100). But in the end, the code ended up an amalgamation of utterly generic rules, which were published as a manual in 1937. Specific regulations on aesthetics were only reactive in nature, allowing municipal councils to deny building permits that threatened to tarnish the aesthetic harmony of the built environment (Oostvogels, 1937: 481–2). But there were hardly any tools to actively promote and guard aspects of urban beauty. According to the model code, suburban municipalities were supposed to fill up with 6 metre-wide two-storey houses with pitched roofs and rear structures of maximum 2.5 metre height. This description is very similar to the generic "land surveyor's section" that, according to many, helped Belgium eventually evolve into the "ugliest country in the world." It strongly seems that, in the translation from suburban "imagination" to "building regulations," things had gone wrong. Arguably, it needed more than some generic rules to actively pursue the imaginaries that the RSAA architects had had in mind to develop the suburbs.

In his 1933 study, "The Problem of Regulating the Height of Buildings," which was republished by installment throughout the 1934 editions of the *Monthly Bulletin of the RSAA*, Belgian architect Victor Horta came to a similar conclusion. His study was a reaction to the appearance of higher and larger buildings, mainly along suburban thoroughfares. Horta concluded that urban aesthetics could never be an absolute and predefined category but rather was dependent on many contingencies that had to be considered case by case (Horta, 1933). In the context of suburban Antwerp, at least some architects had aspired to having a day-to-day position within the provincial authorities and within the SCAA, from which they could have developed a more meaningful and integral approach towards regulating suburban development. Yet they never advanced further than co-writing a generic building code. In a way, the architects ended up contributing to the implementation of rigid building regulations undergirding suburban development that they themselves afterwards despised.

Conclusions

Throughout this chapter, we have shown how the suburban phase of the urbanization of the Antwerp agglomeration at the turn of the twentieth century brought about a shift in the professional position of the architect. In the early twentieth-century suburbs, a new watch of young architects was eager and able to break away from their beaux-arts training at Antwerp's Académie Royale as liberal artists, relying on "formal dogmas" and "lifeless art principles" that they themselves considered out of tune with the concrete desires and development logics of an emerging suburban society. We have argued that this suburban phase encouraged and obliged these architects to engage in concrete transactions with public, technological, economical, and legislative spheres. It forced them to explore other, more relational forms of creativity by engaging with emerging forms of entrepreneurship that came with suburban expansion. A lot of creativity was needed to explore practices beyond the traditional field of the architecture of the individual building and construct an architecture fit for the suburbanization process. These evolutions are in line with reflections that geographers like Gottmann made in the era of massive suburbanization, stating that "the profession of city planner and architect should certainly not renounce their artistic formation; but their training needs to include some competence in many fields they rather despised in the past: the financing of the construction; the economics of the area that they will build; the geography of a wider region around it; and the social needs and habits of the users" (Gottmann, 1962).

In order to have a real and significant impact on the spatial qualities of suburban development, these young architects denounced their purely artistic position and engaged in new and more entrepreneurial modes of city-making. Moderating the public debate and engaging in mutual enterprises with new building agents, real estate agents, and policymakers, these young architects tried to free architectural practice from the clutches of beaux-arts formalism, monumentalism, and abstract thought that had, according to them, still strongly informed architects in building the nineteenth-century city. They were, to paraphrase Jeremy Till, moving away from the individual muse and the contingency of the self as a leitmotiv and instead rooting their profession in an engagement with the intractable contingencies of the real world (Till, 2009; Rorty, 1989). As such, they began acting as spatial agents with *reformative intent*. This methodological shift from liberal artistry to spatial agency went hand in hand with a shift from aspirations of a "particular monumentality" to an embrace of the "shared every day" – a shift also from the particular building to the urban fabric and the gross domestic capital of building in general. Moreover, this creative turn occurred thanks to and in support of suburban development, not in spite of it (Bain, 2013).

In *The Creative Mind*, philosopher and cognitive scientist Margaret Boden identified two senses of creativity in this regard. P-creativity is an idea that is novel "with respect to the individual mind which had the idea," while H-creativity concerns an idea that is "novel with respect to the whole of human history" (Boden, 1990: 32, 8). H-creativity is therefore a historical category that "depends on shared knowledge and shifting intellectual fashions" (32, 8). Arguably, the shift from city to suburbanization at the turn of the twentieth century in Antwerp enabled such a transition from P- to H-creativity in architecture. It was a shift that encouraged architects to become part of the suburban process at large, which inevitably depended on "the working together of many hands and many minds to secure common ends which are quite unattainable by the unaided efforts of isolated individuals" (Unwin, 1924: 16).

NOTES

1 The Royal Society of Antwerp Architects (RSAA) is a translation of the Koninklijke Maatschappij der Bouwmeesters van Antwerpen (KMBA).
2 All translations, unless otherwise indicated, are by the authors.
3 Italics by original author.
4 Among these architects, we find J. Hertogs, W. Van Kuyck, M. De Braey, J. Dries, M. Winders, J. Van Hoof, and P. Van Oenen.
5 In Great Britain, for instance, private building societies managed to build next to one million houses between 1920 and 1936. At the time of publication, it was estimated that their total production approximated 2.25 million houses (Collin, 1938: 72, 272).

REFERENCES

Agentschap Onroerend Erfgoed. 2020. "Portielje – De Braey." Portielje – De Braey: *Inventaris Onroerend Erfgoed*. https://id.erfgoed.net/personen/8383

Awan, N., T. Schneider, and J. Till. 2011. *Spatial Agency: Other Ways of Doing Architecture*. New York: Routledge.

Bain, A. 2013. *Creative Margins: Cultural Production in Canadian Suburbs*. Toronto: University of Toronto Press.

Bellman, H. 1928. *The Silent Revolution*. London: Methuen.

Benevolo, L. 1975. *The Origins of Modern Town Planning*. Cambridge, MA: MIT Press.

Berger, A. 2006. *Drosscape:Wasting Land in Urban America*. New York: Princeton Architectural Press.

Bertels, I. 2018. "Building Contractors in Late Nineteenth-Century Belgium: From Craftsmen to Contractors." In A. Becchi, R. Carvais, and J. Sakarovitch, eds., *L'histoire de la construction: Relevé d'un chantier européen*. Paris: Classiques Garnier, 1139–67.

Boden, M. 1990. *The Creative Mind: Myths and Mechanism*. London: Weidenfeld and Nicolson.

Bridge, G. 2013. "A Transactional Perspective on Space." *International Planning Studies*, 18(3–4): 304–20. https://doi.org/10.1080/13563475.2013.833728

Broes, T. 2018. "The Architect, the Contractor and the Developer: The Rise and Fall of Real-Estate Architecture in Belgium, 1918–1985." In B. Burquel, T. de Roquemaurel, M. Dumont, and G. Joachim, eds., *The Apartment Building*. Vol. 1 of *Real Estate Architecture*. Brussels: REA ASBL, 14–22.

Broes, T., and M. Dehaene. 2017a. "Mastering the Urbanization Process: The Urban Questions of Engineer August Mennes in the Antwerp Agglomeration." *Planning Perspectives*, 32(4): 503–31. https://doi.org/10.1080/02665433.2017.1314793

Broes, T., and M. Dehaene. 2017b. "Real Estate Pioneers on the Metropolitan Frontier: The Works of Jean-Florian Collin and François Amelinckx in Antwerp." *Cities, Communities and Territories*, 33: 89–112. https://doi.org/10.15847/citiescommunitiesterritories.dec2016.033.art06

Bunge, E. 1907. "Lettre de M. Ed. Bunge à Mr. Ernest Bunge du 22 juin 1907." Koninklijke Heemkring Hoghescote, Folder Edouard Bunge.

Burgel, G., ed. 1989. "La ville fragmentée. Le lotissement d'hier et d'aujourd'hui." Thematic Issue, *Villes en parallèle*, 14.

Castells, M. 1975. *La question urbaine*. Paris: Maspero.

Casus. 1911a. "Over de Kunstopvoeding van den Bouwmeester." *De Bouwgids*, 3(1): 18–19.

Casus. 1911b. "Over de Kunstopvoeding van den Bouwmeester." *De Bouwgids*, 3(2): 31–2.

Chafee, R. 1977. "The Teaching of Architecture at the École des Beaux-Arts." In A. Drexler, ed., *The Architecture of the École des Beaux-Arts*. London: Secker and Warburg, 69.

Choay, F. 1965. *L'urbanisme, utopies et réalités: une anthologie*. Paris: Seuil.

Colenbrander, B. 1993. *Stijl: norm en handschrift in de Nederlandse architectuur van de negentiende en twintigste eeuw*. Rotterdam: NAi Publishers.

Collin, J.-F. 1938. *L'épargne immobilière et sa fonction sociale*. Paris: Librairie générale de droit et de jurisprudence.

Corijn, E. 2006. "Urbanity as a Political Project: Towards Post-National European Cities." In L. Kong and J. O'Connor, eds., *Creative Economies, Creative Cities: Asian-European Perspectives*. Dordrecht: Springer, 197–207.

Culot, M., B. Espion, and M. Provost. 2018. *Blaton, une dynastie de constructeurs*. Brussel: Archives d'architecture moderne.

De Bouwgids. 1911. "Aan onze lezers [Editorial]." *De Bouwgids*, 3(1): 1.

De Heem, P. 1921. *La construction des villes et cités-jardins à la conférence de Londres*. Brussels: Goemaere.

Delbeke, A. 1908. *Plechtige aanstellingszitting der commissie tot inrichting der Antwerpsche Agglomeratie gehouden in het Provinciaal bestuur van Antwerpen den 8 februari 1908*. Anvers: Aug. Van Neylen.

De Vos-Van Kleef, D. 1907. *Le nouvel Anvers: cité-jardin*. Antwerp: Nederlandsche Boekhandel.

De Vos-Van Kleef, D. 1912. "De Oude Garde en de Jonge Wacht." *De Bouwgids*, 4(3): 53–6.

De Vos-Van Kleef, D. 1924. "Tuinwijk Vredeburg." *De Bouwgids*, 16(9): 165–72.

D.P.S. 1910. *Exposition Anvers-Extension: principes pour guider le visiteur*. Anvers: n.p.

Duany, A. 2004. "The Beaux-Arts Model." In S.E. Bothwell, A.M. Duany, P.J. Hetzel, S.W. Hurtt, and D. Thadani, eds., *Windsor Forum on Design Education: Toward an Ideal Curriculum to Reform Architectural Education*. Miami: New Urban Press, 133.

Duany, A., E. Plater-Zyberk, and J. Speck. 2000. *Suburban Nation: The Rise of Sprawl and the Decline of the American Dream*. New York: North Point Press.

Entreprises générales de constructions Van Riel & Van den Bergh. 1930. *Entreprises générales de construction Van Riel & Van den Bergh, Société Anonyme, Rue Lamorinière 159, Anvers*. Anvers: Exposition Internationale d'Anvers.

Ford, G.B. 1913. "The City Scientific." *Engineering Record*, 67: 551–2.

Giddens, A. 1984. *The Constitution of Society: Outline of the Theory of Structuration*. Berkeley: University of California Press.

Giddens, A. 1987. *Social Theory and Modern Sociology*. Cambridge: Polity.

Gottdiener, M. 1977. *Planned Sprawl, Private and Public Interests in Suburbia*. Beverly Hills, CA: Sage Publications.

Gottmann, J. 1962. *Economics, Esthetics and Ethics in ModernUrbanization*. New York: Twentieth Century Fund.

GUST (Ghent Urban Studies Team). 2002. *Post, Ex, Sub, Dis: Urban Fragmentations and Constructions*. Rotterdam: 010 Publishers.

Harris, R. 2013. "Suburban Land." In R. Keil, ed., *Suburban Constellations: Governance, Land and Infrastructure in the 21st Century*. Berlin: Jovis Verlag GmbH.

Harvey, D. 1973. *Social Justice and the City*. Baltimore: Johns Hopkins University Press.

Harvey, D. 1985. *TheUrbanizationof Capital*. Oxford: Blackwell.

Harvey, D. 1989. *The Urban Experience*. Baltimore: Johns Hopkins University Press.

Hawkins, H. 2017. *Creativity*. London: Routledge.

Henderson, G. 2004. "'Free' Food, the Local Production of Worth, and the Circuit of Decommodification: A Value Theory of the Surplus." *Environment and Planning D: Society and Space*, 22(4): 485–512. https://doi.org/10.1068/d379

Hickman, L.A., S. Neubert, and K. Reich, eds. 2009. *John Dewey: Between Pragmatism and Constructivism*. New York: Fordham University Press.

Horta, V. 1933. "Le problème de la réglementation de la hauteur des bâtiments." *Bulletin des commissions royales d'art et d'archéologie*, 72: 86–117.

Horta, V., and H. Van de Velde. 1933. *Rationalisation de l'habitation et de l'industrie du bâtiment*. Brussel: Société Belge des urbanistes et architectes modernistes.

Houtart, A. 1938. *Urbanisme en de Wet*. Brussels: Imprimerie Guyot.

Hypotheekbank Vooruitzicht N.V. 1930. *Verslag over het 25 jarig bestaan der maatschappij in feestzitting van 22 februari 1930, 1905–1930*. Antwerp: De algemene drukkerij.

Ingersoll, R. 2006. *Sprawltown: Looking for the City on Its Edges*. New York: Princeton Architectural Press.

Kennes, H., G. Plomteux, R. Steyaert, and S. Van Aerschot-Van Haeverbeeck. 1992. *Bouwen Door De Eeuwen Heen: Inventaris Van Het Cultuurbezit In België: Architectuur. 3nd: Stad Antwerpen, Fusiegemeenten*. Turnhout: Brepols.

KMBA. 1930. "Aan den lezer [Editorial]." 1930. *KMBA: maandschrift der Koninklijke Maatschappij der Bouwmeesters van Antwerpen: Bouwkunst, Stedenbouw, decoratieve kunsten*, 1(1): 1.

KMBA. 1932. "Grooter Antwerpen." *KMBA*, 3(11): 205–9.

KMBA. 1935. "Inrichting der urbanisatie in België." *KMBA*, 6(4): 101–9.

Laureys, D. 2004. "De architectuur in een stroomversnelling. Art-Deco, modernisme en traditionalisme in de provincie Antwerpen." In W. Aerts and D. Laureys, eds., *Bouwen in beeld: de collectie van het Architectuurarchief van de Provincie Antwerpen*. Turnhout: Brepols, 85–129

Lefebvre, H. 2014. *Towards an Architecture of Enjoyment*. Edited by L. Stanek. Translated by R. Bononno. Minneapolis: University of Minnesota Press.

Lerup, L. 2001. *After the City*. Cambridge, MA: MIT Press.

Lombaerde, P. 2008. "Architectuur, stedenbouw en verkeer." In R. Loyen, E. Buyst, and S. Vanfraechem, eds., *Antwerpen in de 20ste eeuw: van Belle-Epoque tot Golden Sixties*. Antwerp: Pandora Publishers, 209–34

Lynch, K. 1966. *The Image of the City*. Cambridge, MA: MIT Press.

Mantziaris, P. 2014. *La ville-paysage – Rudolf Schwarz et la dissolution des villes*. Geneva: Métis Presses.

Mies Van Der Rohe, L. 1923. "Office Building." *G: an avant-garde journal*, 1(July): 102–3.

Migom, S. 2004. "Bouwen van revolutie tot wereldbrand." In W. Aerts and D. Laureys, eds., *Bouwen in beeld: de collectie van het Architectuurarchief van de Provincie Antwerpen*. Turnhout: Brepols, 17–51.

Nevejans, A. 2011. "Kring voor Bouwkunde (1899–)." *ODIS*. Record last modified 27 May 2017. http://www.odis.be/lnk/OR_17501

Oostvogels, J. 1937. *Handleiding voor de bouwverordeningen van de provincie Antwerpen: met toelichtingen*. Antwerp: Nederlandsche boekhandel.

Phelps, N.A. 2015. *Sequel to Suburbia*. Cambridge, MA: MIT Press.

Picon, A. 1992. *L'invention de l'ingénieur moderne: L'École des Ponts et Chaussées, 1747–1851*. Paris: Presses de l'École nationale des ponts et chaussées.

Rabinow, P. 1989. *French Modern, Norms and Forms of the Social Environment*. Cambridge, MA: MIT Press.

Rorty, R. 1989. *Contingency, Irony and Solidarity*. Cambridge: Cambridge University Press.

Schobbens, J. 1910. "Lettre de Joseph Schobbens à Paul De Heem du 1 juin 1910." Provincial Archives Desguinlei, Depot number 506/108, Folder 1910.

Schorske, C.E. 1980. *Fin-de-Siècle Vienna: Politics and Culture*. New York: Knopf.

Sieverts, T. 1997. *Zwischenstadt: zwischen Ort und Welt, Raum und Zeit, Stadt und Land*. Braunschweig: Vieweg.

Sitte, C. 2002 [1909]. *Der Städtebau nach seinen künstlerischen Grundsätzen*. Reprint der 4. Aufl. 1909. Basel: Birkhäuser.

Smets, M. 1995. *Charles Buls, les principes de l'art urbain*. Brussels: Mardaga.

Soja, E. 2000. *Postmetropolis: Critical Studies of Cities and Regions*. Oxford: Blackwell.

Suykerbuyk, C., and G. De Ridder. 1937. "Aanleg en bebouwing der steden." *KMBA*, 8(7): 165–76.

Talen, E. 2005. *New Urbanism and American Planning: The Conflict of Cultures*. New York: Routledge.

Till, J. 2009. *Architecture Depends*. Cambridge, MA: MIT Press.

Tunnard, C., and B. Pushkarev. 1963. *Man-Made America: Chaos or Control? An Inquiry into Selected Problems of Design in theUrbanizedLandscape*. New Haven, CT: Yale University Press.

Unwin, R. 1924. "The Need for a Regional Plan." In International Federation for Town and Country Planning and Garden-Cities, ed., *International Town Planning Conference, Amsterdam 1924*. Amsterdam: De Erven van Munster, 15–35.

Van der Paal, E. 1937. "Hoog-Middel-Laagbouw. Over hoogbouw in't algemeen." *KMBA*, 7(12): 289–95.

Van Loo, A. 2003. *Repertorium van de architectuur in België van 1830 tot heden*. Antwerp: Mercatorfonds.

Van Reeth, F. 1912. "De uitbreiding van Antwerpen. Drama in twee bedrijven." *De Bouwgids*, 4(4): 80–4.

Verwilghen, H. 1938. *Over de gemeentelijke bouwreglementen: rede uitgesproken in de openingsvergadering van de provincieraad van 1 juli 1938*. Hasselt: Drukk. Crollen.

Wittockx, H. 1937. "Het inrichten van een bouwraad door het provinciaal bestuur." *KMBA*, 8(4): 98–100.

14 Creating Suburbs in North America: A Mutual Blind Spot

RICHARD HARRIS

Introduction

Lately, challenging a pervasive "suburban stereotype," some researchers have argued that suburbs can be creative places. The present volume in the "global suburbanisms" series as well has shown that artists can flourish there (see especially part 1 and chapter 7 of this volume) and that suburbs can be a place of vernacular creativity (chapters 3 and 12), a place where creative practices tie suburban communities together (chapters 9, 10, and 11), and a shelter for original craftsmanship and invention too (chapters 8 and 13; Gibson, 2012; Bain, 2013, 2018; Geraghty and Massidda, 2018). To evoke a famous historical example, Thomas Edison established his "invention factory" in suburban Menlo Park to get away from the "worldly distractions" of New York City while remaining within reach of business contacts and suppliers (Morris, 2019: 532; "Thomas Edison," 2020). For Edison, a suburb was the perfect place for the "laboratory of his dreams" (Morris, 2019: 537).

Suburbs can indeed be ideal laboratories of innovation but in a way that even Edison never imagined. When people speak of creativity, they usually think of two types (Hall, 1998).[1] They have praised cultural achievements in art, music, or architecture, associating them with places such as Florence, London, or New York, and they have shown how cities incubate the technical inventions that drive productivity and profits (Hall, 1998; Hietala and Clark, 2013; Van Damme et al., 2018). But there is another way that creativity can be expressed, one in which suburban settings excel: in the design, production, and adaptation of the built environment (see also chapter 13). This claim may come as a surprise to some readers. There is an old stereotype of suburbs as bland places where everywhere, and everyone, looks the same. Such stereotyping has been especially marked in North America. After all, where is the creativity in Levittown? The answer is "everywhere" – in its scale, design, manner of construction and finance, and in the ways its residents have altered the dwellings they moved into (Kelly, 1993). The problem is that the developers, builders, lenders, and homeowners are not usually viewed as being creative. This concluding chapter to this book devoted to suburban creativity makes the argument that they should be.

The case rests on the wide scope that suburbs offer for development. Land is cheaper in the suburbs than in the city; it is commonly undeveloped and available in large tracts; the road network is lower density; altogether, the suburban setting offers affordable opportunities for experimental design and production. The lower density enables a form of home ownership that provides generous scope for individual expression. None of this guarantees creativity, of course. There are innumerable instances of imitative dwellings and suburbs, just as cities have harboured derivative art and copycat businesses. However, in reconsidering how North American suburbs were, literally, made, I will first sketch a narrative of steady innovation, then offer solid grounds for viewing them as important places of creativity, and finally show how historians have recognized this fact in all but name. Innovation has been apparent in all sorts of suburban districts, including industrial clusters, shopping centres, office parks, technopoles, and corporate campuses, but this final chapter will concentrate on residential subdivisions, the most common, familiar, and thoroughly documented of all suburban spaces in North America. It will carry a larger and more general argument about suburbs too: current debates about *urban* creativity should widen their remit. In addition to weighing people's originality once they are in place, we need to think about how those places came to be. But first we need to consider how it is that the creativity involved in the making of suburbs has occupied the blind spot of two groups of scholars.

A Mutual Blind Spot in Sub/urban Studies

In the English-speaking world, there are two groups of scholars who could, and arguably should, have written about the creativity involved in the making of suburbs, but, in part because they have failed to communicate with one another, neither have effectively done so. The first are the self-proclaimed students of creativity. Abstractly, many writers have noted that cities, as concentrations of people, matter in all sorts of ways (Harris, 2021). Within this tradition, one set of authors has shown how urban environments facilitate innovation and economic growth, especially because of the "buzz" of face-to-face interaction (Jacobs, 1961; Hall, 2000; Scott, 2006; Krätke, 2011; Storper, 2013). Another, smaller but eloquent, body of work speaks about how cities foster artistic creativity – both "high" and vernacular – and more generally cultural change (Burgess, 2006; Edensor et al., 2010; Hawkins, 2017). Many have acknowledged that the configuration of urban spaces can matter. Factories benefit from being located in industrial districts. Design elements such as walkability, the juxtaposition of different types and scales of buildings, and the existence of "third spaces" that facilitate interaction make for creative milieu (Stevens, 2016). The absence of one or more of those features can be a suburban disadvantage (Bain, 2018: 270). Charles Landry (2008: xxvii) has argued that the built environment is "the stage, the setting, the container," "the physical preconditions or platform upon which the activities or atmosphere of the city can develop." But, like many others, he takes the actions of the builders and stage-hands for granted.

The second group are those who have examined the making of the built environment. In so doing, they have sometimes implied the presence of creativity, but without stating it in so many words and certainly without reference to those who have explored its other manifestations. A number have acknowledged the inspiration involved in the design and production of specific buildings. Chiefly in the Global North, architectural historians are the prime example. In the South, following the precedent of British architect John Turner, others have written about the ingenuity of rural-urban migrants in building their own homes, often cooperatively, in squatter settlements (Turner and Fichter, 1972).

Turner's ideas merge with the work of two overlapping subdisciplines, all working on the built environment. The first, mostly historical and cultural geographers, interpreted "ordinary landscapes." They insisted that every dwelling is uniquely deserving of attention because each "has its particular builder and each has been lived in by particular individuals and families" (Meinig, 1979: 43). This is the tradition with which I identify, to the extent that I have taken its perspective for granted for much of my career. Its working assumption asks the question, how could the physical places in which we live not be important? A second subdiscipline is in many ways kindred, although mostly non-overlapping. Its declared interest is in everyday buildings and landscapes, and it formed an association, the Vernacular Architecture Forum, in 1980. The forum encouraged the launching of an ambitious project, the *Encyclopedia of Vernacular Architecture of the World*, edited by Paul Oliver (1997), currently being revised under the editorship of Marcel Vellinga. Within and beyond these two orientations in dealing with the built environment, researchers documented the myriad ways in which builders and residents shaped suburban environments to suit their needs. Stefan Muthesius (1982) documented the variety of the once-suburban terraced housing in English cities; Mary Hayward and Charles Belfoure (1999) did likewise for the Baltimore row house. Oliver extended the argument into twentieth-century suburbs, defending and documenting the supposedly bland English suburban semi (Oliver et al., 1981). He complained that "the conspicuousness of the suburbs is only exceeded by the conspicuous neglect which they have received from the architectural press and practice" (Oliver, 1981: 9). Characteristically, Paul Barker (2009: 10, 145) later pointed out that "life is seldom how designers or architects plan it to be"; "suburbia isn't static; it is endlessly adaptable."

Inspired in varying degrees by these traditions, urban and housing historians in North America have studied the development of the suburban homes and landscapes that they know and experience. They have documented innovations in how those spaces were made and remade: in materials, methods, designs, and financing. The present chapter draws extensively on this literature. But, here again, what has been missing is explicit recognition that such innovations were "creative": the term is not used. This point may sound like a minor, semantic issue, and in one sense it is: by any name, these researchers have taken seriously the making of the built environment. But their omission matters because it has helped to ensure that studies of everyday suburbs have remained separate from those of urban creativity, to the detriment of each. They are in each other's blind spot.

The specialized literature that comes closest to bringing the built environment into discussions of creativity is that of planning historians. Apart from mundane matters like zoning, planning – like architecture – engages with questions of design. As such, it has attracted visionaries (Hall, 2002; Pinder, 2005). Documenting those visions, and resulting projects, planning historians have been active in North America since the 1980s, forming an association, the Society for American City and Regional Planning History, in 1986. These scholars have looked beyond self-styled planners to include a wide range of actors who have been involved in shaping the urban landscape (Sies and Silver, 1996: 2). They, too, have helped fill out the narrative developed below. But even here "creativity" has been avoided.

To establish that "creativity" applies to the changing forms and methods of suburban development, it is initially useful to review, in neutral, descriptive terms, what the main changes have been and then to consider the role of the agents, individuals and groups, and of the ideas and practices that they carried through. The challenge in sketching the main changes and identifying their agents is that there is much ground to cover and an abundance of research. It may be that, by itself, this straightforward account will convince even the sceptical reader that creativity has been involved, but it is worthwhile considering the judgements of their historians, and in particular the language they have used. They may not speak of creativity, but they employ synonyms and phrases that, collectively, speak to its presence. Of course, the case can be overstated, and a concluding discussion offers some cautionary comments before making a wider claim.

How North American Suburbs Have Been Transformed

How do North American suburbs of the 2020s differ from those of, say, 1850? The list of changes is long, and only the most common and significant are noted here. Details may be found in many useful surveys and case studies (Jackson, 1985; Fishman, 1987; Weiss, 1987; Ford, 1994; Hayden, 2003; Harris, 2004; Fogelson, 2005; Lane, 2015). They encompass changes not only in the appearance of residential suburbs but in the ways in which they have been created.

The most obvious changes are those that are visible, and the most striking is the scale of development. Cities have grown in population and, as they have sprawled, even more so in size. In the process, a basic development has been the separation of residential areas from all others (Ford, 1994). Industry is gone from the neighbourhood and so, mostly, is the corner store or small commercial strip. Christopher Leinberger (2007: 50) spoke of "the nineteen standard real estate product types that Wall Street knows, understands, and can be traded in large quantities." Several – including mobile home parks and gated communities – were exclusively residential, and none involved mixed use. It has become possible to speak only of residential development.

That said, a significant number of land users have emerged to directly service residential areas and the people who live in them (Ford, 1994). The suburban shopping centre and now the big box centre are the largest aggregations. These include a variety of enterprises that did not exist in 1850 but which now directly serve residential needs: gas stations, garden centres, home improvement stores, realtors' offices, and lately self-storage operations.

Change has happened at all scales. A typical nineteenth-century residential development contained two dozen lots and dwellings; a residential development now involves hundreds, and perhaps thousands (Doucet and Weaver, 1991). This change of scale now makes it possible for land developers to create designs that were previously unthinkable. Instead of fitting into a gridded street pattern, residential subdivisions now include major and minor roads, superblocks that are framed by major arterial streets, as well as cul-de-sacs (Weiss, 1987; Moudon, 1992). Following the "neighbourhood unit" idea popularized by Clarence Perry in the 1920s, many also include a primary school (Silver, 1985). Recent variants, including planned unit developments and the New Urbanist projects that have followed the development of Seaside, Florida, in the 1980s, combine elements of old and new. Regardless, many now take the form of common interest developments, where public spaces such as parks and even roads are owned collectively by the residents rather than by a municipal government (McKenzie, 1994; Ben-Joseph, 2004). Deed restrictions govern what residents are allowed to do to their properties and often who, for example in terms of age, may live there. In the United States, but less so in Canada, many new projects are gated, whether strictly with a guard or symbolically with a signed entrance. In such ways, residential space is organized very differently than in the past.

Underneath, framing and in many ways guiding the current suburban scene, municipal government plays a much larger role than in the nineteenth century. By the early twentieth century, it was becoming a "directive system" (Foglesong, 1986: 5). Especially in the United States, this system prompted the formation of separate suburban jurisdictions (Keating, 1988). Almost everywhere, development is now subject to master plans and zoning regulations that determine patterns of land use, sometimes with great specificity: not just "single family," for example, but "single family on quarter acre lots" (Hirt, 2014). Construction and maintenance is subject to much stricter building codes as well as health regulations. Physical infrastructure has become more extensive (Kennedy, 2011; Tarr, 1996). It was only in the late nineteenth century that residents in middle-class suburbs came to expect piped water and sewers; many working-class districts still lacked these services in the 1920s, and suburbs were more socially diverse than many have acknowledged (Harris, 2004). Today, whether provided by the developer or municipality, and along with paved roads and street lighting, these services are generally mandatory. Once rare, a strong municipal presence is now taken for granted.

Just as subdivisions have changed, so have the dwellings within them, in terms of size, external appearance, and plan. They have grown larger as, for many decades, did the lots on which they were built. In the process, townhouses or plexes (where two or more dwelling units were stacked above each other) became semi-detached and then detached dwellings, in the process acquiring three frontages (Gowans, 1986: 76). Dominant styles changed (Wright, 1981; Clark, 1986; Lane, 2015). Sometimes-exuberant nineteenth-century Gothic evolved into more restrained frontages in the early twentieth century, for example in the form of the foursquare (Wright, 1980: 3). Bungalows, a term denoting a hybrid style imported from India, spread from the west coast across the continent (King, 1984). There were many local variants: most were targeted at the middle class, but the Chicago version evolved from the working-class cottage (Prosser, 1981; Bigott, 2001;). After 1945, on expanded lots, the single-storey ranch style became common, along with side- and back-splits, all of which were new in "elevation, profile, plan and interior furnishings" (Lane, 2015: 5). What they lacked was a porch. Until the 1920s, porches had been standard, especially in warmer regions, but they were soon rendered optional by the popularity of air-conditioning (Cooper, 1998). While these changes were happening, ground plans became more open. Living and dining areas began to flow into each other. Parlours disappeared, and kitchens opened up (Cromley, 1996; Hayden, 2002; Jacobs, 2015). Formality gave way to "casual living," and then, as houses became large enough to offer individual privacy, the "zoned house" emerged (Jacobs, 2015: 3). Then, gathering momentum from the 1970s, a growing disjunction appeared between interiors and exteriors (Brown, 1977). While interiors continued to modernize, a fashion for historicist styles took over (Harris and Dostrovsky, 2008). As lot sizes shrank, Georgian, Gothic, Tudorbethan, and other traditional styles reappeared, often as hybrids. Having evolved from streetscapes of townhouses to landscapes of single-family dwellings, the trend went into slight reverse.

Increasingly, exteriors have accommodated a garage (Jackson, 1976). Beginning at the rear of mansions, and accommodating chauffeurs, garages for automobiles now define the frontages of almost all homes. Initially separate, they became attached to houses via covered passages or as car ports and, by the 1960s, as part of the dwelling. Eventually, a connecting door made it into a room, accommodating storage, a fridge, and a work bench too.

While this evolution was happening, front and backyards (gardens) also changed (Grampp, 2008). Initially, especially in working-class areas, they were used for growing vegetables as well as raising chickens. By the early twentieth century, backyards became utilitarian in a different way, storing fuel. In middle-class suburbs, gardens had always served an aesthetic function, which became generally dominant. In the 1950s, the fashion was minimalist, with lawns connecting properties that, at least out front, lacked fences (Lane, 2015). Lately, the trend has been towards a variety of plantings. Needing more time and money, it is this trend that has supported the growth of garden centres.

A parallel growth has affected "do-it-yourself" (DIY), a term coined in the early 1950s (Harris, 2012). Dwellings have always required maintenance, but in the 1920s more home owners began to undertake this work themselves. After 1945, DIY expanded as more owners undertook renovation projects, while a housing shortage encouraged many to build homes from scratch. This period saw an expansion in the range of construction materials and, notably, the appearance of power tools. As with the gardening trend, DIY supported dedicated retail outlets, in this case home improvement stores.

Interest in gardening and home improvement was associated with the expansion of urban home ownership. In both Canada and the United States, owner-occupiers as a proportion of all households rose from about 30 per cent in 1900 to about 66 per cent by the 1990s, before levelling off. The percentage in suburban areas has always been higher, and it is the homeowner who has the opportunity and financial incentive to invest sweat equity in home and garden. The owner also has more incentive to lobby local government to prevent developments that threaten property values. Altogether, there has been increasing scope for residents to put their stamp on their home and residential environment (Kelly, 1993; Barker, 2009). This trend is reflected on the domestic scene, and also politically, in NIMBY (not-in-my-backyard) initiatives. Indeed, Clarence Perry (1929: 100) himself reckoned that it was home ownership, more than the neighbourhood unit idea for which he is known, that best guaranteed an active local community.

The increasing scale of residential suburbs and the growth of home ownership have reflected changes in how suburbs are made and modified. The basic elements have been the processes of land subdivision, construction, and finance. Typically, businesses have played the dominant role, and most have typically been specialized.

Land subdividers and developers have grown larger. This growth has corresponded to the increasing size of subdivisions and also to the involvement of some of them in multiple projects, sometimes in different cities (Weiss, 1987; Hise, 1997; Rowe, 1991; Harris, 2004: 141–3; Lane, 2015). Although, from the 1920s, most new suburban projects were directed at the middle class, a few were targeted at workers, B.E. Taylor's Brightmoor project in Detroit in the 1920s being a prime example (Loeb, 2001). In boom times, some companies became vertically integrated, taking on construction. Samuel Gross was the largest in late nineteenth-century Chicago. The 1950s saw a resurgence of this practice, the Levitt brothers being the best known (Kelly, 1993). But this business strategy left companies more vulnerable during downturns, one of which left the Levitts bankrupt. Moreover, as Ned Eichler (1982: 114–15) has shown, operating in more than one city posed new challenges. Mostly, developers have remained local and left construction to the builders.

In the nineteenth century, builders operated locally on a small scale, erecting a handful of houses a year and relying on subcontractors. This pattern persisted, and, if anything, the use of subcontractors has increased (Maisel, 1953; Kelly and Associates, 1959; Harris and Buzzelli, 2005). This arrangement is especially true in the renovation business, which became more important from the 1920s (Harris, 2012). That said, some builders grew large, and today several operate nationally across the United States, but not in Canada. Over the course of the twentieth century, they were joined by manufacturers of prefabricated homes. Some, for example Lustron, proved short-lived, but after 1945 mobile homes became common, especially in the United States (Kelly, 1951; Wallis, 1991; Knerr, 2004). In the early and middle decades of the twentieth century, on-site builders resisted such competition by assembling the mail-order kits that customers had purchased. These kits took advantage of balloon (and later platform) framing technology developed in Chicago in the mid-1800s. However, kits, which came with instructions, also enabled competition from owner-builders (Harris, 2012). A few builders, notably B.E. Taylor, capitalized on the willingness of working-class buyers to invest sweat equity (Loeb, 2001). All competed by using prefabricated parts, such as window assemblies and roof trusses, as well as power machinery and tools (Ventre, 1980; Schlesinger and Erlich, 1986; Harris, 2012). In this manner, the advantages of specialization and large-scale production were realized on-site through the use of manufactured parts and specialized subcontractors.

These changes in the making of suburbs have depended on developments in finance. In the nineteenth century, developers required little capital: land was cheap because it was usually unserviced, and the scale of subdivision was small. As operations grew and developers became responsible for the cost of infrastructure, financing changed. Groups of investors formed syndicates (see, for example, Doucet and Weaver, 1991; Fogelson, 2005; Parsons and Harris, 2020). In the 1920s, mortgage bonds became available, which channelled the savings of thousands of investors (Harris, 2013). Sometimes, developers borrowed from the farmers who had sold them the land (see, for example, Eichler, 1982: 20–1). Today, especially when they design conventional products, developers can tap major financial institutions (Leinberger, 2007).

At the same time, the growth in home ownership has depended on changes in consumer finance. Until the 1930s, mortgages were available for under half of the purchase price, taking the form of "balloon" arrangements in which interest was paid but the full amount of the loan came due at the end of the term, commonly five years. Many loans in the United States, and most in Canada, came from individuals. By the 1950s, the current arrangement was becoming dominant, whereby lending institutions offer high-ratio, twenty-to-thirty-year amortizing mortgages. This arrangement became tied into wider financial markets, initially by the creation of a secondary market for mortgages and then by their securitization. A combination of irresponsible lending and poor governmental oversight contributed to the major financial crisis of 2008; since that time, home ownership has dipped in the United States but not in Canada (Immergluck, 2009; Walks, 2019). But in the long run, changes in real estate finance have been closely associated with the growth of home ownership, with all that that has implied for gardening, DIY renovations, and NIMBY politics.

In sum, the shape, appearance, and production of North American suburbs has changed in striking ways over the past two centuries. But to be able to speak of creativity, we need to be able to show that at least some of these changes depended on the initiatives of specific agents, responding with originality to the pressures and opportunities with which they were faced.

The Agents of Change

The above sketch has already mentioned most of the agents who have been involved in the making and remaking of suburbs. But what has been their relative importance, and to what extent have they shaped the course of events, as opposed to responding predictably to wider forces and pressures? That is a large question to which there is no definitive answer. The following is no more than a plausible sketch.

It is clear that, in part, the changing shape of the suburbs has reflected broader technological innovations, as well as trends in economy and society. Robert Fishman (1987: 3) has interpreted the first residential suburbs as "bourgeois utopias," a "middle-class invention" that inevitably reflected this group's ability to realize an aversion to city living. Later, prosperity enabled more people to buy homes, to buy larger homes, and to equip and improve them. The practical application and wide availability of electricity made possible the creation of labour-saving appliances and tools that revolutionized domestic labour and construction (Tobey, 1996). The invention and spread of the automobile made sprawl possible, along with the wide separation of homes from workplaces and stores. And the changing character of the family, including the role of women, provided an incentive to rethink home layouts (Clark, 1986; Cromley, 1996; Hayden, 2002). These trends, and more, created new challenges and opportunities.

How have people and groups responded, and to what extent in original ways? Surveys have noted that "Americans have had to find new housing alternatives" and have "crafted a new manner of living" (Wright, 1981: xviii; Jacobs, 2015: 1), but in general terms many responses were predictable. With rising incomes, it was inevitable that people would buy and improve larger homes, to make them more liveable and to enhance their value, and thereby become housing consumers. The surge of owner-building and DIY activity after 1945 reflected a unique combination of rising incomes, family formation, and a housing shortage. The opening up of interior spaces expressed a trend towards casual living, with changes in women's roles. Supporting these changes, the growth of garden centres, home improvement stores, and realtors was surely inevitable. And perhaps it was even inevitable that the optimism of the 1950s would be expressed in clean, modernist-influenced designs, while growing social uncertainty has recently been reflected in a preference for more comforting historicism (Harris and Dostrovsky, 2008). What matters for the present argument is what people, and entrepreneurs, made of those broad shifts in demand.

Home owners, Meinig's (1979) "particular individuals," now had greater scope to express themselves, and many took advantage of this opportunity. They put their own stamp on standardized dwellings (Kelly, 1993). They redesigned and decorated domestic space, including the garden, to suit themselves and the tastes and needs of family members (Harris, 2012). No two families are the same, and neither were their solutions. They welcomed guidance from salesmen and consumer magazines but, inevitably, did something personal and unique. Influential individuals and groups seized the opportunity to give broad trends a series of foci and directions. There are many examples. The most significant has been Ebenezer Howard's vision of the garden city. Promoted by the Regional Planning Association of America and many professionals, and watered down by innumerable developers, this vision shaped suburban design across the continent (see, for example, Knox, 2011: 94; Hise, 1997; Hall, 2002; McCann, 2017). Recently, variants of New Urbanist design have proved popular and at various scales. Witold Rybczynski (2008) has provided a compelling account of the making of one such subdivision. Such designs took account of changing possibilities and tastes, but did more than merely respond to consumer demand.

In this context, the role of developers, builders, and lenders is ambiguous. They are usually the agents that design and create the suburbs, playing a crucial, potentially creative, role, but most would claim to be simply providing what buyers want. This paradox can be resolved by noting diversity. Some developers acted as pioneers, whether, like J.C. Nicholls, in applying elaborate deed restrictions in Kansas City, Missouri (1920s); or, like Samuel Gross (1880s), Henry Ford (1920s), and the Levitts (1950s), in terms of scale and streamlined production; or, like Joseph Eichler (1950s), in terms of home design (Barrow, 2015; Eichler, 1982; Kelly, 1993; Weiss, 1987; Worley, 1990). Others translated and adapted such ideas, bringing them to new places (Lane, 2015). Given that housing markets are local, translations often involved vision and chutzpah. And then others, like the syndicate that developed Hamilton's Meadowlands (1990s), have applied tested formulae, in that case by combining a big box "power centre" with conventional single-family and townhouse developments (Parsons and Harris, 2020). Like artists, not all developers are original.

As many writers have observed, the same is true of builders. In fact, if individual changes to home building have often been modest, they have also been relentless. Davis (1999: 174) speaks of the "culture of building" being one of "tradition as innovation"; Schlesinger and Erlich (1986: 153) see it as "evolutionary rather than revolutionary"; I have referred to its "continuity in change" (Harris and Buzzelli, 2005: 62). Major changes included the development of balloon framing and the use of power tools (Kelly and Associates, 1959). A succession of new materials – which since 1900 have included tin roof sheeting, concrete blocks, asphalt shingles, gypsum board, latex paint, linoleum, pre-pasted wallpaper, and prefabricated components – demanded new skills while rendering other ones (such as plastering) almost obsolete (Harris, 2012). In the process, these changes, and new materials, made housing more affordable and adaptable, while encouraging new designs. Like developers, most builders employ proven techniques, but over the decades the changes have been "dramatic" (Schlesinger and Erlich, 1986: 159).

Change happened more quickly with building suppliers (Harris, 2012). As late as 1945, most were located in unattractive locations, with grubby yards that catered to male builders and tradesmen. Some still do. But, by 1960, many had opened stores on major thoroughfares and in shopping centres. They employed smartly dressed sales agents and stocked a wider array of goods in new, bright, and clean showrooms that attracted consumers, including many women. Through effective marketing, and by offering training courses, they made DIY accessible to more people. Responding to demand, they reshaped it.

The same may be said of lenders, although here the impact was more sweeping. From 1909, Simon W. Straus made the mortgage bond a valuable tool for the emerging breed of large developers (Weiss, 1987; Harris, 2013). In the 1930s, backed by the US National Association of Real Estate Boards, Frederick Babcock linked new federal guidelines for residential design with mortgage insurance so as to establish the long-term amortizing mortgage as a norm (Weiss, 1987). Canada's then federal finance minister, W.C. Clarke, and from 1946 the Central Mortgage and Housing Corporation, did much the same (Bacher, 1993; Harris, 2004). Along with later revisions, these unprecedented public-private Depression-era initiatives have framed suburban development across the continent ever since.

The examples of innovations could be multiplied. Little has been said about the most original builders because many proved unsuccessful. Upstream work by the manufacturers of tools and building materials has only been hinted at. But hopefully enough has been said to indicate that suburban agents have done much more than simply respond to broad changes in consumer demand. The visibility and significance of what they did depends on our perspective. Contemporaries and social scientists, focused on the present, often deplored stasis and inertia. But historians, surveying decades or centuries, have been struck by the magnitude of change, as their language indicates.

Creativity Hidden in Plain View: The Perspective of Historians

Taking the long view, historians of suburban change have used striking phrases when interpreting its magnitude, significance, and originality. Typically, Peter Hall (2002: 2–3) speaks of the "visionaries," many "utopian," who imagined large designs. Speaking of what actually happened, Larry Ford (1994: 262) notes the "invention and proliferation of new types of special-purpose buildings," while Peter Rowe (1991: 3) sees an overall "territorial transformation." Focusing on middle-class residential suburbs, Robert Fishman (1987: 3) interprets them as "perhaps the most radical rethinking of the relation between residence and the city in the history of domestic architecture." Dramatic stuff!

Accounts of the development process and residential design are no different. Marc Weiss (1987: 2) speaks about how twentieth-century developers were "creating residential subdivisions" in ways that entailed a "transformation of urban land development." For Greg Hise (1997: 20), the "crafting" of "complete communities" required "vision." Looking at his father's development company, Ned Eichler (1982: 65) reckons that a series of technical and managerial innovations required "tremendous effort and ingenuity." His judgement is not mere filial piety. Scholars have routinely viewed land developers as "tenacious visionaries who persist in the face of local resistance, lender scepticism, and bureaucratic delays" (Parsons and Harris, 2020: 3). Similar language has been used in relation to residential design. Howard Davis (1999: 153) talks in general terms about "typological transformations" of dwelling types; Barbara Lane (2015: 221) judges that houses as well as suburbs were "dramatically transformed" after 1945; Clifford Clark (1986: 4) interprets these and other changes as reflections of new "visions" of the family. And of course, in popular discourse since the 1920s, the very notion of the owner-occupied home has been viewed as the American (and sometimes as the Canadian) "dream."

Even the work of builders has been described in generous, even extravagant terms. Douglas Knerr (2004: 2) suggests that Carl Strandlund, designer of the manufactured Lustron house, "intended to become the Henry Ford of housing." Eichler, whose "model was a production line turning out Ford's Model T," was more successful because he adapted it for on-site assembly (Eichler, 1982: 65). The housing industry was notoriously "so competitive" by comparison with others that it is no wonder many historians speak of the "experimentation," "innovations," "advances," and industrial "progress" that came from the "entrepreneurship" of "pioneers" (Kelly, 1951: 12; Maisel, 1953: 6; Kelly and Associates, 1959: xx; Ventre, 1980; Wallis, 1991: vii; Knerr, 2004: 3). Indeed, Loeb (2001) deploys the term "entrepreneurial vernacular." Interestingly, in one of the most detailed accounts of the building industry anywhere, Donna Rilling waxes eloquent in her account of Philadelphia's. In a "competitive" business, she describes builders as "enterprising," "aggressive and ingenious operators," who "maneuver[ed] financial and property instruments in inventive ways" while "reorganizing production within crafts," thereby producing "innovation and change" (Rilling, 2000: vii, 186, 188). This praise in a period – the early nineteenth century – which, as far as the building industry is concerned, is not generally seen as being especially creative!

And finally, the field of real estate finance has included periods of striking originality. Recently, the term "creative finance" has denoted techniques that harmed borrowers, lenders, and others besides, but many innovations proved valuable. To be sure, some were minor or have been superseded. Building suppliers have routinely extended short-term credit to builders, and for a time, they did the same for consumers when the DIY market grew. In time, however, credit cards rendered this service redundant. But other financing methods, major and ingenious, have stood the test of time: mortgage bonds, amortizing mortgages, secondary mortgage markets, and real estate investment trusts. Many writers have recognized each of these to be original and transformational. Dolores Hayden (2003), for example, has noted how the introduction of mortgage insurance "reshaped" homes and suburbs from the 1930s. Perhaps because most of the more important financial arrangements are used by buyers, their importance has been apparent to almost everyone.

It is natural for writers to wax eloquent about the significance of their subject matter. Doubtless, some have overstated the originality and significance of the suburban initiatives that they have described. But those quoted above include leading scholars in the field, and no one has seriously contested the truth of their claims. Although the term "creativity" itself has rarely been used, there is a consensus that the making and remaking of suburbs has consistently involved creative acts.

Concluding Discussion

Accepting that creativity has been involved in the making of suburbs, how should we judge it, and especially by comparison with other fields of human endeavour? Certainly, this concluding chapter should not be read as a straightforward paean to the creativity of suburban building and design. There has been much innovation and adaptation but few instances where large, visionary projects were carried through. Amanda Hurley (2019) has described a few of these "radical suburbs," but they have been the exception not the rule. Ground-breaking ideas – the garden city, New Urbanism – have usually been watered down, with sometimes unfortunate results (Hall, 2002: 3). Some of the most original building forms and methods – the Lustron house being an example – proved unviable. Compromises and failures characterize all fields of human endeavour, and suburban development is no exception.

Compromises should not be surprising. The process of city-building has been around for ten thousand years. If, as Howard Davis (1999) has argued, innovation is normal, it is also true that most new ideas in suburb-building have been incremental. A tradesman discovers an ingenious workaround that saves a few hours per house; a retailer develops an evening course for amateurs in order to promote DIY; a draftsman adds a gable or veneer to a frontage to give the illusion of individuality. Many more inventions remain invisible. The balloon frame is hidden under a brick veneer; the supporting pre-assembled truss lies under the shingle roof. And then those initiatives that are both visible and important become invisible for a different reason: they become common, and we soon take them for granted. Successful innovations were copied again and again. And again. The result has been the often thoughtless repetition of practices and design features. As Barbara Lane (2015: 98) notes, in the 1950s, ranch-style bungalows were at first viewed as "shockingly different" in suburban Philadelphia, but within a few building seasons, they had become normal, and perhaps boringly so. Forty years later, Witold Rybczynski (2008: 19) felt "a shock" when seeing Seaside, Florida, but within two decades "New" Urbanism had become just another suburban design. Such is the fate of successful innovations in any sphere of activity – by definition!

Worse. Once an innovation becomes the norm, it acquires inertia. In 2007, when Christopher Leinberger spoke of the nineteen standard real estate product types, he noted that their success was a barrier to change. But each of those types – including mobile home parks and gated communities – had once been invented. The suburb's problem is that imitations are not only numerous but highly visible; they confront us when we look out of the window or on our way to work. To recognize the inventiveness that they embody, we must take the long view and reflect on their origin.

Another challenge that suburbs face in getting respect from those who value creativity is that they have turned out to be mixed blessings. "Creativity" is supposed to be a boon, but many of the practices now embodied in suburbs harm our health and the environment. They include everything from lead pipes and asbestos insulation to the whole package of car dependency: sprawl and land use segregation. Some financial innovations have been catastrophic, either in conception or in the way they were used: mortgage bonds contributed to one real estate bubble and subprime mortgages to another. But, once again, we must remember that few, if any, inventions do nothing but good, from the steam engine to social media and artificial intelligence. Practical creativity commonly has its downside. If so, an innovative response is usually forthcoming, either to mitigate negative effects or to avoid them altogether. Risky financial schemes may trigger regulation. Similarly with suburbs. Today, attempts are being made to reduce the impact of sprawl by promoting densification and land use mix to reduce car dependency. Regulations and incentives are helping to reduce the carbon footprint of those automobiles. And so it seems reasonable to view the making of North America's suburbs, and surely those elsewhere, as an ongoing example of human adaptability and innovation.

If that is so, there is a larger message. There is a form of creativity that has been overlooked by those who discuss the subject. Everywhere, and not just in the suburbs, the creation and the re-creation of the built environment deserves more academic interest and respect.

NOTE

1 Hall and some other writers also speak about the novel initiatives of
   municipal governments as a third type of creativity. They are distinctively
   urban, not only because they are mostly confined to cities but also because
   they respond to the specific challenges created by the urban environment
   (Harris, 2021). Such initiatives are rarely discussed in the general literature
   on urban creativity.

REFERENCES

Bacher, J. 1993. *Keeping to the Marketplace: The Evolution of Canadian Housing
   Policy*. Montreal and Kingston: McGill-Queen's University Press.
Bain, A. 2013. *Creative Margins: Cultural Production in Canadian Suburbs*.
   Toronto: University of Toronto Press.
Bain, A. 2018. "Suburban Creativity and Innovation." In R. Shearmur,
   C. Carrincazeaux, and D. Doloreux, eds., *Handbook on the Geographies of
   Innovation*. Cheltenham, UK: Edward Elgar, 266–76.
Barker, P. 2009. *The Freedoms of Suburbia*. London: Frances Lincoln.
Barrow, H.B. 2015. *Henry Ford's Plan for the American Suburb*. DeKalb: Northern
   Illinois University Press.
Ben-Joseph, E. 2004. "Land Use and Design Innovations in Private
   Communities." *Land Lines*, 16(4): 8–12. https://www.lincolninst.edu
   /publications/articles/land-use-design-innovations-private-communities
Bigott, J.C. 2001. *From Cottage to Bungalow: Houses and the Working Class in
   Metropolitan Chicago, 1869–1929*. Chicago: University of Chicago Press.
Brown, D.S. 1977. "Suburban Space, Scale and Symbols." *VIA*, 3: 41–7.
Burgess, J. 2006. "Hearing Ordinary Voices: Cultural Studies, Vernacular
   Creativity and Digital Storytelling." *Continuum: Journal of Media and Cultural
   Studies*, 20(2): 201–14. https://doi.org/10.1080/10304310600641737
Clark, C.E. 1986. *The American Family Home, 1800–1960*. Chapel Hill:
   University of North Carolina Press.
Cooper, G. 1998. *Air Conditioning America: Engineers and the Controlled
   Environment, 1900–1960*. Baltimore: Johns Hopkins University Press.
Cromley, E.C. 1996. "Transforming the Food Axis: Houses, Tools, Modes of
   Analysis." *Material History Review*, 44: 8–19.
Davis, H. 1999. *The Culture of Building*. Oxford: Oxford University Press.
Doucet, M., and J. Weaver. 1991. *Housing the North American City*. Montreal
   and Kingston: McGill-Queen's University Press.
Edensor, T., D. Leslie, S. Millington, and N. Rantisi, eds. 2010. *Spaces of
   Vernacular Creativity: Rethinking the Cultural Economy*. New York: Routledge.
Eichler, N. 1982. *The Merchant Builders*. Cambridge, MA: MIT Press.

Fishman, R. 1987. *Bourgeois Utopias: The Rise and Fall of Suburbia*. New York: Basic Books.

Fogelson, R.M. 2005. *Bourgeois Nightmares: Suburbia, 1870–1930*. New Haven, CT: Yale University Press.

Foglesong, R.F. 1986. *Planning the Capitalist City: The Colonial Era to the 1920s*. Princeton, NJ: Princeton University Press.

Ford, L.R. 1994. *Cities and Buildings: Skyscrapers, Skid Rows and Suburbs*. Baltimore: Johns Hopkins University Press.

Geraghty, N.H.D., and A.L. Massidda, eds. 2018. *Creative Spaces: Urban Culture and Marginality in Latin America*. London: Institute of Latin American Studies.

Gibson, C., ed. 2012. *Creativity in Peripheral Places: Redefining the Creative Industries*. London: Routledge.

Gowans, A. 1986. *The Comfortable House: North American Suburban Architecture 1890–1930*. Cambridge, MA: MIT Press.

Grampp, C. 2008. *From Yard to Garden: The Domestication of America's Home Grounds*. Chicago: Center for American Places.

Hall, P. 1998. *Cities inCivilization: Culture, Innovation and Urban Order*. London: Wiedenfeld and Nelson.

Hall, P. 2000. "Creative Cities: Cities and Economic Development." *Urban Studies*, 37(4): 639–49. https://doi.org/10.1080/00420980050003946

Hall, P. 2002. *Cities of Tomorrow: An Intellectual History of Urban Planning and Design in the Twentieth Century*. Cambridge, MA: Blackwell.

Harris, R. 2004. *Creeping Conformity: How Canada Became Suburban, 1900–1960*. Toronto: University of Toronto Press.

Harris, R. 2012. *Building a Market: The Rise of the Home Improvement Industry, 1914–1960*. Chicago: University of Chicago Press.

Harris, R. 2013. "The Rise of Filtering Down: The American Housing Market Transformed, 1915–1929." *Social Science History*, 37(4): 515–49. https://doi.org/10.1017/S0145553200011950

Harris, R. 2021. *How Cities Matter*. Cambridge: Cambridge University Press.

Harris, R., and M. Buzzelli. 2005. "House Building in the Machine Age, 1920s–1970s: Realities and Perceptions of Modernization in North America and Australia." *Business History*, 47(1): 59–85. https://doi.org/10.1080/0007679042000267479

Harris, R., and N. Dostrovsky. 2008. "The Suburban Culture of Building and the Reassuring Revival of Historicist Architecture." *Home Cultures*, 5(2): 167–96. https://doi.org/10.2752/174063108X333173

Hawkins, H. 2017. *Creativity*. New York: Routledge.

Hayden, D. 2002. *Redesigning the American Dream: Gender, Housing and Family Life*. New York: Norton.

Hayden, D. 2003. *Building Suburbia: Green Fields and Urban Growth, 1820–2000.* New York: Pantheon.

Hayward, M.E., and C. Belfoure. 1999. *The Baltimore Rowhouse.* New York: Architectural Press.

Hietala, M., and P. Clark. 2013. "Creative Cities." In P. Clark, ed., *The Oxford Handbook of Cities in World History.* Oxford: Oxford University Press, 720–36

Hirt, S. 2014. *Zoned in theUSA:The Origins and Implications of American Land Use Regulation.* Ithaca, NY: Cornell University Press.

Hise, G. 1997. *Magnetic Los Angeles: Planning the Twentieth-Century Metropolis.* Baltimore, MD: Johns Hopkins University Press.

Hurley, A.K. 2019. *Radical Suburbs: Experimental Living on the Fringes of the American City.* Cleveland, OH: Belt.

Immergluck, D. 2009. *Foreclosed: High-Risk Lending, Deregulation and the Undermining of America's Mortgage Market.* Ithaca, NY: Cornell University Press.

Jackson, J.B. 1976. "The Domestication of the Garage." *Landscape,* 20(2): 10–19.

Jackson, K.T. 1985. *Crabgrass Frontier: The Suburbanization of the United States.* New York: Oxford University Press.

Jacobs, J. 1961. *The Death and Life of Great American Cities.* New York: Vintage.

Jacobs, J.A. 2015. *Detached America: Building Houses inPostwarSuburbia.* Charlottesville: University of Virginia Press.

Keating, A.D. 1988. *Building Chicago: Suburban Developers and the Creation of a Divided Metropolis.* Columbus: Ohio State University Press.

Kelly, B.M. 1951. *The Prefabrication of Houses.* Cambridge, MA: MIT Press.

Kelly, B.M. 1993. *Expanding the American Dream: Building and Rebuilding Levittown.* Albany, NY: SUNY Press.

Kelly, B.M., and Associates. 1959. *Design and the Production of Houses.* New York: McGraw-Hill.

Kennedy, C. 2011. *The Evolution of Great World Cities: Urban Wealth and Economic Growth.* Toronto: Rotman-UTP Publishing.

King, A.D. 1984. *The Bungalow: The Production of a Global Culture.* London: Routledge and Kegan Paul.

Knerr, D. 2004. *Suburban Steel: The Magnificent Failure of the Lustron Corporation, 1945–1951.* Columbus: Ohio State University Press.

Knox, P. 2011. *Cities and Design.* London: Routledge.

Krätke, S. 2011. *The Creative Capital of Cities: Interactive Knowledge Creation and theUrbanizationEconomies of Innovation.* Chichester, UK: Wiley-Blackwell.

Landry, C. 2008. *The Creative City: A Toolkit for Urban Innovators.* London: Earthscan.

Lane, B. 2015. *Houses for a New World: Builders and Buyers in American Suburbs, 1945–1965.* Princeton, NJ: Princeton University Press.

Leinberger, C. 2007. *The Option of Urbanism: Investing in a New American Dream.* Washington, DC: Island Press.

Loeb, C. 2001. *Entrepreneurial Vernacular: Developers' Subdivisions in the 1920s.* Baltimore, MD: Johns Hopkins University Press.

Maisel, S. 1953. *Housebuildingin Transition.* Berkeley: University of California Press.

McCann, L. 2017. *Imagining Uplands: John Olmstead's Masterpiece of Residential Design.* Victoria, BC: Brighton Press.

McKenzie, E. 1994. *Privatopia: Homeowner Associations and the Rise of Residential Private Government.* New Haven, CT: Yale University Press.

Meinig, D.W., ed. 1979. *The Interpretation of Ordinary Landscapers: Geographical Essays.* New York: Oxford University Press.

Morris, E. 2019. *Emerson.* New York: Random House.

Moudon, A.V. 1992. "The Evolution of Twentieth-Century Residential Forms: An American Case Study." In J.W.R. Whitehand and P. Larkham, eds., *Urban Landscapes: International Perspectives.* London: Routledge, 170–206

Muthesius, S. 1982. *The English Terraced House.* New Haven, CT: Yale University Press.

Oliver, P. 1981. "Introduction." In P. Oliver, I. Davis, and I. Bentley, *Dunroamin: The Suburban Semi and Its Enemies.* London: Barrie & Jenkins, 9–26.

Oliver, P., ed. 1997. *Encyclopedia of Vernacular Architecture of the World.* 3 vols. Cambridge: Cambridge University Press.

Oliver, P., I. Davis, and I. Bentley. 1981. *Dunroamin: The Suburban Semi and Its Enemies.* London: Barrie & Jenkins.

Parsons, J., and R. Harris. 2020. "Hometown Advantage: The Making of a Modern Suburb." *Urban Geography,* 41(2): 247–67. https://doi.org/10.1080/02723638.2019.1647756

Perry, C.A. 1929. "City Planning for a Neighborhood Life." *Social Forces,* 8(1): 98–100. https://doi.org/10.2307/2570059

Pinder, D. 2005. *Visions of the City: Utopianism, Power, and Politics in Twentieth-Century Urbanism.* London: Routledge.

Prosser, D.T. 1981. "Chicago and the Bungalow Boom of the 1920s." *Chicago History,* 10(2): 86–95.

Rilling, D.T. 2000. *Making Houses, Crafting Capitalism: Master Builders in Philadelphia, 1790–1850.* Philadelphia: University of Pennsylvania Press.

Rowe, P.G. 1991. *Making a Middle Landscape.* Cambridge, MA: MIT Press.

Rybczynski, W. 2008. *Last Harvest: From Cornfield to New Town.* New York: Scribner.

Schlesinger, T., and M. Erlich. 1986. "Housing: The Industry that Capitalism Didn't Forget." In R.G. Bratt, C. Hartman, and A. Meyerson, eds., *Critical Perspectives on Housing.* Philadelphia: Temple University Press. 139–64

Scott, A.J. 2006. "Creative Cities: Conceptual Issues and Policy Questions." *Journal of Urban Affairs,* 28(1): 1–17. https://doi.org/10.1111/j.0735-2166.2006.00256.x

Sies, M., and C. Silver, eds. 1996. *Planning the Twentieth Century American City*. Baltimore, MD: Johns Hopkins University Press.

Silver, C. 1985. "Neighborhood Planning in Historical Perspective." *Journal of the American Planning Association*, 51(2): 161–74. https://doi.org/10.1080/01944368508976207

Stevens, Q., ed. 2016. *Creative Milieux: How Urban Design Nurtures Creative Clusters*. New York: Routledge.

Storper, M. 2013. *Keys to the City: How Economics, Institutions, Social Interactions, and Politics Shape Development*. Princeton, NJ: Princeton University Press.

Tarr, J.A. 1996. *The Search for the Ultimate Sink: Urban Pollution in Historical Perspective*. Akron, OH: University of Akron Press.

"Thomas Edison and Menlo Park." 2020. Thomas Edison Center at Menlo Park. https://www.menloparkmuseum.org/history/

Tobey, R.C. 1996. *Technology as Freedom: The New Deal and Electrical Modernization of the American Home*. Berkeley: University of California Press.

Turner, J.F.C., and R. Fichter, eds. 1972. *Freedom to Build*. New York: Macmillan.

Van Damme, I., B. de Munck, and A. Miles, eds. 2018. *Cities and Creativity from the Renaissance to the Present*. London: Routledge.

Ventre, F.T. 1980. "On the Blackness of Kettles: Inter-Industry Comparisons in Rates of Technological Change." *Policy Sciences*, 11(3): 309–28.

Walks, A. 2019. "Financial Infrastructures and Suburbanization: From Suburbanization to Value Extraction." In P. Filion and N. Pulver, eds., *Critical Perspectives on Suburban Infrastructure*. Toronto: University of Toronto Press, 88–113.

Wallis, A.D. 1991. *Wheel Estate: The Rise and Decline of Mobile Homes*. New York: Oxford University Press.

Weiss, M. 1987. *The Rise of the Community Builders: The American Real Estate Industry and Urban Land Planning*. New York: Columbia University Press.

Worley, W.S. 1990. *J.C. Nichols and the Shaping of Kansas City: Innovation in Planned Residential Communities*. Columbia: University of Missouri Press.

Wright, G. 1980. *Moralism and the Model Home: Domestic Architecture and Cultural Conflict in Chicago, 1873–1913*. Chicago: University of Chicago Press.

Wright, G. 1981. *Building the American Dream: A Social History of Housing in America*. New York: Pantheon.

Contributors

Alison Bain is professor of the Geographies of Inclusive Cities and co-chair of the Urban Geography section in the Department of Human Geography and Spatial Planning at Utrecht University. She is the author of *Creative Margins: Cultural Production in Canadian Suburbs* (University of Toronto Press, 2013) and co-editor (with Linda Peake) of *Urbanization in a Global Context* (Oxford University Press, 2017; 2022) and (with Julie Podmore) of *The Cultural Infrastructure of Cities* (Agenda Publishing, 2023). She is currently the principal investigator (with Julie Podmore) of "Queering Canadian Suburbs," a pan-Canadian Social Sciences and Humanities Research Council of Canada research project on LGBTQ2S place-making on the periphery of Canada's largest cities.

Sarah Bilston is professor of English literature at Trinity College, Hartford (CT), where she specializes in the British Victorian era. She is the author of *The Awkward Age in Women's Popular Fiction, 1850–1900: Girls and the Transition to Womanhood* (Oxford University Press, 2004) and *The Promise of the Suburbs: A Victorian History in Literature and Culture* (Yale University Press, 2019). She has published widely in leading journals, including *Victorian Literature and Culture, Nineteenth-Century Contexts, ELT*, and *Victorian Review*, and has also published two novels with HarperCollins. She is currently completing a manuscript on the networks of nineteenth-century plant-hunting, "The Hunt for the Lost Orchid," and building a website that explores the richness of a late-Victorian suburb, www.promiseofthesuburbs.com.

Tom Broes is a postdoctoral researcher at the Department of Architecture and Urban Planning, Ghent University, where he teaches research seminars and co-supervises research. In his PhD, *Urbanisons! Urbanising the Antwerp Agglomeration, 1907–1939*, the case examined is used to build an understanding of urbanism and planning as a reasoned and situated societal response to overall urban questions that structure the urbanization process in specific ways in different historical time-spaces. His research contributes to grounding urbanism in an analytical understanding and theory of urbanization. Currently, he is involved in the research project "Construction History, Above and Beyond: What History Can Do for Construction History" (Ghent University, VUB, ULB), in which he develops the research-track "Urbanizing Construction History" that investigates the mutual relationships between different modes of urbanization and construction practices.

Margaret Crawford is professor of architecture at UC Berkeley. Her research focuses on the evolution, uses, and meanings of urban space. She has published numerous articles on immigrant spatial practices, shopping malls, public space, and other issues in the American built environment. Together with John Chase and John Kaliski, she published *Everyday Urbanism* (Monacelli Press, 2008), which examines the daily rituals and cycles that shape urban communities.

Tatiana Debroux holds a PhD in geography and is postdoctoral coordinator and lecturer in urban studies and cultural geography at the Vrije Universiteit Brussel (Cosmopolis) and the Université libre de Bruxelles (DGES-IGEAT). She is also editor-in-chief of the journal *Brussels Studies*. Her research includes work on spatial dimensions of artistic activities (for example, artists' studios and art galleries, arts districts, cultural tourism), historical and contemporary urban dynamics (for example, gentrification processes), narrative cartography (fictional literature as a source for geographers), and historical mapping (geolocation).

Michiel Dehaene is full professor at the Department of Architecture and Urban Planning, Ghent University, where he teaches courses in urban analysis and design. His work focuses on the epistemology of urbanism and the relation between urbanism, ecology, and urbanization. He is engaged in questions of suburban renewal and the sustainable reconversion of low-density urban development. With Chiara Tornaghi, he led an international consortium on the development of an agroecological urbanism.

David Gilbert is professor of urban and historical geography at Royal Holloway, University of London. His work is focused on the histories and geographies of modern London, with particular interests in London's role in imperialism and its significance as a centre for the fashion industry, as well as its suburban cultures. His recent work has concerned religion in suburbia, particularly in a major research project "Making Suburban Faith," which included a series of collaborations between suburban faith communities and artists in West London.

Richard Harris is a Guggenheim Fellow and past president of the Urban History Association (2017–18). Now retired, he taught urban geography at McMaster University. He has written about the history of housing, neighbourhoods, and suburbs in North America and British colonies. His most recent work is *How Cities Matter* (Cambridge, 2020), and he is currently working on *A History of Canadian Neighbourhoods, 1880s–2020s*.

János B. Kocsis is associate professor at the Institute of sustainable Development, Department of Geography and Planning, at Corvinus University of Budapest. He has been analysing the societies and developments of suburbs around Budapest and their interaction with the socio-economic changes within the entire metropolitan area for about twenty years. He has written various articles, book chapters, and research reports, both from a historical perspective of suburbanization in Hungary and as an in-depth scrutiny of the present state of given settlements. His current research focuses on integration of local communities, neighbourhoods, and the challenges posed to previously calm, almost rural settlements by the heavy influx from the central areas of the city and the recent economic transformations of urban periphery.

Johanna Lilius is a senior scientist at the Department of Architecture, Aalto University, Finland. Her background is in geography and planning, and her research has focused on housing and housing policy and development, (strategic) urban planning and development, suburban regeneration, urban cultures and lifestyles, as well as urban entrepreneurs. She is the author of *Reclaiming Cities as Spaces of Middle Class Parenthood* (Palgrave Macmillan), as well as of articles in, for example, *Housing Studies, European Urban and Regional Studies*, and *Gender, Place and Culture*.

Ruth McManus is associate professor in geography and associate dean for teaching and learning in the Faculty of Humanities and Social Sciences at Dublin City University. She is the author of *Dublin 1910–1940: Shaping the City and Suburbs* (2002, new edition 2021) and *Crampton Built* (2008), co-author of *Building Healthy Homes: Dublin Corporation's First Housing Schemes 1880–1925* (2021), and co-editor of *Leaders of the City* (2013). She is particularly interested in the nature of the urban and suburban landscape, and much of her work focuses on the physical and social development of everyday spaces. She is an editor of the *Irish Historic Towns Atlas* and a member of the International Commission on the History of Towns. She is the current president of the Geographical Society of Ireland.

Ulrike Müller is a lecturer in heritage studies at the University of Antwerp and a research associate at the Royal Museums of Fine Arts of Belgium, Brussels, specializing in the history and theory of collecting and museums. She obtained a PhD in art science and history from the Universities of Ghent and Antwerp (2019), with a dissertation on the public role of private art and antique collectors in Brussels, Antwerp, and Ghent during the long nineteenth century (ca. 1780–1914). Her research interests include the history of taste, material culture, public and private collecting, and the art market in Belgium and internationally. She has published, among others, on the accessibility, display, and function of private collections in nineteenth-century Belgian cities; private collectors' interaction with the public cultural life in past and present; and the historical origin and socio-cultural context of several Belgian museums and collections.

Susan Reidy is an urban historian, specializing in the social and planning histories of Australian cities and suburbs. Her PhD (University of Melbourne) examined the historical, cultural, aesthetic, and environmental trajectories of Australia's urban public parks, botanic gardens, and recreation grounds from the colonial period to the twenty-first century. Her fields of research have included the twentieth-century evolution of a suburban community; Australian town planning associations and their crusading members; the role of urban public parks in heritage, remembrance, and environmental restoration; children's playgrounds as physical embodiments of improvement theory; and the influence of new sports on the shaping of public urban landscapes.

Ilja Van Damme is full professor in urban history at the University of Antwerp, where he is a board member of the Centre for Urban History (CUH) and the Urban Studies Institute (USI). His research focuses on processes of urban modernization, with specific attention to consumer, entertainment, and retail history; changes in the public and private spheres; and topics related to the city as a site of creative and socio-cultural interaction. He previously co-edited *Cities and Creativity from the Renaissance to the Present* (Routledge, 2018).

Simon Workman is program director of the English and history program at Carlow College, St. Patrick's. He has published articles, chapters, and reviews on modern Irish writing and culture in a number of different journals and collections, with his work appearing in, among others, the *Irish Studies Review, Irish University Review, Irish Literary Supplement, Poetry Ireland*, and *The Review of English Studies*. He has recently co-edited, with Eoghan Smith, a collection of essays on the cultures of Irish suburbia entitled *Imagining Irish Suburbia in Literature and Culture*, which was published by Palgrave at the beginning of 2019, and is currently editing a special edition of the *Irish Studies Review* on contemporary Irish fiction to appear in 2023.

Index

Page numbers in *italics* refer to figures and tables.

GLOBAL SUBURBANISMS

Series Editor: Roger Keil, York University

Published to date:

Milton Keynes UK
Ingram Content Group UK Ltd.
UKHW012144210424
441411UK00002B/37